Cities and Nature

Series Editors

Peter Newman, Sustainability Policy Institute, Curtin University, Perth, WA, Australia

Cheryl Desha, School of Engineering and Built Environment, Griffith University, Nathan, QLD, Australia

Alessandro Sanches-Pereira, Instituto 17, São Paulo, Brazil

Cities and Nature fosters high-quality multi-disciplinary research addressing the interface between cities and the natural environment. It provides a valuable source of relevant knowledge for researchers, planners and policy-makers. The series welcomes empirically based, cutting-edge and theoretical research in urban geography, urban planning, environmental planning, urban ecology, regional science and economics. It publishes peer-reviewed edited and authored volumes on topics dealing with the urban and the environment nexus, including: spatial dynamics of urban built areas, urban and peri-urban agriculture, urban greening and green infrastructure, environmental planning, urban forests, urban ecology, regional dynamics and landscape fragmentation.

Indexed in Scopus!

Carlos Cobreros · Emanuele Giorgi ·
Tiziano Cattaneo
Editors

Regenerative Design

New Contexts, New Visions, Emerging
Practices and Perspectives

Editors
Carlos Cobreros
School of Architecture, Art and Design
Tecnológico de Monterrey
Querétaro, Mexico

Emanuele Giorgi
School of Architecture, Art and Design
Tecnológico de Monterrey
Chihuahua, Mexico

Tiziano Cattaneo
Department of Civil Engineering
and Architecture
University of Pavia
Pavia, Italy

ISSN 2520-8306 ISSN 2520-8314 (electronic)
Cities and Nature
ISBN 978-3-031-76889-7 ISBN 978-3-031-76890-3 (eBook)
https://doi.org/10.1007/978-3-031-76890-3

© The Editor(s) (if applicable) and The Author(s), under exclusive license to Springer Nature Switzerland AG 2025

This work is subject to copyright. All rights are solely and exclusively licensed by the Publisher, whether the whole or part of the material is concerned, specifically the rights of translation, reprinting, reuse of illustrations, recitation, broadcasting, reproduction on microfilms or in any other physical way, and transmission or information storage and retrieval, electronic adaptation, computer software, or by similar or dissimilar methodology now known or hereafter developed.
The use of general descriptive names, registered names, trademarks, service marks, etc. in this publication does not imply, even in the absence of a specific statement, that such names are exempt from the relevant protective laws and regulations and therefore free for general use.
The publisher, the authors and the editors are safe to assume that the advice and information in this book are believed to be true and accurate at the date of publication. Neither the publisher nor the authors or the editors give a warranty, expressed or implied, with respect to the material contained herein or for any errors or omissions that may have been made. The publisher remains neutral with regard to jurisdictional claims in published maps and institutional affiliations.

This Springer imprint is published by the registered company Springer Nature Switzerland AG
The registered company address is: Gewerbestrasse 11, 6330 Cham, Switzerland

If disposing of this product, please recycle the paper.

For Yumkaax, Stefano Alessandro and Giacomo,
who are going to live this better and possible future.

Acknowledgements

This research on Regenerative Design was performed thanks to the participation and interest of a great number of persons. As authors, we authors are very grateful to all contributors, who participated in this stimulating discussion process, and to all the anonymous reviewers, who provided constructive comments for all the chapters of this book.

We are especially thankful to those Directors and Deans of the of Tecnológico de Monterrey, who encouraged and supported this project, at different scales: Dr. Roberto Iñiguez Flores (National Dean, School of Architecture, Art and Design), Pascual Alcocer (General Director of campus Queretaro), Armando Bahena Vigna (General Director of campus Chihuahua), Alfredo Henry Hidalgo Rasmussen (Associate Dean of Research, Graduate Studies and Continuous Education, School of Architecture, Art and Design), Dr. Diego Alberto Rodríguez Lozano (Associate Dean of Faculty Development, School of Architecture, Art and Design, Tecnológico de Monterrey), Emmanuel A. González Anaya (Central/South Regional Dean, School of Architecture, Art and Design), Diana Susana Urías Borbón (North Regional Dean, School of Architecture, Art and Design), Pablo Hernández Quiñones (Director of Division, School of Architecture, Social Sciences and Humanities, campus Chihuahua).

Moreover, we are particularly grateful to Dra. Aleksandra Krstikj, Dra. Alessandra Cireddu and Mariana Covarrubias-Castro (School of Architecture, Art and Design, Tecnológico de Monterrey) and Dra. Mariana Flores-García (Universidad De La Salle Bajío), who provided valuable comments on methodology and the earlier structure of this book.

Several other designers, students, and professors must be thanked because they make possible very interesting and remarkable initiatives on Regenerative Culture, in particular Dr. Rob Roggema and Dr. María Elena de la Torre Escoto.

Finally, we are especially grateful also to Juliana Pitanguy and Banu Dhayalan from the Springer publishing team, who constantly helped us in the publication process.

Contents

Introduction on the Concept of the Aesthetic of Regenerative Design and Development 1
Tiziano Cattaneo, Emanuele Giorgi, and Carlos Cobreros

New Narratives

Leapfrogging for Regeneration: Contours of a Positive Story for Latin American Cities 19
Rob Roggema

More-Than-Human Connections: Regenerating Through Multispecies Design 39
David Sánchez Ruano

Fostering Regenerative Processes Through Responsible Advanced Design and Circularity 69
Laura Succini, Erik Ciravegna, and Clara Giardina

The Emergence of Practice 95
Delfina Terrado and Martin Bunge

The Importance of Nature

Exploring Urban Ecotones and Citizen-Driven Regenerative Actions 115
María Elena de la Torre, Miguel Ángel Bartorila, and José Javier Alayón González

Citizen Action to Renaturing the Historic City of Crdoba (Spain) 141
Francisco Crespo-García and Ángela Lara-García

State of the Art of Biophilic Design Research in the Latin-Mediterranean Region 167
Lucía Martín López and Rodrigo Durán López

Rural

Sustainable Development of Rural Communities in Mexico 191
Rena Overgaard Porsen, Abril Joana Margarita de León Rincón,
and Irma Beatriz González Hernández

**Regenerative Rural Housing, Case Study: Sierra Gorda Biosphere
Reserve in Querétaro, México** 223
Lervin Benitez Álvarez and Carlos Cobreros

Urban

**Barrio's Regeneration and Reoccupation Process, a Landscape
Approach to a Border City** 255
Diana Cecilia Hernández Fonseca and Héctor Rivero Peña

**Ecovillages and Transition Towns—What Can We Learn
from Them for a Regenerative Future?** 269
Edeltraud Haselsteiner

**Gastrosophy—Fed Cities: The Perspective of Food in Urban
Planning** 295
Jaime Gastalver López-Pazo and Lilian Weikert García

**Building Energy Legislation and Incentives, a Latin American
Vision Towards Regenerative Design** 327
Seyedehniloufar Mousavi and Caribay Godoy-Rangel

Urban-Bioclimatic Aptitude as Nature-Based Solution 353
Marisol Ugalde-Monzalvo

Learning Systems and Experiences

A University focused on Regeneration: What Have We Learned? 399
Federico Llamas, Victoria Haro, Andreea Dani, Ana Lucía Castaño,
and Elizabeth García Rojas

The Birth of an Evolutionary Learning Ecosystem 415
Delfina Terrado

**A Critical View on the Strengths and Challenges of Outdoor
Preschools and Nature-Based Education in Spain** 431
Katia Hueso-Kortekaas

**The Art of the Possible. Imagining and Modelling Regenerative
Cultures Through Participatory Experience** 443
Joanna Crowson, Jorge Gallardo, and Andrew Zionts

Conclusions

There is Hope. Better Possible Futures Through Regenerative Design 469
Carlos Cobreros, Emanuele Giorgi, and Tiziano Cattaneo

Contributors

José Javier Alayón González Departamento de Arquitectura, Pontificia Universidad Javeriana (PUJ), Bogotá, Colombia

Miguel Ángel Bartorila Facultad de Ingeniería, Universidad Autónoma de Querétaro, Querétaro, México

Lervin Benitez Álvarez Craft Vink, Querétaro, Mexico

Martin Bunge Experiencia Raiz, Buenos Aires, Argentina

Ana Lucía Castaño Faculty of Agroecology, Universidad del Medio Ambiente, Valle de Bravo, Mexico

Tiziano Cattaneo Department of Civil Engineering and Architecture, University of Pavia, Pavia, Italy

Erik Ciravegna Alma Mater Studiorum, Università di Bologna, Bologna, Italy

Carlos Cobreros School of Architecture, Art and Design, Tecnologico de Monterrey, Querétaro, Mexico

Francisco Crespo-García AMASCE SCA, Cordoba, Spain

Joanna Crowson Almenara Culturas Regenerativas, Vejer, Cádiz, Spain; Be.Time SCA, Vejer, Cádiz, Spain

Andreea Dani Faculty of Sustainable Architecture, Universidad del Medio Ambiente, Valle de Bravo, Mexico

María Elena de la Torre School of Architecture, Art and Design, Tecnologico de Monterrey, Guadalajara, Mexico

Rodrigo Durán López Faculty of Architecture, Universidad Anáhuac, Ciudad de México, Mexico

Jorge Gallardo Párpado - soñando en el umbral, Vejer, Cádiz, Spain; Be.Time SCA, Vejer, Cádiz, Spain

Elizabeth García Rojas Faculty of Sustainable Architecture, Universidad del Medio Ambiente, Valle de Bravo, Mexico

Jaime Gastalver López-Pazo laPlasita, Sevilla, Spain

Clara Giardina Alma Mater Studiorum, Università di Bologna, Bologna, Italy

Emanuele Giorgi School of Architecture, Art and Design, Tecnologico de Monterrey, Chihuahua, Mexico

Caribay Godoy-Rangel School of Architecture, Art and Design, Tecnologico de Monterrey, Ciudad de México, Mexico

Irma Beatriz González Hernández School of Architecture, Art and Design, Tecnologico de Monterrey, Monterrey, Mexico

Victoria Haro Rector, Universidad del Medio Ambiente, Valle de Bravo, Mexico

Edeltraud Haselsteiner Urbanity—Architecture, Art, Culture and Literature, Vienna, Austria

Diana Cecilia Hernández Fonseca Architecture Department, Universidad Autónoma de Ciudad Juárez (UACJ), Ciudad Juárez, México

Katia Hueso-Kortekaas Department of Mechanical Engineering, ICAI, Comillas Pontifical University, Madrid, Spain

Ángela Lara-García Universidad de Sevilla, Sevilla, Spain

Federico Llamas Chairman, Universidad del Medio Ambiente, Valle de Bravo, Mexico

Abril Joana Margarita de León Rincón School of Social Sciences and Government, Tecnologico de Monterrey, Monterrey, México

Lucía Martín López Escuela de Ingeniería de Fuenlabrada, Universidad Rey Juan Carlos (URJC), Madrid, Spain;
School of Architecture, Art and Design, Tecnologico de Monterrey, Querétaro, Mexico

Seyedehniloufar Mousavi Faculty of Sustainable Resource Management, State University of New York College of Environmental Science and Forestry, Syracuse, USA

Rena Overgaard Porsen School of Architecture, Art and Design, Tecnologico de Monterrey, Monterrey, México

Héctor Rivero Peña Architecture Department, Universidad Autónoma de Ciudad Juárez (UACJ), Ciudad Juárez, México

Rob Roggema School of Architecture, Art and Design, Tecnologico de Monterrey, Monterrey, Mexico

David Sánchez Ruano School of Architecture, Art and Design, Tecnologico de Monterrey, Guadalajara, Mexico

Laura Succini Alma Mater Studiorum, Università di Bologna, Bologna, Italy

Delfina Terrado Iniciativa Murmullo, Buenos Aires, Argentina

Marisol Ugalde-Monzalvo School of Architecture, Art and Design, Tecnologico de Monterrey, Estado de México, Mexico

Lilian Weikert García laPlasita, Sevilla, Spain

Andrew Zionts Almenara Culturas Regenerativas, Vejer, Cádiz, Spain; Be.Time SCA, Vejer, Cádiz, Spain

Introduction on the Concept of the Aesthetic of Regenerative Design and Development

Tiziano Cattaneo, Emanuele Giorgi, and Carlos Cobreros

Abstract While the challenges for the socioecological future are becoming harder and harder, the idea of a "sustainable development" is no more enough: is clear the need for new concepts and strategies within contemporary design, able to propose innovative solutions to regenerate our original environments. While the concept of not only limiting socio-environmental damage but actively restoring and regenerating existing ecosystems is not new in the field of design, its focus on the Latin-Mediterranean context, rich in environmental resources and architectural traditions, has not been extensively explored. In this region, where concerns about socioenvironmental decisions meet debates in practice and academia, a formal collection of experiences, good practices, and theoretical considerations is more than ever needed. A seven-year-long collaboration among Tecnológico de Monterrey (Mexico) and University of Pavia (Italy) set the bases for this book, which aims to provide the initial foundation for this necessary "formal collection". The focuses on ecological design and social design meet in this edited book, with the aim to propose a holistic view on regenerative design in Latin-Mediterranean context.

Keywords Regenerative Design · Sustainable Development · Regenerative Development · Aesthetic

---

T. Cattaneo (✉)
Department of Civil Engineering and Architecture, University of Pavia, Pavia, Italy
e-mail: tiziano.cattaneo@unipv.it

E. Giorgi
School of Architecture, Art and Design, Tecnologico de Monterrey, Chihuahua, Mexico
e-mail: egiorgi@tec.mx

C. Cobreros
School of Architecture, Art and Design, Tecnologico de Monterrey, Querétaro, Mexico
e-mail: ccobreros@tec.mx

© The Author(s), under exclusive license to Springer Nature Switzerland AG 2025
C. Cobreros et al. (eds.), *Regenerative Design*, Cities and Nature,
https://doi.org/10.1007/978-3-031-76890-3_1

1 Environmental Entanglement

In a rapidly urbanizing world, climate change and biodiversity loss are among the greatest risks for the environment and societies around the planet (de Rooij and van Hattum 2022). In its annual report on global risks, the World Economic Forum defines these issues as the risks with the greatest probability and impact for the global economy. The need for action is widely recognized, but the path forward and perspectives for action are not always clear (de Rooij and van Hattum 2022). Keeping clearly in mind this actual situation we can consider the environmental crisis not only as a crisis of the physical and technological environments; but also, as a crisis of the cultural environment—of the modes of acting, operating and living through which society relates to the complexity of environmental systems. As Ghosn and Jazairyv (2018, p. 11) wrote in their book *Geostories*,[1] "People observe daily weather changes, but they do not perceive climate, a phenomenon that is, in its modern definition, a statistically created object of investigation with a long-term assessment period". Furthermore, while weather is experienced locally, comprehending the global effects of climate change necessitates viewing the world as a whole. Thus, the question could be: how do we think about something as intangible and invisible as climate? This is especially challenging as the language of scientific expertise can be misinterpreted, sometimes deliberately, by journalists, politicians, and those working on behalf of industry interests.

In addition to this question, the book addresses other prominent issues, such as how to rethink our environment, including the built environment and beyond. This involves embracing ecological responsibility and understanding that humans are part of nature.

Within this context we cannot avoid considering (one of the recent mantras repeated by many) that there are no 'untouched' places left on Earth and that this is what it means to live in the Anthropocene, the age shaped by human activity at unprecedented pace (Jørgensen et al. 2013).[2] Actually this is true and, in fact we live in a century of architectural activity on a scale unparalleled in history. It could be an obvious information, but it is the factual recognition that architecture is extended and fulfilled in environment. Yet architecture was never a single object; today it is more than ever a form of knowledge that can enable the convergence of physical space and historical time (Andraos 2016, p. 299).[3] Environment then becomes the exemplification, indeed, the very paradigm of a new aesthetic, as defined by Arnold Berleant: the aesthetic of engagement (Berleant 1992, p. 148).[4]

---

[1] Rania Ghosn and EL Hadi Jazairy. *Geostories. Another Architecture for the Environment.* New York, Barcelona: Actar Publishers (2018). Pag. 11.

[2] Jørgensen, F. A., Karlsdóttir, U. B., Mårald, E., Poulsen, B., & Räsänen, T. (2013). Entangled Environments: Historians and Nature in the Nordic Countries. *Historisk Tidsskrift, 92*(1), 9–34. http://www.idunn.no.

[3] Amale Andraos, "What Does Climate Change? (For Architecture)," in *Climates: Architecture and the Planetary Imaginary*, ed. James Graham (Zurich: Lars Müller Publisher, 2016), 299.

[4] A. Berleant, *The Aesthetics of Environment* (Philadelphia: Temple University Press, 1992): 148.

To achieve the aesthetic of engagement, we must consider not only architecture but all the disciplines of the built environment, in order to open the academic disciplinary boundaries, and professional expertise, creating the possibility of connecting, thinking and acting cross scales including communication, experience, interaction and space. With the aim to better design the relationship between people and the environment, we need a variety of knowledge including, for example, philosophy, psychology, sociology, urbanism, landscape science, economics and so on.

Furthermore, we believe that architectural design, if combined with concept of aesthetic of engagement, can guarantee social sustainability and be an answer to the contemporary problems and demands of spaces in our cities. According to this, the concept of aesthetic of engagement entangled with architecture becomes a cultural, free choice of living in relationship with the environment, developing activities and collaborations in shared and common spaces in order to achieve complete sustainability. This requires the ability to engage with the people in the space (buildings or urban spaces) as participants, to see their desires and fears, and then to design contexts and environments to address them (Berleant 1992, pp. 12–13).[5] It doesn't mean design for interaction between buildings; it means to design environments which create a fundamentally social engagement. Collective elements and a functional mix can contribute to recreate new entanglement inside the urban space. This kind of care, applied to the urban and rural spaces brings a scientifically proved improvement of the human wellness and permits a sustainable growth (Giorgi 2020).

Moreover, tradition and innovation promoted in the context of participation and along with the notion of the "scales of environments" (Andraos 2016, p. 301)[6] must be a reference for design, allowing to the confused urban settlements to rediscover their lost identity and memories. While the attention to the urban-scale problems (abandoned areas, third landscape and brown areas, climate change, etc.) become commonplace at global level, we must be aware that these topics must be combined with the social and environmental aspects of the settings. Abandoned areas and suburbs (rural or urban, localized or spread) must become Entangled Environment where communities can recognize themselves (Cattaneo et al. 2022). Our environment design must focus on using holistic, human-centered, and interdisciplinary approaches to create and enable a sustainable eco-system, including experience, communication and place that facilitate interaction of humans with their surroundings. Places are not a static entity, but are constantly changing (Gibbons 2020) thus, it is necessary for prosperity to develop capabilities in the human and communities' components be able to change and adapt constantly in order to regenerate their environment at higher levels of health and well-being rather than degenerate it (du Plessis and Brandon 2015; Mang and Reed 2012).

Having said that, it is possible to develop the concept of Environmental Entanglement that involves crucial aspects that have been too often disregarded or trivialized: such as environmental aesthetics which is an emerging discipline entangled with other fundamental aspects such as Nature, culture, design, arts, and architecture

---

[5] Ibid., 12–13.

[6] Andraos, What Does Climate Change?" (note 2), 301.

among many other. The concept of Environmental Entanglement implies focus on architectural design as a process to create environments, contexts of participation and social engagement. In other words, the concept of Environmental Entanglement implies to look towards to regenerative architecture. Architecture is regenerative when the term architecture incorporates something more than the building. Architecture is the place, the site, the systems, the energy, the building, the fauna, and flora, etc. (Littman 2009). For instance, 'green building' design approach treats nature as a resource rather than a living eco-system(s) and aims only to minimize the net negative impacts on the environment Birkeland (2007). Buildings and cities could increase the ecological base as well as improve the economic and social health of surrounding regions (Birkeland 2007).

In a series of papers and books, the American philosopher Arnold Berleant began to develop a theoretical account of which the central concept was the idea of 'engagement,' later specified as 'aesthetic engagement.' Aesthetic engagement became the central concept of an aesthetic that emerged as an alternative to the aesthetic disinterestedness that was central to traditional aesthetic theory.[7]

Basing on Arnold Berleant's philosophical studies and his definition of "Aesthetic Paradigms for Urban Ecology" four environmental situation as paradigms are given for the conception of Environmental Entanglement.

As Berleant wrote "a sunrise or a sunset where the horizon is unobstructed is the most spectacular light show ever observed. The mediaeval cathedral is a total physical and social environment that brings together not only most of the arts engages all the senses as well. The sailing ship, an artifact that combines the most perfect functionalism with great beauty of object, demands a complete engagement of the experiencer in an environment whose sensory and utilitarian features must be fully integrated. The circus, a highly charged world of fantasy and adventure, surrounds us with an environment thick with sound, sight, and movement, a dream world of the magical, the grotesque, the dangerous". Yet, "the experience of the circus, the cathedral, the sailing ship, and the sunset, while intriguing environments in their own right, represent in clear form dimensions of the experience of the city that are often not clearly recognized and that develop only fortuitously." (Berleant 1992, pp. 61–62).[8]

So, following the philosophical approach of Berleant we can develop the four environmental situations as paradigms for a developed aesthetic of the Environmental Entanglement:

– The sailing ship as functional environment;
– The circus as fantasy world;
– The cathedral as metaphysical environment;
– The sunset as cosmic environment;

---

[7] Arnold Berleant is the founding editor of *Contemporary Aesthetics*, an international on-line journal of contemporary aesthetic theory, research, and application. Journal overview is available at http://www.contempaesthetics.org.

[8] Berleant, "The Aesthetics of Environment." (note 3), 61–62.

Each paradigm stands as a paradigm for a dimension of the city:

- City as functional environment;
- City as imaginative experience: the fantasy world of the city;
- City as monumentality experience: the skyscraper as contemporary cathedral;
- City as a cosmic dimension that move beyond itself through the vision of the horizon.

Arnold Berleant states that these ranges of environmental paradigms are analogies. "As analogies, the paradigms are useful as imaginative sources in reshaping the urban environment into habitats in which people can display their existences. Yet as analogies they prove nothing; the only value of the paradigms lies in the sensitivity with which we use them to draw forth the moral and aesthetic potential of the city" (Berleant 1992, pp. 61–62).[9]

What is important to underline here is that the Berleant's study on the aesthetic of engagement does not has at center of his discourse the concept of regenerative development. Nevertheless, we can note that the aesthetic of engagement brings us to a new perspective of the concept of regenerative development. In fact, the concept of regenerative development has recently been receiving increased attention but has received limited attention if combined with the topic of the aesthetic of the environment.

To make clearer the relationship between aesthetic of the environment and the regenerative development we can connect indeed the four environmental situations earlier stated by recalling the three interrelated ecological principles by Benne and Mang to understand better how the terms scale and scale linkage are defined and used in the paradigm of regenerative sustainability (Benne and Mang 2015).

We can summarize the three interrelated ecological principles as follows:

1. Life is structured as holarchies, ecosystems are made of many smaller systems nested within each other, and they themselves are nested within larger systems.
1. The first ecological principle has direct implications for design: Given the very nature of living systems, defining the holarchy in which a project is nested becomes the essential first step of a regenerative design process, followed by a second step which is understand the dynamic and interdependent relationships between and within scales, or holons. (Benne and Mang 2015).
2. The ability of a holarchy to maintain and develop its integrity and viability in a continually changing environment depends on the ability of the holons that compose it to perform their systemic functions. Living systems are self-creating (autopoietic), self-managing and self-regenerating. These capabilities depend on the subsystems that compose them to perform their functions in a set of reciprocals, co-evolutionary and multidimensional relationships (Benne and Mang 2015).
3. Life is continually produced, repaired, and perpetuated through networks of metabolic exchange (Benne and Mang 2015). The essential characteristic of

---

[9] Ibid., 78.

life is metabolism, the incessant flow and transformation of energy, matter, and information through a network of dynamically interconnected processes. The metabolic network of a living system is what allows it to maintain itself, repair itself and perpetuate itself (Capra 1996; Capra and Luisi 2014).

It is so very clear that regenerative development is a place-based development and design methodology that increases the capabilities necessary for living systems to increase in complexity, diversity, ability to sustain all life, and potential for change to provide future options (i.e. health and well-being) (Gibbons 2020).

On the other hand, aesthetic is a concept already well-known and investigated in architecture and urban design but, to the best of authors' knowledge, aesthetic has never been put together with the concept of regenerative design. So, in this sense, the concept of *aesthetic of the regenerative development* could be a new standpoint through which we can foster new visions on regenerative development as it has been already earlier expressed by many scholars and prominent authors such as Wahl, Benne, Mang, Reed, Lyle among many others.

They have already outlined the main features of regenerative development and design. Our contribution is to introduce a new perspective related to the aesthetic of regenerative development in architectural and urban design. If it's true, as stated by Mang & Reed (2020), that regenerative development relies on human activity as a source of health and regeneration, and on strategies informed by understanding the inner workings of ecosystems (living systems) or the concept of 'interbeing' (Wahl 2016), in order to 'shape' processes that generate new and healthier patterns in a place (Lyle 1984), then we can affirm that the concept of the aesthetic of regenerative development can be seen as a new component to enhance regenerative design in architecture, landscape, and urban design. Regenerative design is, here understood as stated by East, based on a deep understanding of the integral and interdependent nature of living, social and biotic systems (East 2020). However, while 'Regenerative' encompasses social and ecological systems, and the interactions between these systems formulate strategies for sustainability (Cole 2012), regenerative design and development have their roots in the ecological worldview. The main basis of its philosophy emphasizes the importance of 'interrelationships', as stated by du Plessis (2012). Regenerative tools, from the processes of regenerative design and development methodology, have proven to be useful in understanding places from their uniqueness through a comprehensive, participatory, proactive, and multidisciplinary model (Arellano and Cobreros 2022).

This edited book, while not claiming to be exhaustive, represents a step forward in understanding regenerative design and development, particularly in the context of Latin-Mediterranean regions, through the lens of the aesthetic of regenerative development.

1.1 Reasons, Motivations, Structure and Contents of the Book

The book could be considered as result of a 7-year long relationship among scholars and practitioners led by the School of Architecture, Art and Design of Tecnológico de Monterrey, together with international partners.[10] Several scientific seminars, courses and workshops have been organized along this period. Nevertheless, it is not considered by the Editors as final product of a fixed research program, it is instead a new starting point that aims to help scholars and designers to clarify ideas and to find opportunities to go forward with the concept of sustainability. According to its nature, the topic of regenerative culture and regenerative design is open to various interpretations. Accordingly, the contributions included here are also open and inclusive of possible further insights. Moreover, the reasons why this book is important right now are twofold: first, methodologies, processes, and policies to implement regenerative strategies strongly depend on the context of application. In fact, while regenerative culture already plays an important role in the architectural debate in English-speaking countries, the Latin and Mediterranean context still suffers from a lack of interest from a large public. Despite this deficit, this context, rich in culture, creativity, and debates, is showing interesting impulses towards this new way of seeing and interpreting design choices. Despite these thrusts and this cultural richness, little has been written about these trends in the Latin and Mediterranean areas. For this reason, it is urgent to be able to count on a scientific product that allows experts and non-experts to know what is being done in a specific context such as this. Second, it's from the consciousness that we as human being are immersed in a socio-environmental crisis on a planetary scale. We are facing an unprecedented and uncertain future and a fundamental rethinking of the way we develop our cities and landscapes is needed. In fact, sustainability is no longer enough. We need a hopeful response, we can actively create a positive and abundant future through mindful, rooted in a living-systems-based worldview, that requires a change in worldviews from 'mechanistic' to 'ecological'. With the 21st century a new paradigm emerges, and with it, the search for new design development and regenerative design processes, as earlier mentioned they should be based on a methodology that bets on human activity as a source of health and regeneration and on strategies based on understanding the inner workings of ecosystems (living systems). Currently, different approaches are being developed, in continuous evolution, to find meaning and identify appropriate actions and solutions to the global situation. Thus, of the relevance and of the complexity of this "call to act" that the idea of "Regenerative Design in the Latin-Mediterranean context" arises from: while regenerative approaches will be more and more relevant, the socio-environmental situation will become more and more complex to analyze and to intervene.

---

[10] Main partners of this research are the Department of Civil Engineering of the University of Pavia, Italy, and the College of Design and Innovation of Tongji University, China.

Therefore, this book is an edited book which collects thoughts of international experts about the experiences, benefits, and challenges to spread regenerative approaches in contemporary society.

This book is divided into five main parts opened by this chapter along with a concluding chapter developing the topic of regenerative culture and design.

In detail, the five sections have been structured with the aims to provide bases for reasoning on what opportunities the regenerative approaches to architectural design can bring to our territories (rural or urban) and living systems (environment, society, city, and learning). From the chapters here included clearly emerges that regenerative approaches are tendencies which go over the practices for sustainable strategies, since they look for the regeneration of a previous condition of balanced human–environment relation. Moreover, as above mentioned, the focus on Latin-Mediterranean region, rich in culture, environmental resources and urban-architectural tradition is a perfect geographical context to investigate thanks to their actual condition. In fact, even are context far in terms of geographical coordinates they are similar in many features (cultural, environmental, etc.) and both are experiencing a moment of concern and deep debate on how to direct policies and ambitions to be a reference on the international panorama for environmental choices and implementation strategies.

For this reason, the chapters selected for this edited book include research, experiences, and considerations from different types of experts (scholars, professionals, practitioners, educators, activists, etc.) that present innovations from the Latin-Mediterranean context, which sometimes is considered "left behind", but which can nurture further research, thought and applications. Ultimately, the goal of this book is to provide the reader with case studies, analyses, and guidelines resulting from scientific research in academia, practices, and experiences in the Latin-Mediterranean context. These studies, projects, and applications are organized into five sections as follows, discussing from the importance of new narrative facing future environmental challenges, to the appropriate methodologies and approaches for regenerative culture with systemic view. The book closes with outputs, dreams, and hopes of regenerative design: challenges and opportunities to rewrite our alliance with the environment and to move forward our relationship with the socio-environmental capital.

The five sections are:

Section I: New narratives—new visions;
Section II: Nature;
Section III: Rural;
Section IV: Urban;
Section V: Regenerative learning systems, practices, and experiences.

The various chapters offer and propose a possible path to understanding the multiplicity of sustainability approaches as a coherent phenomenon genuinely able to match the complexity of today's global problems. They explore different directions, potentialities, and tools on how these paradigms are evolved in the Latin-Mediterranean context. Different approaches to the topic "Regenerative Design" are being developed to find meaning and identify appropriate actions and solutions to the global situation, but from the uniqueness of the place. In this sense the editors do hope that this book may be a valuable resource for all those professionals, scholars and public agents involved in the response to the challenges we are facing under the unprecedented socio-environmental crisis on a planetary scale, in which we are immersed, providing relevant contributions, recent research findings and case studies' analysis centered on the Latin-Mediterranean context.

Relevant and up-to-date topics, along with the main contributions of this book divided into five parts, are described in detail in the next sections.

1.1.1 Section I: New Narratives—New Visions

The first section introduces an overview on the importance of new narrative for regenerative culture and design. The four chapters of this section discuss how different understandings of regenerative design result in multiple interpretations describing the challenges that our contemporary society has to face in the time of climate change, biodiversity loss, pandemics, social insecurity and injustice. It is very interesting to see how 'narrative' may be the base to an architectural, urban, and landscape design methodology. To frame this new opportunity to overtake the old-fashioned urban planning narrative and develop a novel story of environmental and social inclusiveness and equilibrium, Rob Roggema starts from a fundamental question: how do we, as humanity, want to enter an uncertain future? In view of the claimed need for a fundamental transition towards regenerative cities, the transformation from a mechanical to organic worldview is now extremely necessary. *Roggema* in his chapter introduces the contours of this new narrative for regenerative urban futures by describing the current state of cities: the cities in the global north, many of them are historic, old, and culturally saturated. The appetite, and the capability to discover new pathways for future living are limited. On the contrary the cities in the global south, in general, in recent decades, have grown fast and are still maturing to adapt to changing conditions, new challenges, and disturbances.

With the same emphasis on the relationship between human and environment *David Sánc*hez addresses an important issue for regenerative strategies stating that we, as humans, are a young species and we are still adapting to inhabit our planet. By observing other species in our world it is possible to see how they have been evolving and already know how to cover their needs and regenerate through constant change. Life cycles, rhythms, and patterns in them represent a true language that we can read and respond to. However, human communities have been creating disconnection shaping infrastructure, services, and interactivities sometimes without purpose for the web of life. Thus, a fundamental fact for regenerative strategies is to collaborate with the more-than-human worlds. Biomimicry, biofabrication, biophilia, conservation or restoration can be seen as efforts and tools for regenerative design. *Sánchez* presents interesting results through three case studies developed as part of various courses at the School of Architecture, Art and Design at Tecnológico de Monterrey. The design school provided an open canvas to explore possibilities towards sustainability in the locality. The first project focuses on the design for animals, the second is an exploration of biophilic design in a determined district closer to the campus and the third was developed with organizations that develop urban agroecology in the locality.

Each project developed competencies established by the school but also the sensitivity required towards regeneration. Several tools were implemented that can serve as a guide for teachers and professionals that want to incorporate non-human centered designs. These three examples demonstrate the potential to implement regenerative approaches for design education.

The chapter by *Succini* and coauthors moves from the convergence of the systemic approach of Advanced Design (AD) with the principles of Responsible Innovation (RI), with the aim of promoting more collaborative and ethical practices in complex scenarios. From the overlaps and integrations between the principles of RI and the tools and practices of AD, Responsible Advanced Design (R-AD) arises as a new approach to change and regeneration, especially to address current crises. To refine and test the model, the chapter introduces a selection of case studies, both from the Mediterranean region (Italy) and Latin America (Chile), choosing as an exemplary field of study the so-called Packaging System, which is a multidisciplinary and cross-sectoral system involving different actors and interchange relationships for the development of a complex artifact such as packaging.

The chapter by *Delfina Terrado, Martin Bunge* is an invitation to investigate ways of acting and participation. In the last few years Regeneration and all possible ways of making it visible (agriculture, forestry, economy, social field) has become a trending topic for all those who are sensitive to the context we are living in. There is scientific agreement on the urgency for a shift in our way of intervening, of treating land and air, and water, and people. Within this context the research questions posed by the authors are: how do we build a more committed and conscious way of participating? What does a regenerative practice look like? In this chapter authors explore regeneration from a practice-based perspective, listening to the experiences from three Latin-American cases to explore how the path into the strengthening of their own practices. The reader will go through into these questions, and perhaps will find his/her own questions and movements towards a regenerative practice.

1.1.2 Section II: Nature

Chapters in the second part of the book discuss the multiscale methods, approaches, and principles of contemporary research and practice for regenerative design with nature. *De La Torre* and coauthors investigate a new approach to the relationship between nature and the urban scale through the case study of the urban ecotone of the Río Grande de Santiago Canyon. From the regenerative perspective, alternatives are sought to overcome the relationship between extraction and degradation to reach coevolutionary cooperation.

This perspective should allow the regeneration of the urban and socio environmental marginalization of the northern edge of the metropolitan area of Guadalajara through a relationship of care, cooperation, and appreciation. *De La Torre* and coauthors find two dynamics in tension: first, that of intensive urban development at the expense of ecosystem services and, second, the emergence of citizen initiatives motivated by the search for alternatives to cohabit in respect and care for nature. Eventually, the chapter presents the methodology of regenerative design in the urban ecotone and articulates the narratives of the collectives to trace the route towards possible strategies for a "regenerative evolution" according to the potential of the place. The authors conclude with an important statement: regeneration starts from specific resources: natural relics, inhabitants' awareness of the environment and regenerative actions promoted by social groups. The social agreements strengthen the spaces for exchange with the logic of living systems for health and provide more opportunities for dialog.

In their chapter *Francisco Crespo García and Ángela Lara García*, bring the argument of the challenge of the renaturalisation at the scale of urban environment. In details they investigate a very important scale and condition of Mediterranean cities: the historic city. The chapter introduces an interesting case study of citizen action in the historical urban context of the Axerquía Verde in Córdoba in Spain. Their assumption is both simple and relevant at the same time: nature greenery can cool the environment through evapotranspiration. This ability is useful for cities in hot climates, where heatwaves have become more intense in recent years due to Climate Change and the urban heat island phenomenon. The case study identifies traditional cultural practices related to the use of vegetation as opportunities to promote the renaturalisation of historical cities. This is the case of the *Fiesta de los Patios de Córdoba*, which was inscribed to UNESCO'S Representative List of the Intangible Cultural Heritage of Humanity in 2012. The knowledge of how those initiatives have benne developed could offer explanations that can be of great use to boost similar future initiatives committed to achieving the sustainable green transition for historical cities in Mediterranean region.

The final chapter of section II is a systematic review of the literature on the concepts of Biophilia and Biophilic Design. *Lucía Martín López and Rodrigo Durán López* chapter aims to identify the state of the art of research on these topics in the Latin-Mediterranean context. The study includes the visualization of these aspects through diagrams made with VOSviewer that allow the reader to construct his or her own reflections to establish a debate on the state of the art of the subject. Moreover, despite of some contingent limitations, this study has the great merit to be the first study to consider the territory comprised of the different countries as a region in itself (Latin-Mediterranean Region), and it brings to the light that perhaps this lack of consistency as a Region means that there is a lack of collective work done by researchers on the topics of Biophilia and Biophilic Design. This finding should be an incentive to create a new opportunity for Latino researchers to participate and collaborate with Mediterranean researchers and vice versa.

1.1.3 Section III: Rural

The third part of this book presents possible sustainable regeneration in rural areas. It is well-known that the crisis of rural areas is a planetary problem and despite of obvious geographical differences worldwide it is distinguished by a set of problems that can be summarized as follows: depopulation and ageing of the population, abandonment and decay of small town, difficulty in keeping existing businesses and/or in launching new jobs, intensive agricultural practices to the detriment of biodiversity, pollution, a lack of infrastructures and services for tourism, as well as a shortage of job opportunities for the population, etc.

The two chapters in this section share the perspective of viewing the rural condition as a heritage and a potential foundation for the regeneration of local communities.

The first paper by *Porsen Overgaard, De León Rincón and Hernández González* introduces the application of a model of sustainable regeneration of El Refugio de Cerros Blancos, a small village of the municipality of Mier y Noriega, in Nuevo Leon, Mexico. This model is founded on entrepreneurial outreach based on social development, community leadership and social cohesion, methodologies of Design Thinking and strategic planning in the design processes of community-based architecture and the implementation of follow-up strategies. The second chapter of section III by *Lervin Benítez and Carlos Cobreros* begins with the analysis of the modification of housing as a consequence of the migratory effects in the community of Tilaco located in the Sierra Gorda Biosphere Reserve of Querétaro, presenting problems in its habitability, construction system, adaptation and even loss of identity. The authors generate the contextualization of housing, defined by several scholars as a transcendental space for the development of human beings in the different aspects of their lives. Therefore, the planning, construction and use of housing are important elements in its life cycle. The tools used in their study were regenerative design, based on the connection between humanity and nature, maintaining balance; the Living Building Challenge Certification, which sought to have a positive impact on the community and its cultural fabric; and bioclimatic architecture, analyzing the physical and climatic aspects of the community to make the best use of its resources.

1.1.4 Section IV: Urban

The fourth part of this book presents visions and projects of regenerative approaches in urban contexts addressing various scales and issues. *Hernández Fonseca and Rivero* start from the assumption that nowadays too often "urban development" is understood exclusively as "urbanization" and it is imposed on the territory. On the contrary, they propose adopting the "landscape approach," which allows for defining the intervention process from a broad perspective and incorporating the social, urban, and ecological pre-existing conditions of the place and its complexity. The authors propose rethinking the concept of "barrio" as a place-landscape and as a basis for city regeneration and reoccupation processes in emergency contexts, such as in the border city of Ciudad Juárez.

Edeltraud Haselsteiner's chapter presents a very interesting overview of ten case studies of ecovillages from the European and Mediterranean context (France, Greece, Italy, Austria, Slovenia, Spain, and Turkey). The study starts from the questions: What does regenerative sustainability mean in the context of communities, districts, and municipalities? How is this lived and implemented in eco-communities? In attempt to answer these important questions the author main assumption is that the rapidly advancing climate change and its negative consequences require urgent action. Thus, civil society initiatives (grassroots initiatives) are therefore demanding more radical and immediate changes. Following initial initiatives from the 1960s to the 1970s, which can be categorized more as "back to the country" movements, the idea of exemplifying ecological communal life forms is gaining new impetus due to the increasingly evident climate impacts over the past 10–15 years. Ecovillages operate as a way to empower people to find solutions to pressing environmental and societal problems within social communities and away from technology-centric positions.

López-Pazo and Weikert Garcia chapter presents a project introducing the concept of the Gastrosofia in order to build a new ethical corpus with the goal of introducing the perspective of food in urban planning. On the other edge *Marisol Ugalde-Monzalvo* introduces a study on Nature-based solutions for increase the resilience and sustainability of a city. A model to evaluate the Urban-Bioclimatic aptitude is proposed in this chapter as a Nature-based solution or a *Solution Based on the capacity of Nature*. Bioclimatic applies natural acclimatization strategies so that the building and built environment for themselves collect, accumulate, distribute, or dissipate the energy. These can be applied at different scales and in both new projects and existing built environments. The Urban-Bioclimatic aptitude of a city describes its current state, its limitations, and its potential.

The chapter by *Mousavi and Godoy Rangel* can be read in parallel to the previous one. In fact, the starting point of their chapter is that global climate concerns have largely contributed to developing sustainable solutions throughout different industries. In the building sector, one of the most polluting industries, designers generally seek to reduce negative impacts on the environment by trying to cover the pillars of sustainability. However, this may no longer be enough, and it is necessary to generate positive impacts that can reverse the impact of buildings on climate change. This new approach changes the expression from less bad to better. Positive Carbon/Energy Building (PCB/PEB) is one of the approaches that focuses on creating new relationship between natural energy sources and appropriate construction materials, repairing atmosphere from emissions, as well as increasing well-being and comfort of occupants. In order to apply these strategies in the societies, it is necessary to have a legislative framework sensitive to new needs and aimed at promoting the integration of different measures, not only where energy efficiency measures are required but also the generation of positive energy from renewable sources. This chapter provides a mapping of a regional vision of current sustainable building legislation and incentive plans to motivate citizens in shaping a regenerative future, by focusing on the energy aspect.

1.1.5 Section V: Regenerative Learning Systems, Practices and Experiences

The last section focuses on an essential topic for the regenerative culture and development: learning systems and experiences, regenerative learning, and practices. To boost a regenerative culture, education and experiences are of fundamental importance. *Federico Llamas* and coauthors describe how their activities at Universidad del Medio Ambiente (UMA) in Mexico have integrated regeneration into the design process of their projects, while highlighting some of the main challenges they have encountered. This chapter could be of great interest to who in the field of education aspire to implementation ideas and projects on regenerative development. The chapter by *Delfina Terrado* describes in a testimonial form an innovative experience within Argentina's educational field during the time of the world pandemic. The main intention was to support and co-evolve educational initiatives by creating a collective safe, nurturing and learning space in order to develop an interbeing awareness approach to problem-solving for individuals, communities and institutions. The Evolutionary Learning Ecosystem (ELE) was the term coined to host and potentiate through mutual interaction experiential learning. The revealing effect of ELE and the Circles of Community Response (CRC) was the psychological and social regeneration of the people and initiatives that participated during such uncertain and complex times. This is a detailed account of a real time innovative initiative application that could serve as an example to other regenerative educational projects.

Katia Hueso-Kortekaas chapter offer a critical view on the strengths and the challenges of nature-based education in Spain by experimenting a very interesting learning tool and models. Despite those models not yet being officially recognized by the authorities, they could serve as inspiration to other stakeholders interested in regenerative development in the education systems. *Joanna Crowson, Jorge Gallardo and Andrew Zionts* with the intriguing title *The Art of Possibility* explore narratives and methodologies that, without denying the extent and complexity of the problems and solutions, motivate citizens to act. The results of the authors' qualitative research, undertaken in community projects including a training course for youth workers, regenerative agriculture initiatives, and activities blending art and ecology, show how a focus on shared processing of the emotional impact of the significant biodiversity loss we are experiencing can empower participants to become change agents.

The editors complement the book with a conclusion chapter. The last chapter critically reviews the outputs of the whole research path implanted so far and open up discussion for further detailed studies along with opening up an Hope for our future. By the title *"Hope. Better and possible futures through Regenerative Processes design"* it is quite clear that Carlos Cobreros, Emanuele Giorgi, Tiziano Cattaneo by analyzing a state of the art in which we are facing an unprecedented and uncertain future immersed in a socio-ecological on a planetary scale, thus a fundamental rethinking of the way we develop our cities, landscapes, communities, buildings, services, and products is needed. Although sustainability promises not to affect future humanity, it does not contemplate the possibility of contributing and co-evolving with nature. Sustainability is no longer enough.

A new paradigm based on the idea of regenerative cultures emerges, and with it, the search for new regenerative design and development processes, based on a methodology that bets on human activity as a source of health and regeneration and on strategies based on understanding the inner workings of ecosystems (living systems). In this conclusion chapter, we review the new and different narratives, approaches, projects, initiatives, developed and rooted in a living-systems-based worldview, that requires a change in worldviews from 'mechanistic' to 'ecological', in continuous evolution, to find meaning and identify appropriate actions and solutions to the global situation from local actions.

We do hope this book besides proposing several propositions (and narratives) on how design can follow and implement regenerative devolvement, it makes evident that a shift of paradigms is also needed that starts from a renewed balance between human and nature. Many of the themes here presented have existing applications, others are experimentation in progress, and yet others are just the starting point of new possible investigations or practices. Nothing is definitive and everything is open to be improved, highlighting the importance and multiple goals that need to be pursued together for human health, quality of life, well-being, and so forth, with the aim of providing evidence for a multitude of knowledge and interdisciplinary approaches.

Acknowledgements We thank the external reviewers for providing critical and helpful comments on earlier versions of this chapter and on the book proposal.

References

Adapon Turvey R, Kurissery S (eds) (2019) Sustainability research. IGI Global: Hershey, PA, USA [CR]

Andraos A (2016) What does climate change? (For architecture). In: Graham J (ed) Climates: architecture and the planetary imaginary. Lars Müller Publisher, Zurich

Arellano RML, Cobreros C (2022) De la Arquitectura Sostenible a la Arquitectura Regenerativa, un cambio de paradigma en el contexto mexicano. Perspectivas De La Ciencia y La Tecnología, Universidad Autónoma De Querétaro 5:82–93

Benne B, Mang P (2015) Working regeneratively across scales—insights from nature applied to the built environment. J Clean Prod 109:42–52

Berleant A (1992) The aesthetics of environment. Temple University Press, Philadelphia

Birkeland J (2007) Positive development: Designing for net positive impacts. Environ Des Guide 1–8

Capra F (1996) The web of life: a new synthesis of mind and matter. Flamingo [CR], London, p 336

Capra F, Luisi PL (2014) The systems view of life: a unifying vision. Cambridge University Press

Cattaneo T, Giorgi E, Herrera AMF, del Socorro Aceves Tarango V (2022) Introduction on design for vulnerable communities In: Giorgi E, Cattaneo T, Herrera AMF, del Socorro Aceves Tarango V (eds) Design for vulnerable communities. Springer, Urban Book Series, pp 1–17

Cole R (2012) Regenerative design and development: current theory and practice. Build Res Inform 40(1):1–6 de Rooij B, van Hattum T (2022) Design for regeneration – a nature-based future perspective on cities. In: Roggema, R (eds) Design for regenerative cities and landscapes. contemporary urban design thinking. Springer, Cham. https://doi.org/10.1007/978-3-030-97023-9_6

Du Plessis, C. (2012). Towards a regenerative paradigm for the built environment. Build Res J 40(1):7–22

du Plessis C, Brandon P (2015). An ecological worldview as basis for a regenerative sustainability paradigm for the built environment. J Clean Prod 109

East M (2020) The transition from sustainable to regenerative development. Ecocycles 6(1):106–109

Ghosn R, Jazairy EH (2018) Geostories. Another architecture for the environment. Actar Publishers, New York, Barcelona

Gibbons LV (2020) Regenerative—the new sustainable?. Sustainability 12(13):5483; Goodall J, Abrams D (2021) The book of hope: a survival guide for trying times. Celadon Books

Giorgi E (2020) Technocene. In: E. Giorgi (ed) The co-housing phenomenon. In: Environmental alliance in times of changes. Springer, Urban Book Series, pp 1–26. https://doi.org/10.1007/978-3-030-37097-8_1

Jørgensen FA, Karlsdóttir UB, Mårald E, Poulsen B, Räsänen T (2013). Entangled environments: historians and nature in the Nordic countries. Historisk Tidsskrift 92(1):9–34. http://www.idunn.no

Littman JA (2009) Regenerative architecture: a pathway beyond sustainability. Masters theses, 303

Lyle JT (1984) Designing human ecosystems. Wiley, Hoboken

Mang P, Reed B (2012) Designing from place: a regenerative framework and methodology. Build Res & Inf 40(1). Regenerative design and development. https://doi.org/10.1080/09613218.2012.621341

Mang P, Reed B (2020) Regenerative development and design. In: Loftness V (eds) Sustainable built environments. Encyclopedia of sustainability science and technology series. Springer, New York, NY. https://doi.org/10.1007/978-1-0716-0684-1_303

Wahl DC (2016) Designing Regenerative Cultures. Triarchy Pr

New Narratives

Introduction

Roggema opens the book with a chapter where he sets out a perspective on a novel paradigm of regenerative urban planning, outlining the urgency of its application because of the kind of future problems and the need to respond quickly and profoundly to them. Roggema argues that the maturing cities of the global south, rather than the existing fixed cities of the global north, have the best opportunities to adapt and shift their focus to become regenerative urban regions because of their young populations, more openness to new avenues and the urgency of the problems, which will require fundamental and appropriate policy choices and a new urban narrative. This new urban narrative should include innovative research and experimentation guided by ecological principles, from a platform for exchange, inspiration and experimentation, which will require thinking in three ways: Multiple generations; Multiple scales; and Multiple actors. From this chapter, we could review the limitation of focusing on the easier and more hopeful path, or the great opportunity that lies in the global south, but we cannot lose sight of the imperative need for a radical change of model and paradigm for the cities of the global north, which are the main consumers of raw materials, of energy, the main polluters.

Sánchez-Ruano argues how humans, as a young species still adapting to our planet, have created infrastructures, services and interactivities sometimes without purpose for the web of life, unlike other congeneric species, which have evolved and already know how to meet their needs through constant change. For a regenerative future humanity must collaborate with other species, from a multi-species perspective, from the potential of implementing regenerative approaches in educational practices, capable of transforming knowledge and human agency. If we are willing to co-create with the more-than-human world, the ethics and aesthetics of regeneration must follow a symbiotic dimension, in which the human being understands other forms of life and contributes to the web of life. The ecological transition to regenerative practice involves some challenges, such as reconciling two worldviews between human and non-human ecology. Developing a biophilic and multi-species mind allows us to stimulate our sense of coherence within the world and encourages the collective and individual unconscious to understand the notions of regeneration, environmental policy, and ecological design.

Succini et al. make an interesting contribution to the discussion of more collaborative and ethical practices in complex scenarios, which leads us to the idea of Advanced Responsible Design (ARD) that is deeply related to the concept of regenerative sustainability, with a more holistic vision and approached from the uniqueness of the place, from a local, anticipatory, cooperative and transformative perspective, essential to face the current crises. The model is tested by analysing a series of Packaging System case studies. R-AD model integrated with the circularity framework allows projects to be observed through a filter that focuses on the relationships between the actors in the system, the social and cultural impacts of the product–service and sustainability. Its application makes it possible to map and systematise good practices, design, production and distribution systems, useful for future implementations of the model. An interactive and integrated approach between Advanced Design, Responsible Innovation and Circularity can be a course of action to address the needs of the complex systems around us; to bring to light critical issues and gaps in current product–service systems and to respond to change and improvement of production processes.

Terrado and Bunge, understanding the urgency of a change in our way of intervening, of dealing with land and air, and water and people, explore regeneration from a practice-based perspective, listening to the experiences of three Latin American cases to explore the path towards strengthening their practices. They reveal the refinement of conscious participatory action, but also the immersion in the human soul and the rootedness of the quest for reunion with nature. Regeneration, as a participatory practice, must begin with wonder, opening ourselves to the witnessed situation or phenomena, from the awareness of how we approach things as a moral act, from presence, from empowering conversations within the community of facilitators, sharing situations, through the recognition of emotions. Knowledge and practice emerge through participation and action in the visible world, finding alternative paths, transforming and evolving, with new questions and enquiries, new intimate challenges of what it means to remain active in these less than delicate times, new challenges of how to connect, repair, restore or heal us and the world around us. Regenerative practice is a conscious doing that needs to be embraced with discipline and discernment from our inner ecology to care for and transform our outer ecology.

Leapfrogging for Regeneration: Contours of a Positive Story for Latin American Cities

Rob Roggema

Abstract Given all sorts of global problems, climate change, biodiversity loss, pandemics, social insecurity and injustice, there is a fundamental question on the table: how do we, as humanity want to enter an uncertain future? There is a need for a fundamental transition towards regenerative cities. Transformation from a mechanical to organic worldview. Current state of cities: the cities in the global north, many of them are historic, old, and culturally saturated. The appetite, and the capability to discover new pathways for future living are limited. For cities in the global south, in general, this is different. On a long-term timescale, in recent decades, these urban agglomerations have grown fast. They are still maturing and eager to adapt to changing conditions, new challenges, and disturbances. This opens a new opportunity: to leapfrog the old-fashioned urban planning, instead develop a novel story of environmental and social inclusiveness and novel equilibria. This chapter introduces the contours of this new narrative for regenerative urban futures.

Keywords Narrative · Regenerative · Maturing cities · Future visioning

1 Introduction

The contemporary world is predominantly urban. Cities cover 2% of the earth land surface and house more than half (UN DESA 2014) the human population. The importance of cities is only going to increase in coming decades as another 2.5 billion people move too urban centers (UN 2015). Currently, over 50% of the global population lives in cities and this number is expected to increase to 80% in 2050 (WEF 2022). Activities within urban boundaries consume over 75% of the planet's material resources, more than doubling the material consumption (WEF 2022), inevitably increasing waste generation and global pollution. Moreover, cities account for 71–76% of global carbon emissions, contributing to climate change (UN Habitat, undated).

R. Roggema (✉)
School of Architecture, Art and Design, Tecnologico de Monterrey, Monterrey, Mexico
e-mail: rob.roggema@tec.mx

© The Author(s), under exclusive license to Springer Nature Switzerland AG 2025
C. Cobreros et al. (eds.), *Regenerative Design*, Cities and Nature,
https://doi.org/10.1007/978-3-031-76890-3_2

On a positive, cities attract (talented) people, turning cities in centers of interaction, initiative and accelerating ideas and innovations (Florida 2012), being the place where up to 80% of GDP is generated (Worldbank 2022). The urban is seen as a "vast and variegated whirlpool replete with all the ambivalence of a space full of opportunity, playfulness and liberating potential, while being entwined with spaces of oppression, exclusion and marginalization" (Lefebvre 1991 (1974)). "Cities seem to hold the promise of emancipation and freedom whilst skillfully mastering the whip of repression and domination" (Merrifield and Swyngedouw 1997).

This dominant role of the urban environment is therefore both problematic and a blessing and has implications for both human and natural health. It causes biodiversity loss, and leads to an increase of mental issues, such as mood and anxiety disorders and schizophrenia (Peen et al. 2010), and diseases, such as obesity (Caballero 2007), due to lack of exercise and unhealthy food patterns. The accessibility, proximity and availability of green and natural spaces are seen as an important factor in health problems. Nature and green in the vicinity of residential living, have a positive impact on mental health (Bratman et al. 2012; Callaghan et al. 2021) and/or physical health (Kondo et al. 2018; Ma et al. 2019).

The options to adjust the urban constellation and reroute negative impacts of urban life differ in every city. Contemporary cities all have (had) different development pathways, whether it be their historic growth, or their potential future perspective. In an overly generalized way, three city models can be distinguished: the rusted, pointless, and leapfrogging city. With a decrease of car dominance and rising fuel prices, and changes in demographics, each of these typological cities will have to respond in another way to change. The flexibility, or the room for adaptation is not in every city the same. Especially when external factors influencing urban life become disruptive, such as climate impacts, some cities have better abilities to cope than others.

1. Rusted urbanity: European cities have had their main development stage before the car started to dominate urban environments. Therefore, these cities have often narrow street patterns, denser city cores and mixed use. Over a long period, European cities have established these urban patterns that are difficult to adjust because of their long history, have become path-dependent, and are governed by a public administrative system with a long tradition. Their shrinking and/or ageing population holds on to these old traditions. Though many see future uncertainties rising, the appetite to act is similarly low. The way European cities respond to uncertainties, due to their long and established experience in planning the city, is to deal with what they know happens, or those situations that can be predicted.
2. Pointless urbanity: Cities that grew since the car appeared in human lives, in North America and Australia for instance, are characterized by their dominant infrastructure, low dense lay-out and separation of urban uses, such as living and housing mono-functions relatively far from working places and city centers. Many of the car-dependent cities face the problem of rising fuel prices and the impossibility to keep residential neighborhoods far from urban cores or work-places. American cities have a long tradition in infrastructure planning and have

problems to start planning for anything that is beyond a predicted infrastructural future. In the near future, it is likely that these cities will face a lack of capital to support citizens to move. In many of these cities the opportunity to adapt and change has become invisible.

3. Leapfrogged urbanity: Cities in the global south (Latin America, Africa, eastern Asia) show a relative late development, in which a natural growth patterns are confronted with 'modern' demands and requirements. These cities are generally in a process of maturing. Overcrowded infrastructure, lack of basic utilities, intensive mix of uses are some of the characteristics. In maturing cities, cities that have grown very fast in recent decades, the problems are hard felt, at the same time population is vulnerable, also relatively young. Their (of the cities and their population) growth path is still in front of them, and this offers them a unique position: they still can largely create their future. Radically spoken: if the sewage system is not yet there, it can still be shaped according to future wishes. The absence of state-of-the-art utility systems, safety conditions and healthy living environments, opens the way to leapfrog to a 'future-of-the-art'. In Latin American cities the difficulty (and lack) of planning due to an overwhelming set of problems (safety, inclusion, social equity, basic amenities, healthy and clean environment), have led to a situation that from a Western perspective could be seen as 'behind'. However, this could also be an advantage, because these younger and energetic societies can still discover how to plan for their cities. Especially in disruptive or turbulent conditions, that are highly uncertain, this is a big advantage (Table 1).

The question is how to maximize and capitalize this opportunity of Latin American cities to leapfrog the problems that European and American cities currently face. Given the inertia in European cities and the mechanics of American cities, what do Latin American cities need to do to prevent ending up in these conditions, facing similar inescapable problems of inertia and mechanical problem solving. Instead of going down the same pathway, another turn must be taken, that gives these cities a perspective on long-term livability and well-being for its citizens. Therefore, a new narrative is required.

Table 1 Summary of key elements of distinct urban models (By the author)

City model	Urban fabric	Dominant driver	Population	Planning response
Rusted	Old, fixed, dense	Traditions rule	Aged population	Predicted future
Pointless	Sprawl, big infra	Cars rule	American dream	Not planned for
Leapfrogged	Mixed, vulnerable	Opportunities rule	Young population	Plan for uncertainty

2 Uncertainty

Most urban policies are based on the belief to cover for the majority of future events or disruptions. The bell-curve shows which events occur most often. When these events are catered for in policies and plans, most of the problems are covered. This approach plans for the 'known knowns' (Fig. 1, in orange). Implicitly, this means that the known unknowns (in red) are not planned for and are consciously seen as a calculated risk. Should these events occur, the (negative) effects are dealt with after.

Apart from predicted and predictable uncertainties, there are also unpredictable uncertainties, the unknown unknowns (or the things we don't know that we don't know, in blue), and, even beyond these there are, unimaginable uncertainties (in green), which are impossible to imagine. These deeply uncertain futures are potentially also the most disruptive. How to plan for these types of uncertainties, that we cannot know or cannot even imagine? A mechanistic, rationalized response is useless, as we do not know what the problem is we must solve. The only way out here is to create a situation that the (urban) system is so resilient it can cope with whatever the future beholds and is capable of adjusting very quickly to changing conditions. Systems that are very good at self-(re)organizing themselves have better chances to survive sudden and unprecedented disruptions. They will recover, transform, and adapt. Many systems in nature have this capability and could lead us to reformulating an urban narrative for future urban development.

When cities are inert and shaped by the mechanics of infrastructural calculations, the required freedom to self-organize is very low. Indeed, when we try solving a problem in a mechanistic way while we do not know what that problem is, we are heading for failure. Due to the way many cities in Europe and North America/Australia are planned for, these cities are less good in dealing with unprecedented

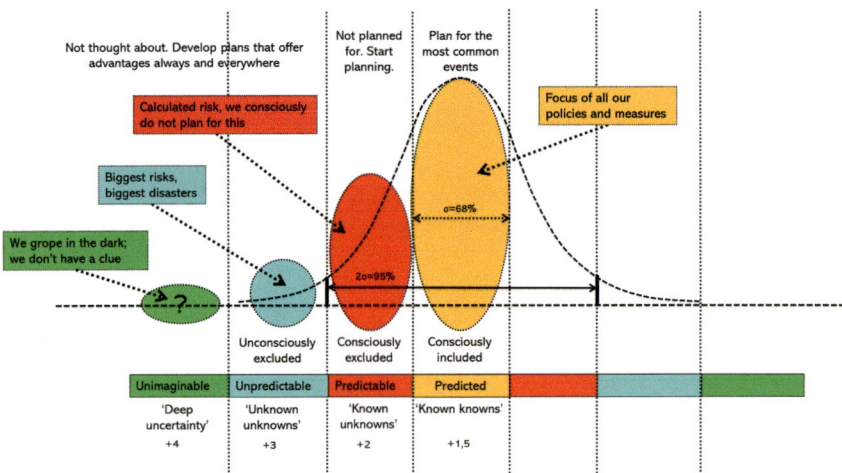

Fig. 1 Types of uncertainties (By the author)

futures, and deep uncertainties. Moreover, they have a traditional culture that doesn't allow them to change their planning systems easily, even when conditions change rapidly. Therefore, it is no surprise that when a disaster happens, these cities suffer and have difficulties to transform.

Notwithstanding the fact that disasters also hit maturing cities, the ones that have been growing rapidly over the last decades, their potential to respond is different. The luck they have is that their planning tradition is not as much established as the European/American style cities, and the combination of a younger population and the high urgency of the types of problems, brings about a window of opportunity. These cities are potentially best placed to develop a novel way to planning for a deep uncertain future. To not end up in the pitfall of the old-school problematic cities, these maturing cities should search for a way to leapfrog the old problems, overcoming and surpassing them by making a basic choice: instead of following current practice to enter a novel pathway?

Current city making has a strong focus on building and constructing, an urban growth that is using resources, and produces waste and pollution, creating unhealthy conditions and food patterns, etc. A single focus on infrastructural arrangements and the supply of resources to the city, leads to diseases and illnesses, degrades the environment, and reduces biodiversity, increases pollution, stress, and surpasses planetary boundaries. This makes current city development extra vulnerable for external stressors, such as climate disasters, or sudden demographic change.

Could the establishment of a regenerative development in maturing cities of Latin America be the answer? The dynamics in these cities are such that they have currently the biggest opportunity to still make this choice. When a new urban narrative of regeneration is created, this could turn the focus on the desired qualities: healthy lives, social interaction, a reciprocal use of resources, connection of culture and nature, and a just and harmonized equilibrium for all people, and living and non-living organisms. Could we gain trust in self-organizing urban processes, inspired by the way ecological processes are self-regulatory, and creating their own resilience and adaptive capacity? Could this help us determining the shape, land use and functioning of urban areas?

Maturing cities are the first ones to explore this pathway to a sustainable urban future and have the best starting position to make progress in this direction.

3 Mechanistic or Organic

For a long time (western) societies have solved urban problems in a rational, mechanistic way. They created artificial water management systems, energy infrastructure, roads and transportation networks, and a logistics system to transport goods, resources all over the world. This has led to a very sophisticated system of narrowly connected economies, that are bound together with individual bits and pieces, that make the machine work. But should one crucial part fail, the whole system is under scrutiny. This makes it a system that functions very well when conditions are stable and change gradually. However, when uncertainties increase and the potential disruptions become larger, the vulnerabilities of such a mechanistic system could lead to severe disruptions. As humanity negotiates the transitional zone between a mechanistic and a whole/living systems worldview, the evolutionary journey of sustainability finds itself, both from a 'politically negotiated version' as a version adopted by the private sector built around efficiency, at 'an evolutionary dead-end' due to 'flawed underlying assumptions, systemic inertia, and the inability [to deal with] complex and living systems (Du Plessis 2012).

The mechanical system works well when we live in a certain world with patches of uncertainty. Problems can be isolated and solved in a linear way, according to beforehand defined agreements. In such a stable context, systems can be global, are homogenous and regulated. Indeed, all systems can follow the same logic because disturbances are rare. This makes the systems easy to control and to calculate whether it performs according to the plan.

However, when the world becomes uncertain with patches of certainty, the globally unified regulations defining homogenous systems no longer work, because the potential problems arising in every place can be different and complex. In this context solving a single problem may cause unforeseen effects and lead to bigger problems. In this case, an organic, self-organizing perspective makes more sense (Fig. 2). Due to the long history of rusted urbanism, urban planning in these cities can be very well predicted. The only form of uncertainty that these cities experience is when something unimaginable occurs. But this doesn't happen often. Most of planning responses are mechanistic. Cities belonging to pointless urbanism are left to market change and though many things in these cities are predictable, due to uncertainties around economic transformations, a realistic future can hardly be imagined. This brings the pointless city in between the mechanistic and organic planning paradigm, reason why they face difficulties in directing their planning efforts. The leapfrogged urbanism must deal with unpredictable and unimaginable changes. To these cities it is clear that an organic planning perspective in which facilitation of changes is more effective than the application of mechanistic planning regulations, is most effective.

The complexity of a mix of interrelated and localized problems requires a multiplicity of solving strategies, that are embedded in local expertise, culture, and methods. The dynamics of such a local system will emerge to a new stable state where a new balance is found. Here, the complexity of problems is not solved in a finite way, but the resilience of the system modifies its components in a way the environment can reorganize and adapt to the changes at hand.

In the 'mechanistic city', often car-dominated, industrial cities in the western world, decision-making occurs in traditional ways, according to the long tradition of constantly reinforced political power balances. Due to this type of decision-making, innovations are relatively difficult, and the adaptability of urban patterns and infrastructure in these cities is relatively low, given the path dependent ways of thinking. This implies that to change course is difficult. These cities are not growing (fast) anymore or even shrink and the population is ageing. Biodiversity is low and nature is close to absent or manufactured. Moreover, due to the investments in existing infrastructure and construction, necessary adaptations to climate change are very

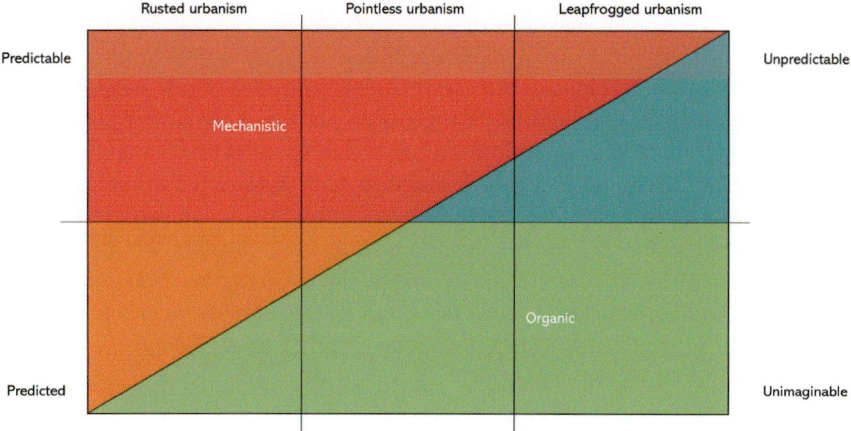

Fig. 2 Three urbanisms dealing with different types of uncertainty (By the author)

costly. This doesn't mean there are no hopeful strategies being developed to increase biodiversity or urban resilience in mechanistic cities, such as reflected in the many initiatives deployed through the MacArthur Foundation (MacArthur Foundation, undated), the Rockefeller Foundation (2023) or the ICLEI (Huseynova et al. 2022), and many more. However, the pace of implementation of these initiatives and the scale at which they are implemented is often hampered by slow administrative and lengthy decision-making processes.

The 'organic city' principles can be easier applied when the direction of the city is still relatively open and can be determined and designed. Decision-making in this case does not replicate decisions from the past and a dynamic culture is more apparent. It is undecided yet how the future shall be shaped, but recent fast growth raised many questions of vulnerable communities, environmental threat and social inclusion and equity. These cities stand at a crossroad and major choices are still relatively open. For Latin American cities, and cities of the global south in general, solving a multitude of problems, many of them related to basic qualities of life, may seem too complex, these maturing cities also have the potential to change course more easily. Contemporary threats, such as climate change impacts and loss of biodiversity can even form a novel driver to leapfrog existing and developing problems. Well-planned adaptation and nature may accelerate and accommodate solutions that improve basic qualities of life simultaneously.

However, when maturing cities 'catch up' with the standards of the cities of the global north, and follow their paradigm of the ever-growing city, they also will import their current problems very soon. They will be (if not already) guided by economically driven urban growth, extract resources and will face difficulties to adapt to change. This 'existing' way of growing urban areas generally reduces biodiversity, makes the city more vulnerable for climate impacts, and is not providing conditions for a better quality of life. The already vulnerable communities will then suffer increasingly and be impacted by climate change more. The expected growth of growth in combination with these higher climatic and other impacts multiplies the vulnerability of the urban population.

In case the maturing cities would divert current pathways of urban development and start following the paradigm of the regenerative city, an alternative future looms. In this paradigm land use is reprioritized, in which space is created and freed for replenishing used resources, allowing nature to self-organize the urban landscape to become more resilient and adaptive, increase biodiversity and self-cleaning the air, the water and the soil. Such a city starts with understanding the local conditions and interconnectedness of ecology, water, soil, topography, local culture, and social coherence. Vulnerable communities would then have easier access to the basic qualities of life: clean water, energy, food, and a healthy environment.

4 Inspirations for a New Urban Narrative

To discover possible components of a narrative for future of the city a selection of potential directions and perspectives are deemed relevant.

4.1 What Future?

The major themes of future urban life and development need to be taken into account. As an example, a recent discussion about the future of cities, brought together visionaries depicting the main urban topics deserving attention in the future (Treviño and Peña 2022).

While the possible future of cities can be scary ('the end of cities, and humanity) or unsettling (continue to live in a permanent crisis), it preferably is desirable and promising, improving the climate, health, inclusion, pluralism, understanding and creativity. To achieve this, a reinvention of the city is required to turn it from one that is feeding crises to a city that mitigates urban impacts through new coalitions and partnerships in the urban arena. The future of cities lies in the reserves of resilience, society, and creativity of urban society in all its corners, even in the least considered. For city designers, access to these reserves starts with understanding the characteristics of each place, and to take care of it in a holistic way, embedding biology, ecology, anthropology, amongst other fields in the architectural and urban profession. Such a universal vision integrates different 'versions' of the city, including the city of well-being and social cohesion, the smart city, and others. In the design of the city the basic needs for its citizens, such as water services, decent parks, security, or public lighting shall be resolved. This also can be used as the entrance point to influence human activities rather than regulating the environment. Therefore, urban districts need to be regenerated, for instance to integrate 'chrono-urbanism', providing proximity of time for its inhabitants to their basic needs on foot or by bike. It allows people to live, work, shop, access a healthy environment, education and culture, and rest in harmony with nearby nature in a short range of distance. Given the urgency of topics of water scarcity and air quality, the reuse of water from treatment, rainwater capture, storage, and use, needs to be integrated in urban planning and design.

4.2 Regeneration

Regenerative design seeks to address the continued degradation of ecosystems by developing the built environment to restore the capacity of ecosystems to function at optimal health for the mutual benefit of both human and non-human lives (Cole 2012). Crucial to regenerative design is a systems-based approach. Rather than being conceived as stand-alone objects, buildings are thought of as nodes in a system, much as organisms form part of an ecosystem (Zari 2015). Regenerative design aims to enable built environments to move into the realm of creating health and wellbeing rather than simply reducing damage (Reed 2007). Therefore, 'A 'regenerative sustainability paradigm' needs to attempt to address the dysfunctional human—nature relationship by entering into a co-creative partnership with nature [and aims] to restore and regenerate the global social–ecological system through a set of localized ecological design and engineering practices rooted in the context and its social–ecological narratives' (Du Plessis 2012).

Regenerative urban development seeks to mimic the circular metabolic systems found in nature (Woo et al. 2014):

- A regenerative city reintroduces treated water into the hydrology cycle, increasingly sources food from urban and peri-urban producers, captures the nutrients from its sewage and waste to be applied to surrounding agricultural land, dramatically reduces its dependence on petroleum products and boosts the deployment of renewable energies in its power, transport and heating sectors.
- Actively work to regenerate the materials and resources the city uses, making the regenerative city a node of production.
- Establish and nurture a symbiotic link between urban areas and their surrounding areas lies at the heart of the regenerative vision. Unlike too many modern cities that heavily depend on their hinterland without giving back anything except waste that cannot be reabsorbed, regenerative cities have a mutually beneficial relationship with their peri-urban and rural territory.

To realize this, comprehensive political, financial and technological strategies should be initiated for an environmentally enhancing, restorative relationship between cities and the ecosystems from which they draw resources for their sustenance (Girardet 2010). The urban society then gives more back to its surrounding and internal ecosystems than it extracts, forming a ReciproCity (Roggema 2019), which allows for the regeneration of resources, ecology, a clean environment, and healthy citizens and species.

4.3 Goodness

Throughout time, city designers, sociologists, spatial economists, authoritarian leaders, sociocratic leaders, all have tried to define what a good city is. The most recent addition to this is the rise of comprehensive eco-cities such as Masdar, BIG in worlds plan, and most recently the line in Saudi Arabia. Though the objectives are respectable, these, often, large urban developments have a centralistic top-down appearance, not always sensitive to the desires of their (future) residents. In responding to these ambitious eco-cities, Elizabeth Farrelly points out some of the core aspects of a good city (Farrelly 2022):

> The best cities, even those set out on a simple grid, are enablers not enforcers. They offer option and enticement as much as constraint, negative spaces as well as positive. A good city enables gathering, but it accommodates melancholy and refusal as readily as shared celebration and joy. Above all, it must offer a sense of freedom. Within the bounds of etiquette and law, the city should liberate your truest self. Because democracy is desire—based we must create not only the physical forms and systems of sustainable cities but also the metaphysical qualities—beauty and choice, intricacy and walkability, social inclusion and genuine citizen engagement—to make them objects of desire. To survive and undo climate change with civilization in any way intact, we must somehow combine design solutions, which are inevitably top-down (implying control), with genuine participation. There are many ways to do this—limit building size to enable citizen builders, encourage co-ops, supply land, and educate and engage the populace instead of excluding and pacifying them. But our efforts to listen and respond to people, and to nature, must be genuine and they must be immediate. If we are to fly free as well as stay safe, we must meet top-down with bottom-up, then equalize these forces to sit side by side.

Learning from concepts that have been proposed to plan for a sustainable or self-organizing city, when these are brought together in a mutual reinforcing way, this could lead to an alternative perspective on conceiving the 'best city' (Roggema 2012). Each of these city-models (the planned, sustainable, or self-organizing city) have their downsides but when these concepts are integrated, they could lead to a stronger and more sustainable proposition. Such a perspective on the city is more adaptable to change hence makes the urban environment safer and more resilient. Moreover, it could open up the urban planning arena for inhabitants to shape their own, sustainable environment.

The advantage of planned cities is they generally have a very robust and generous amount of green and water structures, which can be used to increase the resilience of the entire city. If this space is used for implementation of ecological principles the main advantage of the sustainable city model can be capitalized and become fully integrated with the urban design, adding to the beauty and livability of the city. Finally, when in the design process for these areas the self-organizing power of communities and societies to adjust their environment to their liking and requirements is put at the forefront, clean regenerative interventions can be implemented that are supported by the residents. Integration of the best parts of the aforementioned city-models, are only possible when traditional constraints of city planning are circumvented so the innovation to 'plan for the unplanned' can count on a refreshed mindsets of local policy makers.

To start planning the unplanned the following elements are distinguished (Roggema 2012):

1. Basic conditions: Start with the existing conditions (landscape first, e.g. landscape ecology water system, elevation and soil conditions, both current and potential) and the inevitable (the planning decisions that already have been taken and cannot be reversed).
2. No-regret: Add no regret measures and other measures that are always beneficial, no matter what the future might bring. For instance, to free the riverbed for a natural flow of the water is always an advantage, no matter what future climate disaster might occur.
3. Unplanned space: Add or create spatial reserves to accommodate the unknown future. This requires keeping the option open for a free development in most of the urban landscape. The spatial layout can only evolve when there is no control on the long-term or in detail. This allows for a self-organizing spatial and community power which creates a strong resilient urban system. Self-organization is a continuous process of adjusting the configuration and must therefore be well monitored, allowing for the best possible emergence, and evolving spatial and social process.

4.4 Speculation

One step further on the stated above, ways to genuinely involve residents and the population, developing speculative futures and visions is suggested to broaden the scope of attractive spatial possibilities. This way many different futures can be envisioned and understood by all and are used to mold the most desirable version of an unpredictable future. Approaching the design and development of the city as a set of uncharted speculations could help to widen the scope of options and therefore open the way to solutions for crucial problems that stay without reach in conventional planning. This way, best times are still to come (Hoffman 2022):

> Approaching city making as a process with guaranteed results perpetuates the idea that the future can be forecasted and controlled. That kind of thinking is dangerous. Summer temperatures will continue to spike higher than existing cooling systems have been designed to handle. Internet access and storage demands are already outpacing available tools. Even with the best research and foresight, urban life will continue to morph in ways beyond prediction. The past is not a succulent template for what lies ahead. Clinging to threads of certainty ultimately limits our imagination about what cities can become. Presenting futures as predictable requires grounding them in today's logic. Yet the confines of conventional wisdom often turn envisioning alternative trajectories into impossible tasks. Placing boundaries around collective imagination makes long-standing issues appear increasingly intractable and dystopian futures more inevitable by the day. Celebrating the space between fantasy and reality builds the resilience this century requires. If we want to aim toward less-dystopian destinations, we have to get creative. Our survival on this planet depends on creating nimble responses to accelerating scales, scopes, and speeds of change. By creating containers for collective imagination of what the future can bring, speculative futures helps

us create those responses together. Assuming that devastation is the entirety of what's ahead is limited thinking. What if the best times are still to come? We owe it to ourselves to ask.

Therefore, the conception of speculative spatial futures can be helpful to include unprecedented change and unforeseen conditions.

4.5 Phantasy

In many science fiction movies images of the future are envisioned. Technologies, gadgets, and other features that are seen as impossible have become reality in decades after. For example, the mobile phone, tablets, self-driving cars, holograms, and many other ingredients of these movies we now know in real life and have become common aspects of our lives (Cavendish 2020). A similar development of envisioning the future of the city can be witnessed. Phantasy is used to create an imaginative city as a (sneak)preview of what might be the future. Such a phantasy, that is bound to become real is Wakanda, a fictional country located in sub-Saharan Africa, home to the superhero Black Panther. In the story Wakanda is portrayed as the most technologically advanced nation on the planet.

The capital city, Birnin Zana is envisioned as a pedestrian friendly, car-free environment in which small buslike shuttles cover longer distances. Streets becoming focused on people rather than cars, where people are just able to walk in the middle of them without the fear of being run over. Wakanda's architecture comes in all shapes, sizes, with tall and midrise towers, and human-scale urbanism. The buildings are regional in their expression, not only being organic, but resembling its place and culture, and incorporating traditional African elements, such as thatch roofs and hanging gardens on some of their tallest structures. The city has a strong annotation with emerging ecocities. The question is whether this image of the city inspires the creation of a new type of environment or whether future cities are just reinforcing what is already there. Birnin Zana presents an alternative vision of what future cities could look like and be run in Africa (Malkin 2018). This phantasy city has, in reality, led to the initiative of the Senegalese-American singer Akon to build "Akon City", a solar powered "real-life" Wakanda on 2,000 acres of coastal land in Senegal (Akinwotu 2020). The plans depict gleaming skyscrapers, shopping malls, music studios and eco-friendly tourist resorts. They envisage that "Akoin", a cryptocurrency would be the central currency.

Deliberating how future cities would or should be, the use of the power that these 'phantastic', novel urban concepts bring could inspire city-making in the real world. Would the design of the city as a Marvel-story, create an appealing version of the city that seems to be unreal, just as the technologies and gadgets were seen in science fiction movies? The city becomes then a phantasy first, and (parts) will later evolve into reality. This approach could incentivize regenerative urban development when novel technologies are impregnated with a regenerative vocabulary. Moving a little bit away from the traditional urbanists who design within their professional frames, involving creatives to think about urban futures could overcome the gap between path dependent city making and the urgencies cities of the future will have to deal with.

5 A New Narrative

Composing a narrative for future regenerative cities needs to consider the breadth of perspectives, as mentioned before. To respond to current and upcoming global and local problems impacting life of urban dwellers, inclusiveness and environmental resilience shall be integrated. Moreover, social, and environmental sustainability must be positioned within the dynamic world the city is. A true regenerative city can only emerge when problems and positives of urban life are bridged with experiments and excitement.

5.1 The Two-Cities Paradox

The views on how to care for the best possible city may cause a paradox between the *sustaineers* and the *experimenteers*.

On the one hand side, the objective to create the most sustainable city possible, emphasizes good and healthy living conditions, clean and safe environmental qualities, nature and green surroundings, and a social, inclusive way of living. In this ideal world, there is limited disruptions, and whenever they occur the social and environmental coherence will diminish its impact. This perspective, led by people we might call the 'sustaineers', contains a moral right to know what is best, what is true and promises a better world, in harmony with nature. At the same time, this perspective is close to boredom, and because it often refers to historic knowledge and convictions, there is little space for experimentation, as this implies a certain risk. By striving for ecological welldoing, pointing at the harmony and the laws of nature it prevents crossing limits and searching for the edgy solutions. The DNA of the 'sustaineers' is in the core conservative and non-experimental because every experiment beholds a potential risk.

On the other side, the city is the place where new ideas emerge and due to the proximity of people interactions and novel solutions for problems that not even have been recognized emerge. The urban environment offers continuously new twists to stay attractive and seductive. This is also the reason why people always want to live in cities, even if living conditions are not good. In the city new ways of doing things and experimentation is embraced. Artists, creatives, entrepreneurs, and inventors are looking for an environment to shape the world of tomorrow, finding answers that may or may not be successful in dealing with fuzzy urban conditions. This perspective, led by people we might call 'experimenteers' aims for cutting edge concepts to discover a new vision for the city. This is a city of excitement and experimentation, full of inventions and imagination. A dynamic environment containing seductive energy, where new worlds are created, and the belief in new technologies and advancements is strong. Here, the world of tomorrow is created which may cause a certain level of addiction to experimentation. However, by entering an unknown world there is no guarantee to stay within ecological or planetary boundaries. The inventions created here can never prove beforehand they will be successful, nor better for humankind and nature, hence are risky by nature.

These perspectives of 'sustaineers' and 'experimenteers' may not be new; they are also not commonly used in urban planning and design. In practice, groups representing these viewpoints often fight without making the effort to understand and make use of each other's unique competences. This can be seen as a missed opportunity for improving the quality of urban planning and design towards a more regenerative city. It withholds urban residents from entering a future city that is exciting and livable.

The paradox (Fig. 3) is that when more sustainability is pushed by the sustaineers, the experimenteers get less interested, as they see a boring world full of status quo and harmony. Instead, they are driven by shock and awe, and want to amaze the world with their novelties. So, while the world needs their innovative solutions, they could not be less interested in contributing. At the same time, the need for these innovative solutions (to save the world) is urgent, but it is seen as alien to environmentalists. Experimenteers could bring yet unknown solutions to counteract dangerous climate change and biodiversity loss, but due to their unpredictable effects, these solutions are seen as a dangerous and risky pathway. However, without innovations there will be no novel solutions while at the same time these are urgently needed. Sustaineers and experimenteers have silently agreed to a self-fulfilling and self-enforcing to disagree on the way forward.

It seems hardly possible to bridge these two versions of the city. A suggested and preliminary set of ideas to design the eco-cathedric city as a Marvel story could be ignited as follows:

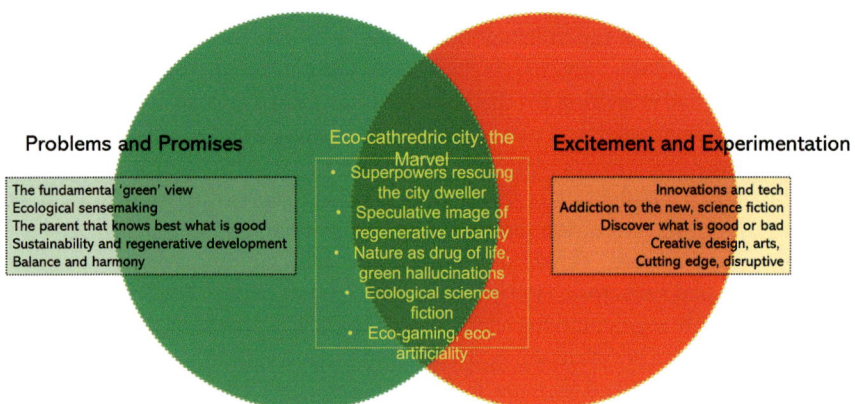

Fig. 3 Paradox of two cities (by the author)

- Attracting electric minds and superpowers to come to the rescue of the urban dweller.
- Designing the city as a cultural ecology of phantasy and the sublime.
- Co-creating a speculative image of the regeneration of urbanity.
- Using urban dynamic as the engine for a dazzling transformation, in which nature is the drug for life, residents gaze of excitement and experience green hallucinations.
- Envisioning ecology as science fiction and the object of futuring science. Conceiving a story of the eco-cathedric city, the marvel.
- Making use of new media and let games, artificial design, and new technologies be guided by ecological principles.

To be able to create inclusive, caring and regenerative urban and rural precincts, cities and landscapes it requires cutting edge and cross-disciplinary research. Nature-driven urbanism must be linked with those shock and awe disruptive experiments, which must be applied in concrete urban contexts (streets, precincts and city-regions), be tested and improved, and evolve into mainstream urbanism.

Cities with a long history (rusted and pointless cities), are the future victims of climate disaster, social unrest, injustice, and unhealthy environments. Despite promising examples promoted by organizations such as the Rockefeller Foundation and ICLEI, they eventually should 'dedevelop'. The maturing cities of Latin America have the opportunity to leapfrog those cities' mistakes and problems. A future eco-cathedric city could be shaped along three directions, time, scale, and governance:

1. The timescale.

Balancing the ecological, climatic, and geological timescales (long term) with urban timescales, which are shorter and related to the human lifetimes and time-horizons (days–years) or political timeframes (4–5 years), and the disruptive timescale (here and now, the disaster does not wait, a brilliant idea appears suddenly) (Roggema 2015).

Mainstream urbanism is concerned with the human and political timescale of days, years, decades. The (eco-)cathedric city emphasizes to integrate and guide developments according to ecological timeframes of one to many (human) generations (Krznaric 2020). Such a city is driven by the dynamics of ecology, and therefore capable of establishing resilience both for disturbances from environmental/climatic impacts and social vulnerabilities. This city aims to start design from the characteristics of local/regional ecosystems, its bio-regional qualities, and opportunities. The basics are inspired by the features and processes in nature, water, elevation, soil, and the metabolism of resource flows. This timescale of ecology, evolution, emergence can be a relatively long process.

2. The spatial scale.

Connect ecological, landscape, geomorphological scales (regions, bioregion, urban region) with the acute visible intervention scale, that is hyperlocal. Specific interventions aim to transform the current urban system towards greater ecological resilience: eco-acupuncture (Ryan 2013; Roggema 2017; Hemingway and De

Castro Mazarro 2022). The design of a city as a blueprint is useless as it only fixes the boundaries of change, limiting adaptive capacities. Designing the eco-cathedric city means definition of the objective as a long-term process using the principles of eco-acupuncture to enforce tipping points towards that end goal. The spatial scales of ecology are fractal, from the miniscule to the biospheric scale. In spatial planning fractal thinking could be applied to constantly moving up and down spatial scales while designing, allowing the small spatial interventions to influence the larger scales.

3. Governance.

The current governance model is solely driven by (a select group of) human beings. Natural elements, hyperlocal or disengaged people, or creatives are often not engaged nor involved in decisions taken for the whole natural/cultural society. In an ecocracy (Prasetyanti 2017; Takebu Good Governance 2021), the voice and vote of nature can be given a position in the governance process, for instance to include the rights of nature (Cullinan and Falstrom 2008). In such a governance model, every decision asks the question how to improve the resilience of the natural systems of ecology, water food, soil, landscape, as well as vulnerable communities, on the long term alongside the current decision-making constellations. What would it mean if natural elements, such as the Whanganui River in New Zealand, are seen as a co-deciding 'person', with rights and a voice (Tanasescu 2017) in the governing process? Any decision taken would then ideally contribute to a socio-ecological equilibrium for future generations of humans, animals, and plant species.

6 Conclusion

In this chapter a broad range of ideas and concepts have been discussed, and this explains its general nature. Though it may seem idealistic at times, it opens up a perspective towards a novel urban regenerative planning paradigm, without the danger of being lingered in planning approaches that repeat past practices that have resulted in some of the problems the cities of today face.

Regenerative city making is urgently needed because of the type of future problems and the ability to respond quickly and profoundly. Not every city-type can transform as easily. Cities of the global north, being often dominated by a path-dependent long history, or being strongly based on car-use may have more difficulties due to a combination of factors: established way of decision-making, repeating former approaches, and the inability to change due to lack of resources to transform from an oil-based transport model to a mixed-use, denser city.

The maturing cities of the global south have the best opportunities to adapt to and change their focus to become regenerative urban regions. This is caused by the young population, more open for new pathways and the urgency of problems, requiring fundamental political choices. These maturing cities have the chance to leapfrog existing fixed cities, preventing them from copying their problems.

This however requires the search for a new urban narrative. In many cities in the global north one can find a separation of people striving for a green, sustainable future and the ones that want to make use of the urban opportunities and explore uncharted territory, with all the unpredictable consequences. For a sustainable future, given the magnitude and urgency of problems uncharted territory must be explored, while this may not immediately (or guaranteed) lead to a sustainable future. There, sustainability advocates become troubled, because there is a risk of supporting these novel explorations may end in a less sustainable futures than envisioned. On the other hand, the explorers of this uncharted territory are not overly excited by the green movement, as they experience this as a world that limits their minds. This is the paradox of two cities, which shall be overcome if a regenerative development is aimed for. To bridge this gap, a narrative for future regenerative cities should incorporate elements of both, including cutting-edge research and experimentation guided by ecological principles.

The explorers and sustainability advocates need to be connected by a shared interest. No matter how impossible this seems, their minds must be mixed, and future visioning shall be loaded with regenerative vocabulary. If the creatives are interested and being challenged in developing stories, games, artificial intelligence, movies which are incentivized with an eye on a regenerative future this paradox might be dismantled. A new narrative of an eco-cathedric city will create a platform for exchange, inspiration, and experiment.

Such an eco-cathredic city requires thinking in three ways:

- Multiple generations: the city is not designed, developed, and build for the current inhabitants, but for many generations to come.
- Multiple scales: the city is not designed as a blueprint, which can be realized as it is drawn on the drawing board. The eco-cathedric city is a long-term, large scale imaginative phantasy, which is realized by acting on the smallest scale possible, whilst influencing all larger scales.
- Multiple actors: not only the aristocracy and current bureaucracy of established leaders should decide on the future of the city, but a voice shall be given to other groups, such as hyperlocals, creatives, and vulnerable communities but a voice should also be given to non-human organisms, such as rivers, animals, and other forms of nature.

This chapter uncovered the very first ideas of a future regenerative city in the Latin American context to successfully merges the sustainability and creative sector. However, many questions remain food for future research. Firstly, the LatAm region is large and contains many good examples of urban planning for regeneration, such as the most famous one of Curitiba. These could be further investigated and serve as models for other cities in the regions. Secondly, the directions for solutions in this chapter require further design research how to design, plan and implement these novel urban models in specific contexts. Thirdly, the process of implementation of these ideas by motivating sustainability experts to think more innovatively, or who are the strategic allies for implementation can be further investigated. Lastly, further design work is suggested on how inspiration drawn from science fiction and comic books can be used in realistic urban planning.

References

Akinwotu E (2020) Music mogul Akon going ahead with futuristic 'Akon City' in Senegal. The Guardian. https://www.theguardian.com/world/2020/sep/01/music-mogul-akon-going-ahead-with-futuristic-akon-city-in-senegal
Bratman GN, Hamilton JP, Daily GC (2012) The impacts of nature experience on human cognitive function and mental health. Ann N Y Acad Sci 1249(1):118–136
Caballero B (2007) The global epidemic of obesity: an overview. Epidemiol Rev 29(1):1–5
Callaghan A, McCombe G, Harrold A, McMeel C, Mills G, Moore-Cherry N, Cullen W (2021) The impact of green spaces on mental health in urban settings: a scoping review. J Ment Health 30(2):179–193
Cavendish L (2020) Welcome to the future: 11 ideas that went from science fiction to reality. All about Space magazine. https://www.space.com/science-fiction-turned-reality.html
Cole R (2012) Regenerative design and development: current theory and practice. Build. Res. Inf. 40(1):1–6
Cullinan C, Falstrom A (2008) If nature had rights. Orion Mag 27:26–31
Du Plessis C (2012) Towards a regenerative paradigm for the built environment. Build Res & Inf 40(1):7–22. https://doi.org/10.1080/09613218.2012.628548
Farrelly E (2022) Futuristic cities like Saudi Arabia's the line have one goal. The Saturday Paper. https://www.thesaturdaypaper.com.au/life/cities/2022/10/15/futur...the-line-have-one-goal?cid=3c71730767c169ee25f2f49aa7714163#mtr
Florida RL (2012) The rise of the creative class, revisited. Basic Books, New York
Girardet H (2010) Regenerative cities. World Future Council, Hamburg
Hemingway JM, De Castro Mazarro A (2022) Pinning down urban acupuncture: from a planning practice to a sustainable urban transformation model? Plan Theory Pract 23(2):305–309. https://doi.org/10.1080/14649357.2022.2037383
Hoffman J (2022) Speculative futures. Design approaches to navigate change, foster resilience, and co-create the cities we need. North Atlantic Books, Berkeley, CA
Huseynova A, Bizzotto M, Vital Estrada V (2022) Resilient cities, thriving cities. The evolution of urban resilience. ICLEI, Bonn
Kondo MC, Fluehr JM, McKeon T, Branas CC (2018) Urban green space and its impact on human health. Int J Environ Res Public Health 15(3):445
Krznaric R (2020) The good ancestor: how to think long-term in a short-term world. The Experiment, New York
Lefebvre H (1991) The production of space. Blackwell Publishers, Oxford
MacArthur Foundation (undated) MacArthur Foundation. https://www.macfound.org
Ma B, Zhou T, Lei S, Wen Y, Htun TT (2019) Effects of urban green spaces on residents' well-being. Environ Dev Sustain 21(6):2793–2809
Malkin M (2018) The real-life possibilities of *Black Panther*'s Wakanda, according to urbanists and city planners. Architectural Digest. https://www.architecturaldigest.com/story/the-real-life-possibilities-of-black-panthers-wakanda-according-to-urbanists-and-city-planners
Merrifield A, Swyngedouw E (eds) (1997) The urbanization of injustice. New York University Press, New York
Peen J, Schoevers RA, Beekman AT, Dekker J (2010) The current status of urban-rural differences in psychiatric disorders. Acta Psychiatr Scand 121(2):84–93. https://doi.org/10.1111/j.1600-0447.2009.01438.x

Prasetyanti R (2017) Ecocracy: ecology based democracy pursuing local goals of sustainable development in Indonesia. Jurnal Kebijakan & Administrasi Publik 21:1–13

Reed B (2007) Shifting from 'sustainability' to regeneration. Build Res Inf 35(6):674–680

Rockefeller Foundation (2023) 100 resilient cities. https://www.rockefellerfoundation.org/100-resilient-cities/

Roggema R (2012) Swarming landscapes: the art of designing for climate adaptation. Springer, Dordrecht, p 266

Roggema R (2015) Three urbanisms in one city: accommodating the paces of change. J Environ Prot 6:946–956. https://doi.org/10.4236/jep.2015.69084

Roggema R (2017) The future of sustainable urbanism: society-based, complexity-led, and landscape-driven. Sustainability 9(8):1442

Roggema R (2019) ReciproCity, giving instead of taking. Inaugural lecture. Hanze University of Applied Sciences, Groningen

Ryan C (2013) Eco-acupuncture: designing and facilitating pathways for urban transformation, for a resilient low-carbon future. J Clean Prod 50:189–199

Takebu Good Governance (2021) "Democracy" から Eco-cracy (Eco-cracy) へ. Environ Symb 37(1):70–76 (original: 建部好治. (2021) 「民主主義」から生態系主主義 (Eco-cracy) へ. 環境共生, 37(1):70–76)

Tanasescu M (2017) When a river is a person: From Ecuador to New Zealand, nature gets its day in court. The Conversation. https://theconversation.com/when-a-river-is-a-person-from-ecuador-to-new-zealand-nature-gets-its-day-in-court-79278

Treviño R, Peña C (2022) 8 reflexiones sobre la arquitectura y el futuro de las ciudades. Conecta. https://conecta.tec.mx/es/noticias/nacional/educacion/8-reflexiones-sobre-la-arquitectura-y-el-futuro-de-las-ciudades.

UN (2015) World urbanization prospects: the 2014 revision, (ST/ESA/SER.A/366). United Nations, Department of Economic and Social Affairs, Population Division, New York

UN DESA (2014) World urbanization prospects. https://www.un.org/development/desa/en/news/population/world-urbanization-prospects.html

UN Habitat (undated) Climate change. https://unhabitat.org/topic/climate-change

WEF (2022) This chart shows the impact rising urbanization will have on the world. Geneva: World Economic Forum. https://www.weforum.org/agenda/2022/04/global-urbanization-material-consumption/

Woo F, Wortmann J, Schurig S, Leidreiter A (2014) Regenerative urban development: a roadmap to the city we need. Hamburg: ©World Future Council

Worldbank (2022) Urban development. https://www.worldbank.org/en/topic/urbandevelopment/overview

Zari MP (2015) Ecosystem services analysis: Mimicking ecosystem services for regenerative urban design. Int J Sustain Built Environ 4(1):145–157

More-Than-Human Connections: Regenerating Through Multispecies Design

David Sánchez Ruano

Abstract We humans are a young species; some argue that we are still adapting to our planet (Thomas 2017). When we consider our fellow species, we can see that they are still evolving, but already know how to cover their needs by regenerating through constant change. Life cycles, rhythms, and the patterns within them represent a true language that we can read and respond to (Illetterati and Michelini 2013). Human communities have been creating disconnection by shaping infrastructure, services, and interactivities, often without thought for the web of life. A fundamental part of regenerative strategies is collaboration with these more-than-human worlds (O'Gorman and Gaynor 2020). Biomimicry, biofabrication, biophilia, conservation and restoration can be used as efforts and tools for regenerative design. Three case studies with multispecies perspectives will highlight and identify co-evolutionary principles and values. These case studies demonstrate our potential to implement regenerative approaches in design education.

Keywords More-than-human · Regenerative design · Multispecies · Patterns · Design education

1 Introduction: Rethinking the Human

The concept of regeneration is now considered an evolving step in our human culture and can be seen beyond sustainability efforts (Reed 2007). Architecture and design disciplines are now being employed with great emphasis to alleviate not only the environmental issues humans have caused, but also to change our social behavior for a better future. In order to explore the concept of regeneration, we must reconsider our posture as species in the so-called anthropocene. If we examine the concepts of restoration, resilience, conservation, etc., we can identify not only a systemic need to survive as species, but also a more-than-human (O'Gorman and Gaynor 2020) recognition of reciprocity. The change of worldview that regeneration necessitates

D. Sánchez Ruano (✉)
School of Architecture, Art and Design, Tecnologico de Monterrey, Guadalajara, Mexico
e-mail: david.sanchezr@tec.mx

may lie in post-humanism, but to access it, we must first recognize our identity as a young species or even a symbiotic agent (Margulis 2000). Other species have worked to inhabit this planet for milennia, which has provided the means for us as humans to born and to thrive. According to Allen (2015) and related theorists of evolutionary biology, hominids developed habitat skills and body features depending on the climate conditions of our planet. Several species have been adapting, hybridizing, and even thriving under the influence of us as humans (Thomas 2017). Regenerative design gives us answers to identify the best way to become resilient once we intervene in the places and spaces of other living systems.

There are three sections to this article. The first offers an overview of the concepts that moved forward the notion of regeneration. It is divided in two subsections: One which explains the urgent need to understand nature dynamics towards regeneration, and one which defines the related disciplines that birthed the regenerative culture. The following section explains how multispecies studies and posthumanism are influencing regenerative design methodologies. The last section explores, in detail, the academic exercises and tools utilized within them which could be very useful in the implementation of regenerative projects. These tools and examples provide an opportunity for those in creative disciplines to transform their teaching approach and create a culture of care for the living world, and facilitate wisdom along the way. As regenerative strategies advance as a practice, there is a need to provide methodological approaches, multidisciplinary efforts and to implement them urgently. In the context of education, we can ignite that regenerative wisdom.

2 Preceding Regeneration

2.1 *Life Cycles, Rhythms and Patterns*

Constant change and fluctuations are part of the dynamics of living systems. Life cycles, rhythms, and the patterns within represent a language we can read, a truth that nature shares for all living things (Illetterati and Michelini 2013). Human communities have been disrupting and disconnecting by creating infrastructure, services, and interactivities sometimes without purpose and lacking consciousness. Seeking regeneration requires the acquisition, or the re-establishment, of the language that nature displays.

When we define life cycles or the cycle of life, we create an image of time. We humans look at temporalities by observing the stars, the seasons through the migration of animals, or the flowering of plants. These observations of natural cycles were represented in ancient expressions of language and survival actions. Our ancestors passed on the knowledge of readings landscape and timescapes (Adam 1998) to respect synchronicity and ecological rhythms. Some expressions, such as the Iroquois' seventh generation or the Kogi's golden thread of time, remain as living examples of following nature's dynamics.

Nowadays the rhythm of our lives is punctuated by a 24-h clock to respond to a productive economic system. Our recent mastery of time has disconnected people from the rhythms of nature (Richardson 2017). Regenerative design strategies respond to cycles, sequences, and successions which redefine planetary ethics so that our hearts, minds, and bodies respond to the Earth's rhythms and requests. This will surely guide us toward good design, architecture, and livable cities that integrate biorhythms. Initiatives such as the slow food (Petrini 2007) or degrowth movements (Demaria et al. 2013) can be seen as ways to stop degeneration culture.

Relating to natural time and cycles entails learning to read the pattern language of the more-than-human world. The more we know about other species, the more we engage with them and change towards regeneration. Those patterns are found in the forms, functions, and processes we translate to our spoken language, algorithms, music, built environments, and technologies. Designing within nature and seeing how our creativity pleases life gives a deep ecological response, and a deep ecological worldview manifests a Gaian Hierarchy (Sánchez Ruano 2016). Metabolizing and attuning to the self-regulating patterns of Gaia is the regenerative prime.

2.2 Rooted Efforts and Tools for Regenerative Design

Sustainability efforts require us to move beyond the perpetuation of current operation conditions and the problems that we face. Focusing on regeneration instead of sustainability opens actionable opportunities for what we can create and modify in our planet. In this way, biomimicry, biophilia, conservation, and restoration can be seen as efforts and tools for regenerative design. Wahl (2016) interprets these disciplines as levels to reach regeneration. In any case, if we are developing projects at any of these levels, it is important to recognize how we can direct them and deepen their effect.

Two paired concepts preceded the idea of regeneration: Conservation and restoration. Efforts in integrating conservation biology and restoration ecology have been made (Wiens and Hobbs 2015) since the beginning of the past century. Both disciplines are intertwined with the idea of regenerative design, along with the notion of preservation and rehabilitation of landscapes and wildlife around the globe. Their origins were born beyond economic and material values (McClintock 2009) and changed to a more ethical posture (Rolston III 2011) which questions human use. These disciplines were usually carried out by government agencies, versus rewilding non-managed areas, which were left for non-human dynamics. For example, in Brazil, several species of forest have been regulated to become productive sources, but conservation targets have been poorly developed (Pacheco et al. 2018). In Mexico, the Yucatan peninsula has been untouched for centuries, but the introduction of new train systems established efforts of conservation and then restoration. With an imminent growth of population, indigenous injustice and disarticulated touristic projects have increased (Camargo and Vázquez-Maguirre 2021). The United States National Park administration transcended frontiers, becoming a model for other countries (Kingsland 2002), but these kinds of natural reserves require constant implementation.

The biophilic level that precedes a regenerative strategy refers to the way in which we value, love, and have nostalgia for nature. The value lays on the ethical dimension. Aldo Leopold said 'we can only be ethical in relation to something we can see, feel, understand, love and otherwise have faith in' (Leopold 1989). The love comes from how close we are to a landscape and how express feelings for it, or for a species of plants or animals, and if we feel this closeness, we feel the need to regenerate what we have lost. Most examples of biophilic projects integrate plants and other elements like water and earthly material, and connect them with those of landscape architecture. Some others related to interior design incorporate outdoor elements such as light, air, plants, and colors which aim for a healthy environment (Söderlund 2019). Forest and horticultural therapies are encouraging people to spend more time outdoors in an attempt to combat stress and even physical illness (Hansen et al. 2017). These incorporations require a practical but sensitive understanding of nature. Biophilic design, then, is a great asset to explore along with regeneration as it touches the spiritual and purposeful act of being constantly close with nature.

Biophilic design intentionally seeks to create interspecies relationships, which increase human well-being. In nature we can find secrets that can inspire us to design and also help us reflect on our human creative gifts and value, this is how biophilia has a close intersection between biomimicry (Malcolm and Sanchez Ruano 2015). Being mentored by nature's genius can help us navigate toward deeper levels of consciousness. Unlocking other species' wisdom requires new methodologies which can guide us to implement regenerative design. Biomimicry can be seen as one of the levels, or steps, towards regeneration, but we certainly can have a deeper focus on studying the biological aspects of it. The aesthetic aspect can be explored in the metaphor, analogy, and emulation of other species, always referring to a purpose, a fruitful function that promotes life and regenerates it.

Many examples of bioarchitecture, biomimetic design and bioinspired engineering display an array of structures, forms, functions, and processes that require a close biological study of a species creating a bond and following the principles of life. Most developments use computational tools and analytic methods with a multidisciplinary approach (Goel et al. 2013). Biomimetic projects such as vehicles, buildings, or objects manifest a living meaning, but through regenerative strategies we can provide a pathway towards a metabolic environment.

Another contemporary aspect approached though regenerative design is the use of biomaterials, or biofabrication, with the assistance of living organisms to heal or clean damaged land or water bodies. Biofabrication connects to the physical aspect of building or designing through the bio-utilization of living matter (Gazit 2016). Synthetic biology, hybridization, transplantation, and related artifices that we humans caused are not a case of anthropogenic isolated solutions, and require ethical responses (Gutmann 2011). For example, creating a garden, restoring a reef by 3D-printing, or even composting or reforesting through the use of technologies expresses the interrelationship between the human and the non-human. Nevertheless, is true intent and reciprocity that is valued.

The development of initiatives beyond conservation, such as "rewilding" (Monbiot 2013) are also gaining attention recently. This notion tries to resist the urge to control nature and allows it to find its own way. For Olson (2012), reviving the wild means "understanding and unlearning our conditioning, the cultural programming that determines how we see and interact with the world." For Higgins (2013), reclaiming the wild is a gift that gives us aesthetic pleasure, relaxation, restoration, tranquility, and an authentic childhood. This return to a wild state, or the "unlearning process," encourages our civilization to become uncivilized, which means truly reconnecting with nature. Rewilding theories and practices (Hawkins et al. 2022) are indeed a synonym of regenerative design strategies, but it might imply a more profound philosophy to leave nature untouched or allow it to take over.

All examples mentioned above represent rooted efforts towards regenerative design strategies. The most widely used, conservation and restoration, requires a strong bond. The same goes for biophilic design, biomimicry, and biofabrication. The methods, tools, and implemented examples following can strengthen the effectiveness of regeneration. A missing link is the understanding of multispecies design and architecture, which is described in the following section.

3 Multispecies Perspectives

3.1 The Empathetic Regeneration

Symbiosis is the mightiest force that promotes evolution. Different from competition, this concept is so powerful that it opens a myriad of possibilities for communion and conviviality with other species and matter. From our cells to stardust, we seek a creative reciprocity. Such synergy generates novelty, as expressed by Margulis (1991). Looking to regenerate, then, requires association with all the living beings, and teaming up with them. If we are willing to create the right conditions of life through a new natural park, ecological corridor, or housing development with hanging gardens, for example, we need to rethink more-than-human relationships.

Teaming up with species of plants requires a finite sensitivity, a deep or neo-naturalistic empathy. Biologists, ecologists, and biotechnologists, along with architects, urbanists, and designers, need the same use of analytic tools Intersections of ethology (Beaujouan et al. 2021). As we consciously establish to design for or with nature, we are required to see other species, with a higher degree of empathy (DiZerega 1995), and to see them not only as individuals but also as groups. Bees or molds, for example, function as a whole, not as individuals. By changing our perspective in this way, we enter into a dialogue in which interspecies relationships need to be understood (Kirksey and Helmreich 2010). In this aspect, environmental humanities have gained attention by bringing the arts and sciences together (Emmett and Nye 2017). The discipline has manifested several ways to generate agency and consciousness, erasing the dichotomy of nature versus humanity.

In his writings, Arne Naess discusses how strengthening a "sense of place" revitalizes the self's internal relationship with the environment (Drengson 2005). With respect to exploring a sense of place, educator Richard Louv identifies a problem: "We can't protect something we don't love, we can't love what we don't know, and we can't know what we don't see, hear, or feel. We have lost the connection with our natural history, being important to recognize where we come from for our conscience" (Louv 2012). This deep ecological worldview can also help us constantly seek a holistic connection with the more-than-human elements.

One of the main issues regarding mass extinction, and of course biodiversity loss, is that it diminishes planetary health. Non-human equity and reciprocity is what is needed (Franks et al. 2020). This can make fundamental shifts beyond human-centeredness, and in consequence, regeneration. Without this worldview, it will be impossible to keep our planet alive. How might we create significant technologies and infrastructures that integrate non-human collaborations? Or, in a way, to serve non-human interest at its best?

3.2 A Post-human Answer

The previous questions can be answered not just by human and some species' efforts, but through multispecies collaborations. Multispecies projects bring reciprocity (Metcalfe 2015), a co-evolutionary factor which must be integrated in the philosophy of regeneration.

Multispecies relationality is always becoming-with (van Dooren et al. 2016). This kind of mutuality not only requires a reconsideration of how we develop design and architecture along with other disciplines, but how we question our human world. Indigenous knowledge, reciprocal ecologies, and ethnographies and ethologies (Lestel et al. 2006) require immersive correlation to understand all the kingdoms of life, including minerals and landscapes. By aiming to study "meanings, interests and affects," as Van Dooren remarks on Lestel, we make it possible to change our naturalistic worldview.

Being immersed in the life of others requires us to become symbiotic and create regenerative narratives of art, design, or architecture. There are many ways to express the words, behaviors, and affections of the non-human: In drawings, poems, performance, art installations, and, of course, interactive objects, buildings or naturalistic landscapes (Kirksey 2014). Tsing (2010) defines this way of inquiry as the art of noticing. Multispecies studies require multidisciplinary studies.

Since Haraway has manifested the phrase "staying with the trouble" (Haraway 2016), the tools we have to build are collaborative, moving from the arts and sciences. As we understand our assemblage in our own bodies (Deleuze and Guattari 2013) and the interrelation with human technologies such as cyborgs and AI (Nath and Manna 2021), we enter into the realms of posthumanism, where the dilemmas of control and transcending together appear. The post-humanism (Forlano 2017) studies are a fundamental partner for regenerative strategies to collaborate with the more-than-human worlds. Cultivating a mutual flourishing, then, will require us to move forward the agency of regeneration.

3.3 Methodological Approaches

The incorporation of the multispecies factor in regenerative design requires various methodological approaches. Depending on the discipline, it is important that the objectives of a project or a brief have a clear multidisciplinary approach to solve more-than.human problems. The nature of regenerative projects refers to methodologies that seek landscape or habitat restoration, user-centered design, experience design, or even social design. Within the methodologies are tools that are complementary or adapted. For example, a persona map or an ethogram are more directed at individuals, while tools that refer to psychometric evaluations or mapping of urban areas are directed to populations or ecosystems. The following section shows examples of tools applied sequentially, mainly using the design-thinking method.

4 Incorporating the More-Than-Human Factor

4.1 Applied Examples

Academic examples shown in this chapter contain features that demonstrate both methodologies and ethical approaches (de la Bellacasa 2017), and these must be shown in every regenerative design project. The following three examples were developed as part of various courses at the School of Architecture, Art and Design at Tecnológico de Monterrey. The design school provided an open canvas to explore possibilities towards sustainability in the locality. The first project focused on design for animals, the second explored biophilic design in a determined district closer to the campus, and the third was created with organizations that develop urban agroecology in the locality.

Each project reached competencies established by the school and helped students develop the sensitivity needed for regeneration. The students implemented several tools which can serve as a guide for teachers and professionals who want to incorporate non-human centered designs into their practices.

4.1.1 Nature-Centered Design

During the last four decades, design disciplines have acquired a new role in solving environmental problems (Papanek 1995). Nevertheless, there is a human-centered instinct to develop better processes, services, experiences, and technologies that sometimes make us extractive, selfish and less human (Norman 2005; Pashia 2023). What if we go beyond human centralism (post-humanism)? What if our center is to design for another species or a natural ecosystem? (Ryn 2013; Sánchez Ruano 2019; Veselova and Gaziulusoy 2022).

There are domesticated species, free species that interact with humans, and species that are in danger of extinction. As described in previous sections, multispecies design requires direct interaction. In this way, user experience design (UXD) today involves an opportunity for innovation and the implementation of interactive technologies to improve the quality of life of the animal and plant species with which we share this world. As expressed by Ruge and Mancini (2022), "animal studies invites us to use tools to try to get into their shells, legs, fins, and claws, because it is only then that we, as fellow human animals, can begin to understand, protect, and positively influence their world".

This first example was implemented at the experience design course conducted for the third-year undergraduate level. This course was originally intended to create human-centered designs with the use of UXD tools. The course was then adapted to implement the idea of nature-centered design, where students referred to an open brief/challenge to choose an animal to design for. A set of animal-centered design (AXD) tools were developed, and the following specific objectives were assessed in order to develop a product to promote interaction with the chosen animal and their context, thus improving their quality of life:

- Identify the need for the species through user analysis tools.
- Integrate aspects of interaction with interfaces and mechanical attributes.
- Consider aesthetics and functionality linked to a rewarding experience.
- Identify the benefits of the various actors and contexts.
- Present functional prototype.

The objectives opened the possibility of not only giving agency to multispecies studies but also to use established tools and design new ones. The products were presented in the following stages:

1. Ethnographic stage

The design brief is given along with an introduction of examples (Fig. 1). The first assignment is to choose the animal, making sure the students have access to observe it directly. The students' own households, local zoos, neighborhood parks, university campuses, or suburban edges are recommended.

A recognition of ethnographic research to identify users, habits and behaviors is part of the first steps of design methods (Laurel 2003; Plattner et al. 2013; Stickdorn and Schneider 2012). In this case, biologizing the tools created a scope to develop empathy towards the animal selected.

Tools used:

1. Multispecies design cards: Metcalfe (2015) developed a set of tools to "facilitate the recognition of animals as clients, identify human-animal interactions

Fig. 1 Refigio. Set for lonely bees. Credit: Maliarts and Nesting brick for Swifts Credit. RSPB UK

as designed experiences and the view of man-made systems as further extensions of ecological systems," resulting in this set of cards. Using them during a session, answering the questions, and reflecting on the statements provides a high sensitivity towards the chosen animals, helping users become emphatic.

2. Photo Safari: Photo safaris are tasks in which users photograph subjects relevant to the project, serving as cultural probes (Gaver et al. 1999). This provides clear information about how the user performs tasks or unknown information about the user's habits. Spending time in the animal's habitat is required. The insights obtained from the pictures can feed the tools in the next stages.
3. Biological Empathy Map: Serving to analyze interviews, notes, documents, pictures, or other means of information, empathy maps help to categorize insights (Gray 2018). In this case, the biological empathy map developed by the author (Sánchez Ruano 2016) is used to obtain taxonomic information of species and relevant stories that confirm the affiliation with the animal as a relative.
4. Animal Persona Template: Persona templates were created to organize all the necessary data about the user including personality, interests, skills, and goals (Goodwin 2009). By adapting this template with the information collected from the animal, the students evaluate the ethical threat and begin to see the animal as a person.

2. Usability (Ethological) analysis stage

User experience (UX) is a discipline that nowadays mainly responds to the digital needs of society (Nunnally and Farkas 2016). Ergonomics, easy interfaces, and pleasant experiences are some of its main features. At this stage, students are encouraged to identify steps, emotions, and responses in order to confirm usability sequences.

Ethology, as described in previous sections, is widely used in multispecies studies. The biological analysis generated from it provides a deep understanding of behaviors including communication, habits, mating, defense, and other related ways of animal individuals or groups. This knowledge of species gives us deep insights into their interactions with individuals or groups, and seeing this in their immediate context helps us to conceptualize better experiences for them.

Tools used:

1. Ethograms: Widely used in zoology to study animals over time, these sheets or tables help to monitor behaviors, health, stability of a group/individuals, and to track social interactions, needs and emotions. This scientific tool can be very detailed and may be used in the wild or in labs (*Observational Ethogram Activity*, n.d.). The instruction for this tool was to identify some formats and then customize your own.
2. Scenario Templates: This tool helps to previsualize how the users will perform by interacting with the new product or space. The components or sequences created affect the overall behavior. Usually, it is illustrated with pictures and storyboards (Kumar 2012). Thinking about what the chosen animal experiences and how we want to help them is represented in the story we tell.
3. User Journey Map: Customer Journey maps are widely used in marketing and service design (Kalbach 2016). They are visual representations or timelines that map actions, touchpoints, and the pains and gains we experience when we interact with spaces, objects, information, and people. By using this map, we can manifest what we observed in the ethograms, persona, and scenarios, creating a good sequence of the experience we want to generate. This tool can be used before and during the conceptualization and after interaction with the prototype.

3. Definition of experiences and conceptualization stage

Seeking to solve the needs of the animal species by conceptualizing not objects, but good experiences, is the aim of this creative phase. After using the previous tool, the students were ready to draw ideas and create digital renders and rapid models.

Tools used:

1. 2D sketching and 3D sketching: Drawing is a fundamental tool of conceptualization. Digitally or otherwise, the forms, functions, and experiences represented in a product or space need to be shown. Here, we recommended drawing the animal's context with some concepts.
2. Mock-ups: Creating volume with the right dimensions is needed to observe the aesthetic, functionality, and configuration of the concept. Some mock-ups or models can be 3D printed or made with representative materials. Creating a 1:1 scale is best, if possible, but they can also be done in another.

4. Functional and mechanical verification stage

The requirement to test the prototype in the animal context is a fundamental one. A latent improvement is part of this assignment by, for example, modifying dimensions, changing materials, or adding new experiential features.

Tools used:

1. Prototyping: Aimed to test, prototypes are an essential part of the design process. Fidelity of materials, size, colors, and ergonomics are required. Use by the animal for a few days is recommended.
2. New ethogram: The same ethogram seen in the usability stage may be implemented. New observations can be recorded as positive or negative as the animal uses the prototype.
3. New User Journey Map. Another iteration of the User Journey Map is suggested. In this way, new amendments can be added to generate a better experience. A second or more prototypes may be tested.

5. Presentation stage

Along with the physical prototype, photographs of models and renders were displayed. Here it was required to make a display with the animal figure (3D or 2D) and some of the project's features, such as the food (nuts, grass, or apple) or part of the context, such as a tree trunk, lawn, or home space.

In order to identify the research phases and tools used, see Table 1.

Some of the most relevant products developed during this course were:

- Device to alleviate anxiety for horses.
- Snack feeder for pigs.
- Hanging nesting skeleton for urban birds.
- Nut dispenser for park squirrels.
- Safe reproductive housing for golden blue macaws (See Fig. 2).
- Playful feeder for ringtail lemurs (See Fig. 3).
- Toy for lonely zoo gorillas.
- Bamboo nesting for inseparable parakeets.
- Safe hiding space for house rabbits.

Table 1 Nature-centred research tools

Nature-centred design phase	Tools
1. Ethnographic	1. Multispecies design cards
	2. Photo safari
	3. Biological emphaty map
	4. Animal persona template
2. Usability (ethological)	1. Ethograms
	2. Scenario templates
	3. User journey maps
3. Ideation	1. 2D and 3D sketching
	2. Mock-ups
4. Prototyping	1. Prototypes
	2. New ethogram
	3. New user journey map
5. Communication	1. Presenting display

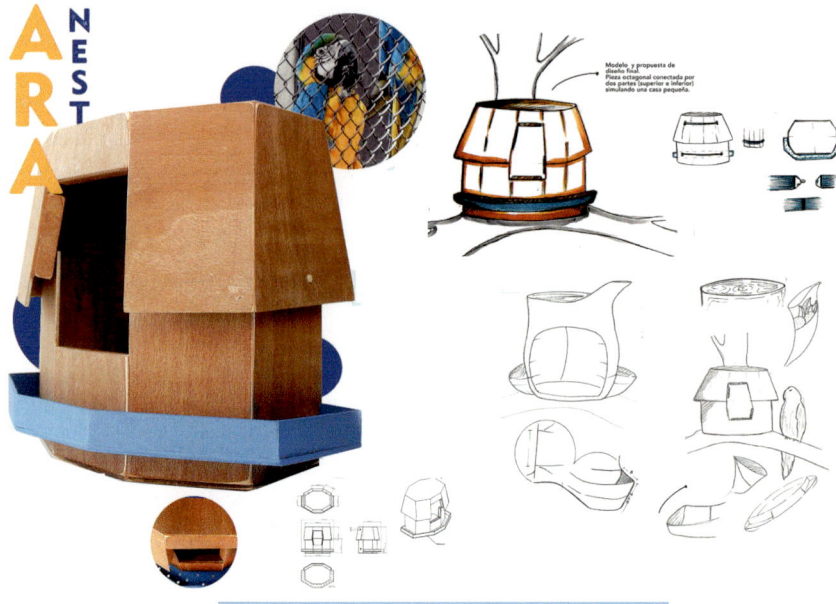

Fig. 2 Safe reproductive housing for golden blue macaws (in Spanish). Credit: Dulce Chavez

Fig. 3 Playful feeder for ringtail lemurs (in Spanish). Credit. Lourdes Rubio

- Hanging recycled feeder for hummingbirds.

The students' final submissions reached the following competencies:

1. Communicate a final design proposal through two and/or three-dimensional representations with professional quality.
2. Systematically expose the design process and tools used.
3. Identify innovation attributes and socio-environmental achievements.
4. Integrate user-centered design elements and user experiences effectively and clearly.

Some general observations can be taken into consideration by creating Animal Centred Design. The most important was that the tools generated through this academic exercise can serve as a continuation and basis for designers, architects, and urbanists in the definition of multispecies design. Another aspect is the urgent need to establish this kind of interdisciplinary academic exercise so it can become part of future professional practice and be made more accessible to the general public. Ultimately, we must recognize that the creation of spaces and facilities for humans with the consideration of other-than-human beings can be an easy step towards regeneration.

4.1.2 The Biophilic District

During the last decades, efforts to create sustainability have focused on raising the well-being and health of citizens. The value that nature offers within urban areas has been key in uncertain times, such as Covid-19 pandemic (Kolandai et al. 2023). It is proven that having access to open natural spaces and interactions with them improves our health, both physical and mental (Rugel et al. 2019). Biophilia refers to the existence of an innate biological connection between human beings and the outside world; an innate and hereditary human inclination towards natural processes and systems (Wilson 1984) There is an imminent need for natural contact that generates emotion, meaning, compassion, and beauty, and provides connection with the living (Lumber et al. 2017). As individuals and as a collective, we seek the transcendence to rediscover nature, and thus to be part of it in everything around us. This attachment has been biologically codified because it is a tool with the ability to improve the physical, emotional, and intellectual aptitude of the human being during their development as a species (Kellert 2012).

Ecopsychological strategies can help us regenerate our relationship with the living world. When our psyche recovers the sensitivity to appreciate nature, we can in turn improve the psyche of nature (Roszak 2001). When we recognize the aesthetic, emotional attractions, and reverential qualities that nature and all its complexities have, we feel the need to reconnect with nature. For example, Pearce et al. (2018) identify three main pathways provided by the natural environment that contribute to mental health benefits: (i) directly through the restorative effect of nature, and then in two indirect ways: (ii) providing opportunities for positive social contact;

and (iii) providing opportunities for physical activity. Much evidence of these benefits emphasizes the "restorative" effect (Kaplan and Kaplan 1989; Ulrich 1981), meaning the reduction of stress and anxiety and improvements in mood, well-being, concentration, and attention.

Nature itself guides us to look for a biophilic design today. The phenomenon of creating urban centers, hospitals, and spaces to grow food and to exercise reveals a tendency to create nurturing and healing spaces (Beatley 2014). The incorporation of natural elements that benefit physical and emotional health begin to permeate throughout our neighborhoods (Cobreros et al. 2021). Features such as outdoor views, natural light, nature patterns, spatial diversity, fractals, plant patterns, or simulations are designed to be restorers of health (Ebrahimpour 2020; Ortega-Andeane et al. 2019). The phenomenon of biophilic cities is expanding, and represents great opportunities to create a regenerative culture.

This second example was implemented at Design Project IV, a course conducted for the fourth-year undergraduate level. The intention of the course was to sensitize the students towards special populations and for them to commit to the community, incorporate functional systems, and address issues related to sustainable development in social aspects.

The brief was named "Biophilic Design: Living Cities" to gain input for the research project "Biophilic Design Strategies for a Regenerative Urban Development," part of Observatorio de Ciudades initiative (Observatorio de Ciudades n.d.), aimed to search innovative approaches and biophilic strategies for urban areas based on interdisciplinary and inter-institutional efforts. Here, industrial design, architecture, urban planning, and psychology joined forces. The project was focused on the urban area influenced by Tecnológico de Monterrey, Campus Guadalajara, and Campus Querétaro in Mexico, as a pilot project. This second exercise was adapted to incorporate a research project focusing on the idea of biophilic design, and to develop urban equipment. In this case, a given space within the city was explored with a series of psychometric, geographic, and design tools.

The following specific objectives were assessed in order to develop a product, installation, or space which would ignite the creation and expansion of a biophilic district project:

Course objectives:

- Be sensitive to special populations and have the ability to negotiate and compromise with public and private institutions.
- Be aware and sensitive of the implications and requirements necessary for special groups responding to universal and inclusive design.
- Apply advanced knowledge of manufacturing and materials such as molded and composite plastics, among others.
- Apply techniques to carry out the complete and formal compendium of the document and graphic/descriptive memory with an entire ergonomic and usability study.

Research project objectives:

- Evaluate physical environments and psychological aspects from the biophilic point of view by generating a tool and adapting metrics, which can be tested and replicated both in other areas of the city and other cities in Mexico.
- Develop strategies and strategic actions that guide the possibility of including these communities within the Biophilic Cities Network (*Cities in the Network*, n.d.).
- Generate a vision and master plan for a biological community.
- Incorporate the biophilic aspect in the initiative of the current Tec District and other possible districts in the future.
- Reaching conclusions on the use of analysis and evaluation tools, as well as on the biophilic design approach for regenerative urban development.

Connecting the objectives opened the possibility not only to develop biophilic designs but also to use established tools and design new ones for the research project. This project opened windows to generate biophilic strategies that resulted further in a regenerative strategy for the district and its inhabitants. The products developed were presented the following phases:

Phase 1. Global Context, Biophilia and Regenerative Cultures

By approaching the notion of the anthropocene (Weber 2019), Gaia theory (Lovelock 1979), and related deep ecological concepts (Sessions 1995), biophilia and regeneration are contained and represent a change of worldview in the student and any person who is willing to create agency or communion with the environment. In this way, studying the roots of these concepts and their authors represents one of the prime exercises for creating reconnection with nature, along with immersive experiences, to have a sense of place.

Tools:

1. Readings: Understanding biophilic design and regeneration requires a theoretical background. Identifying biophilic patterns and the psychological aspect of nature connections also requires a deeper understanding of our close relationship with the more-than-human world.
2. Immersions: Photographic tour through Streetview or in person through the area under study. The aim is to observe and appreciate the history of the place, the essence of the place, and the spatial characteristics of the place at the height of the eyes of the passerby. Each team chooses a number of images sufficient to capture the essence of the place in their assigned sub-zone.

Phase 2: Mapping the biophilic district

Within the ethnographic aspects to analyze within population context, it is necessary to define the right places to intervene in the city. This process is defined here as urban acupuncture (Lerner 2016), requiring an analysis of various points within the city that have the potential to be regenerated but also to be attractive elements for users.

There are three fundamental tools to gather information to precisely develop a biophilic regenerative project. These are geostatistical tools, identification of biophilic patterns, and psychometric reviews. Using them systematically will result in maps and data that can secure the location of interventions, reaching more people and ecosystems.

Tools:

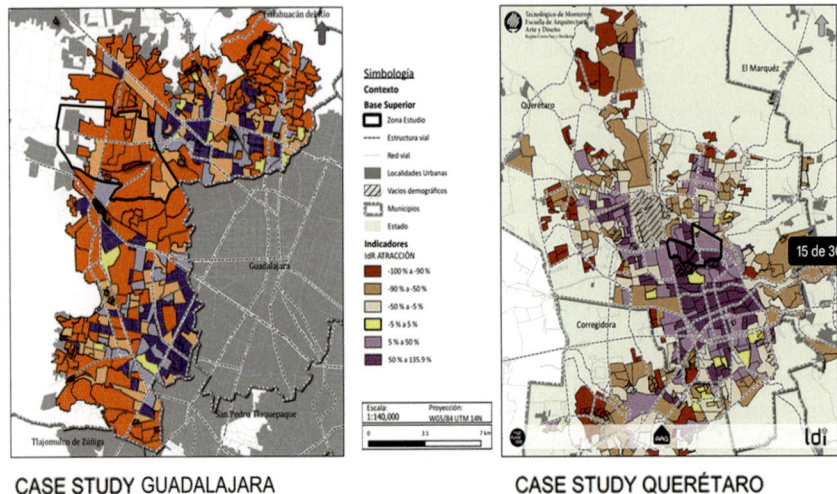

Fig. 4 Example of using AGEBs in different cities

1. Geostatistical tools: Geographical Information Systems (GIS) provide us with elements to obtain statistics and data, most of the time contained in various databases studied from a state, municipality, or locality. This data can be filtered, correlated, and projected according to a given specific area. In the case of a district, a polygon is determined, which contains, among other data, the number of inhabitants, their ages, sexes, and socioeconomic levels (see Fig. 4).

    For this, a spatial analysis is carried out on a metropolitan scale, at the level of AGEB (Basic Geostatistical Area). Each indicator element generates an average value and a cartogram that represents the different percentages of closeness to the average value. The physical and behavioral conditions of each AGEB help to assess the presence of natural elements. The district, or polygon, analysis validates the importance of articulating georeferenced data with behavioral results based on population densities associated with specific activities, and helps identify where to locate the urban equipment or installations. We divide this analysis into two sections:

    a. AGEB demography: Deals with the urban, spatial, socio-economic, and demographic analysis from the blocks of the subzone in order to identify current demographic conditions.
    b. Subzone elements: According to the physical aspects, the number of parks, services, constructions, etc. must be determined and connected with the following psychometric study in order to obtain finer data.

2. Identification of Biophilic Design Patterns. In order to identify the current conditions of the intervention area, 14 elements or patterns are analyzed (Ryan et al. 2014) to identify current conditions and the emphasis on solving the district

problems. This part should be concluded with observations and highlighted insights should always be in relation to the central theme of the course: "Biophilic Design. Living Cities". The identification of biophilic patterns leads to better urban planning processes of buildings, cities, installations, and interior designs which can support healthy and sustainable communities. The biophilic design patterns present three categories:

- *Nature in Space*, which implies a direct experience with nature in seven patterns: The visual connection with nature, the non-visual connection with nature, non-rhythmic sensory stimuli, thermal convergence and air flow, the presence of water, dynamic and diffuse light, and direct connection with natural systems.
- *Natural analogies* allow direct reinterpretation or generation of experiences with nature through forms and biomorphic patterns, connection material with nature, and complexity and order from analogous natural structures identified in the environment.
- *Experiences from space and place*, such as the perspective on the environment, the sensation of refuge and mystery, or the sensation of risk or danger (Terrapin Bright Green 2014).

The following activities must be assigned:

a. Tracking patterns: Using Google Earth and Streetview, the most notable natural and urban elements in the sub-area of work will be identified. So willtheir potential and risks, which must be exposed.
b. Defined biophilic patterns: The degree of biophilia is evaluated and will be represented on a plan with views from Streetview, highlighting representative sections of the main streets and public spaces. It is advisable to use the images/photographs taken during the immersive activity. The identification of patterns, both in plan and from Streetview, must be done in color-code.
c. Worksheets: Integrated references in the analyzed path as well as images and a categorization of biophilic design patterns.

3. Psychometric Reviews: Psychometric concepts were implemented to read and validate biophilic aspects of the exercises. The following reviews and data collection based on Cobreros et al. (2023) should be performed to identify biophilic activities:

a. Access/connectivity with nature scales: The study of the perception, attitudes and connection between the human being and nature has been considered crucial because the future of the environment depends, to a large extent, on human decisions and, of course, on regeneration. Diverse scales must be developed to determine and understand feelings, ideas, beliefs, and, in general, the psychological relationship with the natural world. The scales can be classified according to whether or not they measure general constructs, such as "nature," and specific scales have focused on attitudes towards nature or specific nature issues. Further classifications respond to ethnographies

and demographics and whether they are intended to be applied to adults or specifically to children.
  b. Psychological well-being: The analytical cross-sectional design can be done using a Google form to be shared via email to participants with the following sections: The Scale of Connectivity with Nature (1), 14 items with Likert-type response options, the mood rating scale (2), 16 items on a 5-point Likert-type scale, the Satisfaction with Life scale (3), 5-item instrument; the Perceived Stress Scale (4), 14 items on a 5-point Likert-type scale to assess the level of stress posed by different life situations, and the World Health Organization Quality of Life Scale in its brief version WHOQOL-BREF (5) with 26 items.

  The most significant results shown in the study indicate that there is a significant relationship between connectivity with positive nature and feelings of joy. In addition, this affects quality of life in general, as well as specific dimensions of physical health, psychological health, and social relationships. Feelings such as sadness/depression, anxiety, and anger/hostility might be found. A psychometric tool can determine deviations, focus, or a lack of access to nature.

Phase 3. Biophilic experiences and living spaces

Developing the interventions requires the previous data analysis. As the design process advances, the student focuses their creativity into the development of design concepts. Most of these concepts require feedback to track whether or not the proposals integrate the biophilic and regenerative features. Maps, plans, and main psychometric insights must be collected before proceeding with sketches, mock-ups, and prototypes.

Tools used:

1. 2D sketching and 3D sketching: Drawing is a fundamental tool for conceptualization. Digitally or otherwise, the forms, functions, and biophilic experiences must be represented in the interventions done on the spot selected.
2. Mock-ups: Creating volume with right dimensions is needed to observe the aesthetic, functionality, and configuration of the concept. Some mock-ups or models can be 3D-printed or made with representative materials. Creating 1:1 scale is best, if possible, but they can also be done in another.
3. Prototyping: 1:1 street-sized prototypes can be difficult to develop, so1:4 scale prototypes may be created instead. Fidelity of materials, colors, and biophilic features isrequired (trees, plants, etc.). Human scale needs to be represented.

These are some examples developed by students during the project:

– Circle of natural coexistence, a ground-level bench (see Fig. 5).
– Street refuge to keep trees alive and people covered from climate conditions.
– Agroecological units to have contemplative and productive spaces (see Fig. 6).
– Skyline pavilion to manifest contemplation and freedom.
– Street planters to keep streets fresh.
– Fresh bus-stop bench to keep plants and people protected.
– Secure park hammocks to enjoy the trees.

Fig. 5 Circle of natural coexistence (in Spanish). Credit: Daniela Reyes and Alejandra Ramirez

Fig. 6 Agroecological Circle (in Spanish). Credit: Daniel Jiménez and Daniela Urteaga

Phase 4. Living City Communication

Throughout the project the students developed a new language to interpret their creations and interactions. They confirmed that sustainability, biophilia, and regeneration are needed. Presenting their projects with all the details, including the biophilic patterns, psychometric data, and geographical maps, provided professional competencies by reaching multidiscipline.

Tools:

1. Biophilic maps: Infographics that integrate collected data with district maps, biophilic patterns identified, and the impact reached in terms of well-being and health.
2. Reflections: A final report with pictures, research questions, main discoveries, and observations can provide a good comprehension of the competencies reached.

In order to identify the research phases and tools used, see Table 2.

The students became competent in the following skills:

- Ability to solve social needs based on knowledge of basic technology and mechanisms.
- Identifying tools to design for communities or special populations.
- Ability to analyze data from various disciplines with a comprehensive structure.
- Communicating the design of the project and composing/developing quality products.

This research represents a response to our need for healthy, living, and dynamic spaces. The comfort, strength, calm, happiness, and wonder that nature provides develops our need to pursue that kind of biophilic response in everything we build. This kind of systematic approach focuses on a context that can be regenerated. The data collected can show us the way in which inhabitants are predisposed to biophilic patterns, and those psychological aspects define a transdisciplinary perspective. Identifying these components creates areas of opportunity for construction and actions

Table 2 Biophilic/regenerative research tools

Biophilic/regenerative phase	Tools
1. Contextualization	1. Readings
	2. Immersions
2. Mapping	1. Geostatistics
	2. Biophilic patterns identification
	3. Psychometric reviews
3. Experience ideation	1. 2D and 3D sketching
	2. Mock-ups
	3. Prototypes
4. Communication	1. Biophilic maps
	2. Reflections

that change the dynamic of the city and its infrastructure towards healthier, more inclusive, and sustainable environments. All of this will change human behavior. Connectivity and access to nature reflect a very important aspect in the structure of cities, where the direct connection with nature is not only associated with positive and negative emotions, but also represents the means for manufacturing new experiences with positive impacts. These benefits lead us to embrace biophilic approaches to our cities for regenerative urban development.

4.1.3 Urban Agroecologies

Regeneration is about social action. The participation of people, academia, governments, and corporations in local initiatives to manifest wellbeing and change is vital (Smith and Iversen 2018). As the culture of sustainability and its goals blends with regenerative initiatives, together they will bring value in users, environments, collectives, social enterprises, services, and related systems in order to bring prosperity. Key issues in understanding nurturing living systems are food security (Rhodes 2014) and ecoliteracy (Kahn 2010). We seek less degradation, less depletion of materials, and more social equity (Wever and Vogtländer 2021). Nowadays, food production in and around cities is unseen, part of the urban fabric. Urban and peri-urban horticulture (Brinkley 2012) play an important role in diversifying urban diets and providing environmental services. They can also function as therapeutic (Harris 2017), educational, and social tools. Organic farming, food services, and allotments among the city have emerged recently as megatrends (Davies and Bowman 2016).

Today, the pursuit of the Sustainable Development Goals (SDG) (UN 2016), the increase of vegetarianism and veganism as lifestyles (Jallinoja et al. 2018), and the awareness of reducing waste (Hollander et al. 2017) represent ways of responsible consumption and circular production. These aspects have become vital to generate resilience in an age of inequality and rapid depletion of resources. The SDG 12 seeks to promote green industries, efficient resource management, clean production, energy efficiency in industry, waste reduction and pollution, and environmental sustainability policies in industrial production and consumption as well as the circular economy approach.

Creating urban agro-ecological food systems through co-designing responds to a call to engage sensibly and responsibly with local and global efforts (Berthet et al. 2016; Duru et al. 2015). There is a need for transitions in our current food systems. There is no one-size-fits-all approach to food systems (re)design, and this requires a variety of actions (Parker Krieg and Toivanen 2021). For example, it is necessary to know of plant and fungi species' wisdom and their direct relationship with the soil.

The third example interrelates the notion of social regeneration within the city, specifically configuring projects in terms of agroecologies and social interactions with urban gardens. These represented the need for a reciprocal relationship not just with the soil but also with urban flora, fauna, and fungi.

The research context was the course "Design for Sustainability and Social Innovation." This module is offered for the students for the last semester of their undergraduate studies in Industrial Design at the School of Architecture, Art and Design, campus Guadalajara. The main objective of the course is to understand and visualize co-design as an integrative tool for generating changes in the transition towards a most sustainable society, emphasizing the sociocultural component.

The brief given to the students was to co-design strategies with peri-urban agroecological initiatives. Each initiative presented a series of challenges related to the SDG 12 Sustainable production and consumption. The expected output was to prototype products, services, or systems that generate sustainable behavior towards agroecological practices. The three local initiatives were:

(1) Parque Agroecologico Zapopan, a decentralized agro-educational urban space supported by the municipality of Zapopan, Jalisco. This park reaches both the poor and wealthy neighborhoods by bringing courses, allotment spaces, and events to people of all ages.
(2) Ecocentro Providencia, a greenhouse/green space in the city center created in the back of an abandoned municipal administrative office. The space belongs to a federal organization, but depends on the municipal administration through Health Ministry of Mexico (Secretaria de Salud).
(3) Entreflores Creative Community, a private peri-urban horticulture space focused on the production of vegetables and edible flowers for restaurants. They also offer educational visits, courses, and volunteer experiences.

The following specific objectives were assessed in order to co-design a product or services to deliver agroecological actions:

- Develop critical thinking and reflection skills to actively participate in the co-creation and co-implementation of strategies with high social impact.
- Use collaborative methodologies that involve the participation of different actors of a system throughout the SDGs
- Develop systematic evaluations of the environmental consequences associated with the products, services or systems developed

The methodology to develop the product implied a mixed-method scheme for a co-design process. It aimed to deliver both social innovation and sustainable practice, resulting in the following stages:

1. Context-immersion stage.

    An ethnographic visit to the site of the initiative was conducted before the students were sorted into groups. Once in groups, each initiative presented their objectives, showed their facilities, and directed a conversation.

    Tools used:

    1. Improvised Conversations: This ethnographic tool creates a more naturalistic data collection (Swain and King 2022). As the questions are developed with a

topic in mind, they bring forward the concept of study, the output of insights, and identification of needs, which will later be recorded.

2. SDG's and agroecological goals stage.

   On this visit, the students planned to discuss the #12 SDG goal and the impact on the city, the initiative mission, and the value of design.

   Tools used:

   1. Bibliographic research: State-of-the-art materials on the topic (agroecology, anthropocene issues, SDGs, and regeneration) were assigned to students to deepen their understanding and help them gather information for future interviews.
   2. Interviews: After forming teams, a second ethnographic visit was planned, further structuring directed questions. The aim was to obtain a list of needs and insights.

3. Co-Design stage.

   Back in the classroom, each group analyzed their materials in order to begin generating ideas to solve the problems detected.

   Tools used:

   1. Idea development: Using the insights gathered and sometimes the needs expressed by the agroecological, the groups developed a series of proposals.
   2. Participatory development of ideas: The students conducted a visit to the site to present their ideas and upgrade them with the opinion of volunteers and leaders. The aim was to identify the viability of the ideas and to teach the participatory ingredient required in co-design.
   3. Prototyping: 1:1 street-sized prototype can be difficult to develop, so 1:4 scale prototype may be created instead. Fidelity of materials, colors, and biophilic features is required (trees, plants, etc.). Human scale needs to be represented.

4. Initiatives Review stage.

   A final presentation to demonstrate the usability and viability was conducted and documented. The aim was to identify any modifications needed. The representative team of each agroecological initiative provided feedback.

   In order to identify the research phases and tools used, see Table 3.

This resulted in the following projects:

- Living expositive educational unit: Mobile boards to promote urban agroecology.
- Edible flowers transportation system: A box to maintain fresh edible flowers.
- Flower-recycling community service: Initiative to recycle dry flowers and convert them into pigments.
- Rotary hydroponics rack: A vertical horticulture system to save space inside greenhouses (see Fig. 7).
- Solar dehydrator based on the principle of a hot-air chimney (see Fig. 8).
- Plant-growing didactic toolkit: Helping devices to plant a seed with an effective growing distance.

Table 3 Urban agroecology research tools

Urban agroecology phase	Tools
1. Context immersions	1. Improvised conversations
2. SDGs and agroecological goals	1. Literature review
	2. Interviews
3. Co-design	1. Idea development
	2. Participatory development
	3. Prototypes
4. Review	1. Initiatives colective Review

Fig. 7 Rotary hydroponics rack (in Spanish). Credit Mia Rodriguez

Fig. 8 Solar dehydrator. Credit: Regina Arellano (Sánchez Ruano et al. 2019)

The students became competent in the following skills:

- Understanding and visualizing co-design as an integrating tool regeneration.
- Implementating collaborative methodologies that contemplate interdisciplinary work as a trigger for social innovation.
- Applying systems-based thinking in order to understand and address global issues.

The benefits that agroecology brings to the urban populations are vital. When governments promote policies to help communities, groups, and institutions implement social aids for food security, the design academy becomes an important player to co-design solutions with stakeholders. Social design interventions then promote sustainable consumption and production. Through these types of projects we can conclude that the work is easy when the efforts of academia, local initiatives, citizens, and SDGs come together. The sensibilization towards urban and peri-urban agroecology becomes essential to maintain resilience, well-being, and social collaboration, which are part of the regenerative culture we humans have with soil and plants as living symbiotic ecosystems.

5 Conclusion: The Symbiotic Regeneration

Through this work we observe that the inclusion of regenerative design methodologies in educational practices is capable of transforming knowledge and human agency, and the benefit transcends boundaries. The examples displayed develop multispecies efforts, vital ideas which need to be highlighted in the epistemology of regenerative design. If we are willing to co-create with the more-than-human world, the ethic and aesthetic of regeneration need to follow a symbiotic dimension, where the human understands other lifeforms and contributes to the web of life.

The use of analytic tools to develop multispecies projects, as presented through this chapter, serves as an incentive for seeking new methods for regenerative design. Incorporating these dimensions will enhance agency beyond the social and economic aspects of regenerative design. In this case, education can be seen as the base upon which we can build other systems.

The regenerative approach is evolving and covers an increasingly broad spectrum of sustainability concerns, extending beyond the traditional aspects of design to address the nature of thinking and interactivity required to design and engage in a multispecies realm. Despite the growing interest, the ecological transition to regenerative practice brings some challenges. We are beginning to reconcile the two worldviews of human and non-human ecology on small scales and in scattered places, so this work is not yet widely recognized as part of a broader evolution. What is needed now is to bring awareness and intention to these ideas as they emerge as the new pattern. Design educators, then, should then help to redefine an education that resonates with regenerative culture and moves beyond the anthropocene to the symbiocene.

Developing a biophilic and multispecies mind allows us to stimulate our sense of coherence within the world, and encourages the collective and individual unconscious to understand the notions of regeneration, environmental politics, and ecological design. Human-created animal shelters, coral reef restorations, urban agriculture, nature reserves, and even ethnographic reconciliation with tribes are examples of spaces and actions that must be established in society. Intentions like collecting rainwater, growing food locally, or trying zero-waste lifestyles illustrate that we are becoming aware of our context as humans. Facilitating the health and well-being of bees, mangroves, marine creatures, or entire ecosystems, just to name a few, with a closer relationship in search of reciprocity, is the greatest gift we can give to our living planet.

References

Adam PB (1998) Timescapes of modernity: the environment and invisible hazards
Albrecht GA (2019) Earth emotions. In: Earth emotions. Cornell University Press. https://www.degruyter.com/document/doi/10.1515/9781501715242/html
Allen JS (2015) Home: how habitat made us human. Hachette UK
Beatley T (2014) Imagining biophilic cities. Low Carb Cities. https://doi.org/10.4324/9781315766003-15
Beaujouan J, Cromer D, Boivin X (2021) Review: from human–animal relation practice research to the development of the livestock farmer's activity: an ergonomics–applied ethology interaction. Animal 15(12):100395. https://doi.org/10.1016/j.animal.2021.100395
de la Bellacasa MP (2017). Matters of care: speculative ethics in more than human worlds 41
Berthet ETA, Barnaud C, Girard N, Labatut J, Martin G (2016) How to foster agroecological innovations? A comparison of participatory design methods. J Environ Plan Manag 59(2):280–301. https://doi.org/10.1080/09640568.2015.1009627
Brinkley C (2012) Evaluating the benefits of peri-urban agriculture. J Plan Lit 27(3):259–269. https://doi.org/10.1177/0885412211435172
Camargo BA, Vázquez-Maguirre M (2021) Humanism, dignity and indigenous justice: the Mayan Train megaproject, Mexico. J Sustain Tour 29(2–3):372–391. https://doi.org/10.1080/09669582.2020.1758707
Cities in the Network (n.d.) Biophilic cities. https://www.biophiliccities.org/partner-cities. Accessed 28 Oct 2021
Cobreros C, Flores-García M, Biondi S, Maya M, Ontiveros-Ortíz EN (2021) Re-signifying domestic space in times of confinement based on biophilic design. In: Montoya MA, Krstikj A, Rehner J, Lemus-Delgado D (eds) COVID-19 and cities: experiences, responses, and uncertainties. Springer International Publishing, pp 127–146. https://doi.org/10.1007/978-3-030-84134-8_8
Cobreros C, Medoza-Ruvalcaba N, Flores-García M, Roggema R (2023). Improving psychological well-being in urban university districts through biophilic design: two cases in Mexico. Sustainability 15(7), Article 7. https://doi.org/10.3390/su15075703
Davies FT, Bowman JE (2016). Horticulture, food security, and the challenge of feeding the world. https://pubag.nal.usda.gov/catalog/5634560
Deleuze G, Guattari F (2013). A thousand plateaus: capitalism and schizophrenia. Bloomsbury Academic

Demaria F, Schneider F, Sekulova F, Martinez-Alier J (2013) What is degrowth? From an activist slogan to a social movement. Environ Values 22(2):191–215. https://doi.org/10.3197/096327 113X13581561725194

DiZerega G (1995) Empathy, society, nature, and the relational self: deep ecology and liberal modernity. Soc Theory Pract 21(2):239–269

Drengson PA (ed) (2005) An example of a place: Tvergastein. In: The selected works of Arne Naess. Springer Netherlands, pp 2604–2625. http://link.springer.com/chapter/10.1007/978-1-4020-4519-6_116

Duru M, Therond O, Fares M (2015). Designing agroecological transitions; a review. Agron Sustain Dev. https://doi.org/10.1007/s13593-015-0318-x

Ebrahimpour M (2020) Proposing a framework of biophilic design principles in hot and arid climate of Iran by using grounded theory. Civ Environ Eng 16(1):71–78. https://doi.org/10.2478/cee-2020-0008

Emmett RS, Nye DE (2017). The environmental humanities: a critical introduction

Estrategias de diseño biofílico para un desarrollo urbano regenerativo en México (n.d.) Observatorio de Ciudades. https://observatoriodeciudades.mx/ecologia/estrategias-de-diseno-biofilico-para-un-desarrollo-urbano-regenerativo-en-mexico/. Accessed 12 Jan 2023

Forlano L (2017) Posthumanism and design. She Ji: J des Econ Innov 3(1):16–29. https://doi.org/10.1016/j.sheji.2017.08.001

Franks B, Webb C, Gagliano M, Smuts B (2020) Conventional science will not do justice to nonhuman interests: a fresh approach is required. Anim Sentence 4(27). https://doi.org/10.51291/2377-7478.1552

Gaver B, Dunne T, Pacenti E (1999). Design: Cultural probes. Interact: New Vis Hum-Comput Interact 6(1):21–29. https://doi.org/10.1145/291224.291235

Gazit M (2016) Living matter: biomaterials for design and architecture. Thesis, Massachusetts Institute of Technology. https://dspace.mit.edu/handle/1721.1/107310

Goel AK, McAdams DA, Stone RB (eds) (2013) Biologically inspired design: computational methods and tools

Goodwin K (2009) Designing for the digital age: how to create human-centered products and services. Wiley. Wiley.Com. https://www.wiley.com/en-us/Designing+for+the+Digital+Age%3A+How+to+Create+Human+Centered+Products+and+Services-p-9780470229101

Gray D (2018) Updated empathy map canvas. The XPLANE collection. https://medium.com/the-xplane-collection/updated-empathy-map-canvas-46df22df3c8a. Accessed 21 July 2018

Gutmann A (2011) The ethics of synthetic biology: guiding principles for emerging technologies. Hast Cent Rep 41(4):17–22. https://doi.org/10.1002/j.1552-146X.2011.tb00118.x

Hansen MM, Jones R, Tocchini K (2017) Shinrin-Yoku (forest bathing) and nature therapy: a state-of-the-art review. Int J Environ Res Public Health 14(8), Article 8. https://doi.org/10.3390/ijerph14080851

Haraway DJ (2016) Staying with the trouble: making kin in the chthulucene

Harris H (2017) The social dimensions of therapeutic horticulture. Health Soc Care Community 25(4):1328–1336. https://doi.org/10.1111/hsc.12433

Higgins DL (2013) Claim your wildness: and let nature nurture your health and well-being, 1st edn. Vivid Publishing

Hollander MC, Bakker CA, Hultink EJ (2017) Product design in a circular economy: development of a typology of key concepts and terms. J Ind Ecol 21(3):517–525. https://doi.org/10.1111/jiec.12610

Hawkins S, Convery I, Carver S, Beyers R (2022) Routledge handbook of rewilding. Taylor & Francis

Illetterati L, Michelini F (2013). Purposiveness: teleology between nature and mind. Walter de Gruyter

Jallinoja P, Vinnari M, Niva M (2018) Veganism and plant-based eating: analysis of interplay between discursive strategies and lifestyle political consumerism. Oxf Handb Polit Consum 11:157. https://doi.org/10.1093/oxfordhb/9780190629038.013.52

Kahn RV (2010) Critical pedagogy, ecoliteracy, & planetary crisis: the ecopedagogy movement. Peter Lang

Kalbach J (2016) Mapping experiences: a complete guide to creating value through journeys, blueprints, and diagrams

Kaplan R, Kaplan S (1989) The experience of nature: a psychological perspective. CUP Archive

Kellert SR (2012) Birthright: people and nature in the modern world. Yale University Press

Kingsland S (2002) Designing nature reserves: adapting ecology to real-world problems. Endeavour 26(1):9–14. https://doi.org/10.1016/S0160-9327(00)01396-X

Kirksey E (2014). The multispecies salon. Duke University Press

Kirksey SE, Helmreich S (2010) The emergence of multispecies ethnography. Cult Anthropol 25(4):545–576. https://doi.org/10.1111/j.1548-1360.2010.01069.x

Kolandai K, Milne B, McLay J, von Randow M, Lay-Yee R (2023) Anthropause appreciation, biophilia, and ecophilosophical contemplations amidst a global pandemic. J Environ Psychol 85:101943. https://doi.org/10.1016/j.jenvp.2022.101943

Kumar V (2012). 101 design methods: a structured approach for driving innovation in your organization

Laurel B (2003). Design research: methods and perspectives. MIT Press

Leopold A (1989) A sand county almanac, and sketches here and there. Oxford University Press

Lerner, J (2016). Urban Acupuncture.

Lestel D, Brunois F, Gaunet F (2006) Etho-ethnology and ethno-ethology. Soc Sci Inf 45(2):155–177. https://doi.org/10.1177/0539018406063633

Louv R (2012) The nature principle: reconnecting with life in a virtual age: human restoration and the end of nature-deficit disorder (reprint). Algonquin Books of Chapel Hill

Lovelock J (1979) Gaia a new look at life on Earth. Oxford University Press

Lumber R, Richardson M, Sheffield D (2017) Beyond knowing nature: contact, emotion, compassion, meaning, and beauty are pathways to nature connection. PLoS ONE 12(5):e0177186. https://doi.org/10.1371/journal.pone.0177186

Malcolm J, Sanchez Ruano D (2015) Using nature to inspire design values, issues & ethics. In: LearnxDesign 3rd international conference for design education researchers, Chicago, June 2015

Margulis L (1991) Symbiosis as a source of evolutionary innovation: speciation and morphogenesis. MIT Press

Margulis L (2000) Symbiotic planet: a new look at evolution. Basic Books

McClintock J (2009). Nature's kindred spirits: Aldo Leopold, Joseph Wood Krutch, Edward Abbey, Annie Dillard, and Gary Snyder. University of Wisconsin Press. http://site.ebrary.com/lib/alltitles/docDetail.action?docID=10364102

Metcalfe D (2015) Multispecies design. Doctoral, University of the Arts London in collaboration with Falmouth University. http://repository.falmouth.ac.uk/3223/

Monbiot G (2013) Feral: searching for enchantment on the frontiers of rewilding. Allen Lane

Nath R, Manna R (2021) From posthumanism to ethics of artificial intelligence. AI & Soc. https://doi.org/10.1007/s00146-021-01274-1

Norman D (2005) Human-centered design considered harmful. Interactions 12(4):14–19. https://doi.org/10.1145/1070960.1070976

Nunnally B, Farkas D (2016) UX research: practical techniques for designing better products. O'Reilly Media, Inc.

Observational ethogram activity (n.d.) Wisconsin National Primate Research Center. https://primate.wisc.edu/outreach/observational-ethogram-activity/ Accessed 11 Jan 2023

O'Gorman E, Gaynor A (2020) More-than-human histories. Environ Hist 25(4):711–735. https://doi.org/10.1093/envhis/emaa027

Olson M (2012) Unlearn, rewild. New Society Publishers

Ortega-Andeane P, Vázquez OG, García AM (2019) Factores ambientales y estrés en salas de espera: Comparación en cuidadores primarios informales de pacientes con Cáncer. Revista Latinoamericana de Medicina Conductual / Latin American J Behav Med 9(1), Article 1

Pacheco AA, Neves ACO, Fernandes GW (2018) Uneven conservation efforts compromise Brazil to meet the target 11 of convention on biological diversity. Perspect Ecol Conserv 16(1):43–48. https://doi.org/10.1016/j.pecon.2017.12.001

Papanek V (1995) The green imperative: ecology and ethics in design and architecture. Thames & Hudson

Parker Krieg C, Toivanen R (2021) Situating sustainability: a handbook of contexts and concepts. Helsinki University Press. http://0-search.ebscohost.com.biblioteca-ils.tec.mx/login.aspx%3fdirect%3dtrue%26db%3de001mww%26AN%3d3412587%26lang%3des%26site%3deds-live%26scope%3dsite

Pashia M (2023) Human-centered learning may be missing the point. getting smart. https://www.gettingsmart.com/2023/01/02/human-centered-learning-may-be-missing-the-point/. Accessed 2 Jan 2023

Pearce J, Cherrie M, Shortt N, Deary I, Ward Thompson C (2018) Life course of place: a longitudinal study of mental health and place. Trans Inst Br Geogr 43(4):555–572. https://doi.org/10.1111/tran.12246

Petrini C (2007) Slow food nation: why our food should be good, clean, and fair. Rizzoli Ex Libris

Plattner H, Meinel C, Leifer L (eds) (2013) Design thinking: understand - improve – apply, 2011 edn. Springer

Reed B (2007) Shifting from 'sustainability' to regeneration. Build Res & Inf 35(6):674–680. https://doi.org/10.1080/09613210701475753

Rhodes CJ (2014) Soil erosion, climate change and global food security: challenges and strategies. Sci Prog 97(2):97–153. https://doi.org/10.3184/003685014X13994567941465

Richardson BJ (2017) Time and environmental law: telling nature's time. Cambridge University Press

Rolston H (2011). A new environmental ethics: the next millennium for life on earth

Roszak T (2001) The voice of the earth: an exploration of ecopsychology. Phanes Press.

Ruge, L, Mancini, C (2022) An ethics toolkit to support animal-centered research and design. Front Vet Sci 9. http://oro.open.ac.uk/85307/

Rugel EJ, Carpiano RM, Henderson SB, Brauer M (2019) Exposure to natural space, sense of community belonging, and adverse mental health outcomes across an urban region. Environ Res 171:365–377. https://doi.org/10.1016/j.envres.2019.01.034

Ryan CO, Browning WD, Clancy JO, Andrews SL, Kallianpurkar NB (2014) Biophilic design patterns: emerging nature-based parameters for health and well-being in the built environment. Int J Archit Res: Archnet-IJAR 8(2):62–76

der Ryn SV (2013) Design for and empathic world: reconnecting people, nature, and self. Island Press

Sánchez Ruano D (2016) Symbiotic design practice—discovery—The University of Dundee Research Portal (n.d.). https://discovery.dundee.ac.uk/en/studentTheses/symbiotic-design-practice. Accessed 29 May 2018

Sánchez Ruano D (2019) Nature-centered design. Exploring the path to design as nature. Des J 22(1):2225–2229. https://doi.org/10.1080/14606925.2019.1595016

Sánchez Ruano D, León R, Flores R (2019) Responses in urban and periurban horticulture: social design interventions to promote sustainable consumption and production. In: Sense and sensibility 2019 conference

Sessions G (ed) (1995) Deep ecology for the twenty-first century. Shambhala Publications Inc.

Smith RC, Iversen OS (2018) Participatory design for sustainable social change. Des Stud 59:9–36. https://doi.org/10.1016/j.destud.2018.05.005

Söderlund J (2019) The emergence of biophilic design. Springer International Publishing. https://doi.org/10.1007/978-3-030-29813-5

Stickdorn M, Schneider J (2012) This is service design thinking: basics, tools, cases, 1st edn. Wiley

Swain J, King B (2022) Using informal conversations in qualitative research. Int J Qual Methods 21:16094069221085056. https://doi.org/10.1177/16094069221085056

Terrapin Bright Green (2014, September 12) 14 patterns of biophilic design. http://www.terrapinbrightgreen.com/reports/14-patterns-of-biophilic-design/

Thomas CD (2017) Inheritors of the earth: how nature is thriving in an age of extinction

Tsing A (2010) Arts of inclusion, or how to love a mushroom. Manoa 22(2):191–203

Ulrich RS (1981) Natural versus urban scenes: some psychophysiological effects. Environ Behav 13(5):523–556. https://doi.org/10.1177/0013916581135001

van Dooren T, Kirksey E, Münster U (2016) Multispecies studies cultivating arts of attentiveness. Environ Hum 8(1):1–23. https://doi.org/10.1215/22011919-3527695

Veselova E, Gaziulusoy İ (2022) Bioinclusive collaborative and participatory design: a conceptual framework and a research agenda. Des Cult 14(2):149–183. https://doi.org/10.1080/17547075.2021.2019455

Wahl DC (2016). Designing regenerative cultures. Triarchy Press Ltd.

Weber A (2019) Enlivenment: toward a poetics for the anthropocene. The MIT Press

Wever R, Vogtländer J (2021) Design for the value of sustainability. In: van den Hoven J, Vermaas PE, van de Poel I (eds) Handbook of ethics, values, and technological design: sources, theory, values and application domains. Springer Netherlands, pp 1–31. https://doi.org/10.1007/978-94-007-6994-6_20-1

Wiens JA, Hobbs RJ (2015) Integrating conservation and restoration in a changing world. Bioscience 65(3):302–312. https://doi.org/10.1093/biosci/biu235

Wilson EO (1984) Biophilia. Harvard University Press. 14 patterns of biophilic design. http://www.terrapinbrightgreen.com/reports/14-patterns-of-biophilic-design/. Accessed 12 Sept 2014

Fostering Regenerative Processes Through Responsible Advanced Design and Circularity

Laura Succini, Erik Ciravegna, and Clara Giardina

Abstract This contribution moves from the convergence of the systemic approach of Advanced Design (AD) with the principles of Responsible Innovation (RI), with the aim of promoting more collaborative and ethical practices in complex scenarios. Responsible Innovation originates from the broadening of the scope of the so-called 'Responsible Research and Innovation' (RRI), established by the European Community. From the overlaps and integrations between the principles of RI and the tools and practices of AD, a new approach to change and re-generation emerges, particularly to address current crises, which we have called Responsible Advanced Design (R-AD). We may relate R-AD deeply to the concept of the so-called Regenerative Sustainability, which takes more account of the particularities of different contexts and brings sustainability to a more holistic view. To include both the three dimensions of Sustainable Development—environmental quality, economic prosperity, and social equity—and key concepts such as People, Planet, Prosperity, Peace, and Partnership, as promoted by the SDGs, circularity and its 9Rs strategies were then considered to be integrated into the R-AD approach. This has led to an extension of the R-AD approach that, therefore, can guide anticipatory, cooperative, and transformative design, which is essential to address current crises from a local and global perspective and to promote regenerative processes in particularly challenging sectors. This revisited approach was thus applied to the analysis of a selection of case studies, both from the Mediterranean region (Italy) and Latin America (Chile), choosing as an exemplary field of study the so-called Packaging System, which is a multidisciplinary and cross-sectoral system involving different actors and interchange relationships for the development of a complex artifact such as packaging.

L. Succini (✉) · E. Ciravegna · C. Giardina
Alma Mater Studiorum, Università di Bologna, Bologna, Italy
e-mail: laura.succini@unibo.it

E. Ciravegna
e-mail: erik.ciravegna@unibo.it

C. Giardina
e-mail: clara.giardina@unibo.it

Keywords Advanced design · Responsible innovation · Collaborative processes · Ethics · Packaging system

1 Responsible Advanced Design as a New Approach to Change and Regeneration

1.1 The Advanced Design Approach

Within the so-called "new galaxy of design" (Maffei 2021, p. 19), in recent years, numerous theoretical reflections have emerged on how design practices and approaches offer different perspectives and possible solutions to respond to current crises, with the intention of sustaining change and regeneration and to activate more inclusive forms of innovation capable of considering both the environment and what is part of it among the stakeholders (Carayannis et al. 2012). Among this multitude of approaches, this paper focuses on Advanced Design (AD), framed as part of Design of Processes and seen as a systems approach that relies on multi-stakeholder collaborative practices that can drive transformative change within the realities in which it operates.

As early as 2009, AD began to raise questions about "the relationship between the evolution of the time factor and design-driven innovation" (Celi 2010, cited in Celaschi et al. 2014, p. 22), bringing to light how the concept of anticipation and future studies (Zamenopoulos and Alexiou 2007; Poli 2010, 2019) can be drivers to support an "advanced design approach" capable of responding to the complexity of the ever-changing innovation dimension. The maturation of this methodology, as emerges from the reference literature (Celi 2015; Formia 2017; Iñiguez Flores et al. 2019; Celaschi et al. 2019) leads it not only to open the vision of industrial design to the concepts of anticipation but to propose itself as a discipline (approach) that intervenes in the complexity of relationships related to the changing context (Celaschi and Celi 2015), and the processes of innovation conditioned by environmental and social factors and lack of resources.

In fact, even for AD, sustainability (of processes, relationships, innovation, etc.) is a point of interest and investigation. Through the factors that characterize it, it intervenes on two levels of the design process: the first aimed at the community, trying to change people's behavior by involving them in the value chain through the application of its principles; the second aimed at the designers themselves, leading them to reflect on what the impact of their work might be in the long term.

Celaschi (2015, 2016) and Iñiguez Flores et al. (2014) have defined attributes and areas of intervention that characterize AD in comparison to other approaches and that support their application in real-world contexts. The analysis of the contributions and experiments carried out in recent years by the Advanced Design Unit

(ADU)[1] at the University of Bologna, and the Latin Network for the Development of Design Processes,[2] has led to delve into AD on a theoretical level, creating a synoptic framework of the elements that characterize it and highlighting the gaps on which we need to intervene to make design approaches more responsible, ethical and collaborative and in line with the Sustainable Development Goals (SDGs) from UN's Agenda 2030[3] and the Green Deal of the European Commission[4] meant to outline interventions to transform policies on energy, climate and circularity.

The above-mentioned synoptic framework considers four integrated and interconnected macro-fields of action (Succini 2022, 2023) and constitutes a system of application for the various contexts in which AD operates. These macro-fields are defined as follows.

1. *Time, Future, and Anticipation.* AD by its very nature is an approach that connects past, present, and future, and inserts the time factor as a field of action (Celaschi et al. 2018) within the design process, seeking to "anticipate change, and to make it accessible through the materialization of processes, paths, artefacts and conjectures" (Celi and Morrison 2019, p. 807). AD is based on the ability to imagine possible, probable, potential futures (Celaschi et al. 2019) and applies to the context reacting to changes through continuous innovation.

2. *Innovation of Collective Sharing Processes.* People, the community, and the territory (productive, institutional, etc.) are the actors and active agents for the development of new design forms capable of responding to current challenges. Therefore, the AD through co-design processes and collective intelligence practices seeks to involve all the actors of the territorial eco-system along the value chain (Celaschi et al. 2014) thinking that "the collective, social and shared construction of the vision is the key to the success" (Celaschi et al. 2014, p. 28) of more sustainable products, services, systems.

3. *Transformative Innovation.* It is based specifically on three principles: (i) adaptivity (Mozota 2006; Iñiguez Flores et al. 2014), through which AD supports change and activates collective forms of learning, and allows design to read critical issues into opportunities through the anticipation of practices; (ii) sustainable transformation, leading designers to reflect on what their responsibilities are in transforming the way individuals and communities act in projects; (iii) cultural change, through iterative and continuous experimentation, creating forms of design that start from the future and lead to changes in the present (Celaschi et al. 2019).

4. *Knowledge Innovation.* The sharing of knowledge and the activation of collective knowledge constitute forms of innovation (Celaschi et al. 2019) characterized by: (i) interdisciplinarity and cross-fertilization between both academic and non-academic competencies; (ii) forms of "Reflection-in action" (Schon 1983) in

---

[1] https://adu.unibo.it/.

[2] https://www.forumdesignprocess.org/dgdw22/past-editions/.

[3] https://sdgs.un.org/goals.

[4] https://commission.europa.eu/strategy-and-policy/priorities-2019-2024/european-green-deal_en.

which design is read as a reflective practice; (iii) "learning by doing" training systems and by the collective (Iñiguez Flores et al. 2014) and co-produced learning processes.

1.2 The Emergence of Responsible Advanced Design

Through an integrated and cooperative approach, AD aims to outline possible futures towards which to move. However, it must integrate with other systems and approaches, not typical of design, to adapt to the current emergencies, such as lack of resources, climate change and technological acceleration, and the demand for systems and forms of innovation that are more responsible, ethical, and capable of involving all players in the territorial ecosystem in a more inclusive manner.

AD has thus opened to the principles and concepts of Responsible Research and Innovation (RRI), formalized in 2011 in the context of the European Community's development policies and included—since 2013—within the Horizon 2020 projects. RRI is an approach that anticipates and assesses the potential implications and expectations of society regarding research and innovation, integrating in the projects some identifying concepts that can be summarized in six pillars: Ethics, Open Access, Gender Equality, Public Engagement, Science Education, and Governance (Bezzi 2021). RRI implies that societal actors (researchers, citizens, policy makers, companies, institutions, agencies, etc.) collaborate throughout the research and innovation processes to better align the results with society's values, needs and expectations European Commission 2013; European Commission and von Schomberg 2011, pp. 8–10; Schomberg 2013).

In addition to the emergence of RRI within European Community policies, the concept of Responsible Innovation (RI) has also been established in academic research (Jakobsen et al. 2019) and experimented in different application contexts, such as education, manufacturing, social, etc.

According to Stilgoe et al. (2013) anticipation, reflexivity, inclusivity, and responsiveness are the four dimensions most frequently used as the framework for further implementation of RI. In their contribution, as well as in subsequent academic research and discussions, it is emphasized that these same dimensions coexist in an integrated way in the process.

Further key dimensions of RI of particular interest for this reflection are Care (Burget et al. 2017), Time, and Transparency. Such dimensions, together with anticipation, reflexivity, inclusiveness, and responsiveness, complete the reference system that aims to enhance forms of RI within the complexities of territories. Taken together, these seven principles, when integrated with AD, can provide the framework for establishing the approach of Responsible Advanced Design (R-AD) (Succini and Ciravegna 2022; Succini 2022).

This systemic inclusion allows the AD approach to incorporate a co-construction of the future strongly characterized by a multidimensional quintuple-helix (Carayannis et al. 2012; Iaione and Nictolis 2016) collaborative system, as well as inclusive and ethical design, production, learning and decision-making processes to innovate products, services, organizations, and processes cooperatively and responsibly. Also, the seven dimensions of RI assumed as drivers of the R-AD approach make AD more permeable to the intrinsic values of the context, more sensitive to assessing the impact of design actions throughout the process (from conception to implementation), and more attentive to considering and listening to a wider sample of stakeholders.

The resultant seven dimensions of the R-AD approach then are:

1. *Anticipation.* AD interprets anticipation as "the exploration of the infinite possibilities open to the imagination (where the design process) contributes to building the future" (Celaschi et al. 2018, p. 131). RI uses it to predict the effects of a given project proposal while also assessing its risks (Long et al. 2020) and surfacing potential impacts to respond with sustainable solutions (Deppeler and Aikens 2020).

   Features: ethical and inclusive public engagement, open access of design practices and the dissemination of knowledge, ethics in the design process and in the involvement of people.

   Tools: to make this process closer to reality and adaptable in the long run, it is necessary to activate co-design practices by developing collective anticipatory imagination and problem solving; it is also necessary to use tools to measure the social, economic, environmental, and cultural impacts not only of products/ services but also of the design process.

2. *Reflexivity.* Within AD, the designer is not seen as a sole authorial figure, but as a professional capable of activating an open design space in which users and community can be heard and can bring their input into the process of conceiving and developing ideas. In RI "responsibility makes reflexivity a public matter" (Wynne 2011, cited in Stilgoe et al. 2013, p. 1571) bringing to light the role of the individual as a social engagement in the project (Long et al. 2020).

   Features: principles of transparency about actions, processes, and results; user involvement through responsible forms of knowledge and learning; peer to peer innovation education; enabling non-stereotypical project thinking.

   Tools: use of tacit knowledge, experiential, interactive and collaborative learning systems, internal to the realities involved or open to the whole community; the development of multi-level formats such as seminars, training days, peer-to-peer meetings or academies guided by design practices can make these systems more permeable and more collaborative.

3. *Inclusion.* It could be meant both as the "continuous involvement of society" (Grimpe et al. 2014, p. 2969) and as an open process capable of accommodating all the groups of actors that are part of the territorial ecosystem. In order not to fall back on an already pre-established information process, it is necessary to activate collaborative forms of actor involvement by working on both micro and macro scales and by bringing out the contribution of both the individual and the community.

Features: encouraging ethical, transparent, and universally accessible forms of co-creation; involving interdisciplinary expertise; respecting diversity and incorporating gender equity policies in the various design, production, communication, and training processes.

Tools: use of co-design practices that follow the responsible-driven characteristics of the model, application of responsible design concepts not only at the product/service conception stage but throughout the value chain.

4. *Responsiveness.* This dimension is interpreted in the sense of the ability to adapt to the territorial capital present in each context and to react fluidly to the sudden changes we constantly experience, thus being able to scale and vary the project according to what we are facing.

    Features: responding actively and ethically to the transformation of the planet and the lack of resources; activating and maintaining a collaborative network between actors.

    Tools: within this dimension, AD is the discipline that through scenario design and prototype development can think of different ways to anticipate problems and is able to find solutions to meet ecological, economic challenges.

5. *Care.* It is "a peculiarly human activity that includes everything we do to maintain, preserve or adjust our world, with the goal of living in it under optimal conditions" (Tronto and Fisher 1990, p. 3), cited in Succini and Ciravegna 2022). This concept relates to caring for self, others, community, and planet (Escobar 2018) and thus caring for the future with the aim of rethinking the way we innovate and the principles and policies to achieve it.

    Features: building new sustainable and inclusive behaviors and processes; and anticipating the needs of others and the planet.

    Tools: the use of design semi-finished products, a typical tool of AD, leads to having a set of possible paths to follow as a basis for building new designs more in line with future demands from past experiences (Celaschi et al. 2019, 2014); in addition, the use of responsible-driven impact monitoring systems can help to keep products alive longer.

6. *Transparency.* It pervades and characterizes many of the other dimensions. The principles of transparency are incorporated within the project actions, starting as early as the preliminary stages (team definition) to the long-term verification of the viability of a project.

    Features: making the project, production, distribution, actions, peculiarities, and limitations visible and accessible. Involving the various actors through open and inclusive dialogue systems. Informing in an ethical manner.

    Tools: the tools are the same as those described in the other dimensions enhanced by principles of transparency.

7. *Time.* Different levels of time and its relationship to different spheres are considered in the actions and tools implemented by R-AD: the time of impact in relation to change, which usually takes place over a long period of time and must be compared with the accelerated or slow pace of what surrounds us (humans, non-humans, environment, etc.); the time of people; the time of a productive process;

the time to make a relationship cohesive; the indefinite time of certain actions; the time of adaptation; the time in which innovation processes can change; etc.

Features: respecting the time of people and the environment; activating forms of collaboration and knowledge between the actors of the inclusive territorial ecosystem; considering time as a project variable (Celaschi 2016; Celaschi et al. 2018, 2019).

Tools: imagination (through scenarios and prototypes that imagine possible, probable, and potential futures).

2 Responsible Advanced Design and Circularity for Regenerative Processes

2.1 Regenerative Sustainability and Circularity as Key Factors for Design Practices

As seen so far, R-AD is an approach that arises from the overlaps and integrations between the principles of RI and the tools and practices of AD. This approach relates deeply to the concept of Regenerative Sustainability (Gibbons 2020; Wahl 2016), which takes more account of the particularities of different contexts and brings sustainability to a more holistic view (Gibbons et al. 2018; Ibrahim and Ahmed 2022). According to Gibbons (2020), "regenerative sustainability focuses on transforming the worldviews, paradigms, and thinking underlying manifested reality and, thus, (un)sustainability" (Gibbons 2020, p. 4).

Regenerative Sustainability is rooted in various design practices and theoretical perspectives that over time have promoted key concepts such as responsibility, care or the virtuous relationship between humans and nature, traceable to authors such as Papanek (1971) or Margolin and Margolin (2002), to name a few. In addition, it has its foundation in design approaches such as Ecological Design (Ryn and Cowan 2007), Regenerative Development and Design (Mang and Haggard 2016; Mang and Reed 2013) or, more specifically, Regenerative Design for Sustainable Development (Lyle 1994). These approaches have allowed Regenerative Sustainability to evolve, leading to the development of a more systemic perspective, capable of working across multiple domains and scales of territories and actors.

Regenerative Sustainability integrates complex systems in a deeper and more balanced way and sees ecosystems as networks of dynamic processes in connection with each other, adapting not only to human needs but also to those of other living things, where the role of community is crucial in developing processes of behavioral change, relationships and coping with ongoing social, economic, and environmental crises. The system of interaction between people, environment and the natural system that fosters a form of cooperative, co-evolutionary relationships to achieve co-benefits (Cole 2012; Mang and Reed 2013; Ibrahim and Ahmed 2022) "encourages all system components to evolve, regenerate, and produce more than they consume to improve both the ecosystem's health and people's quality of life" (Ibrahim and Ahmed 2022, p. 5).

The need to go 'beyond sustainability' that underlies Regenerative Sustainability has fostered the development of several projects and experiments in which regenerative processes have been applied to different fields, including architectural design, urban and regional planning (Zhang et al. 2015). Moreover, it is also of particular interest in the field of production, and in the development of product systems that have a strong impact on the territory in all its complexity, from an environmental, cultural, and social point of view.

Based on the point of view of Jenkin and Pedersen Zari (2009) who see Regenerative Design as a linear system of production flows that is transformed into a cyclic system, it is arguable that the concept of 'regenerative process' in the field of our interest (product system) can be linked to the concept of Circularity.

In this perspective, it is useful to recall the definition of Circular Economy according to the Ellen Macarthur Foundation,[5] which refers to it as a systemic approach to economic development designed to benefit business, society, and the environment: in contrast to the linear economy model, circularity "is restorative and regenerative by design" (Ellen MacArthur Foundation n.d.). Our approach to regenerative design thus draws on this notion of circularity, which aims to iteratively build and rebuild the overall health of the system in which it operates.

In recent years, the concept of circular economy has been investigated, particularly for its operational vision applicable to the world of production that can align with the parameters of sustainable development (Ghisellini et al. 2016; Kirchherr et al. 2017). In defining the principles and practices characterizing and operationalizing the circular economy in different contexts, the so-called 9Rs framework has emerged (European Commission 2008; Zhu et al. 2010; Reh 2013; Sihvonen and Ritola 2015; Buren et al. 2016; Potting et al. 2017). According to a schematization by Kirchherr et al. (Kirchherr et al. 2017, p. 224), adapted from Potting et al. (2017, p. 5), in this framework each "R" refers to a specific circularity strategy:

- *R0 Refuse*. Make product redundant by abandoning its function or by offering the same function with a radically different product.
- *R1 Rethink*. Make product use more intensive (e.g., by sharing product).
- *R2 Reduce*. Increase efficiency in product manufacture or use by consuming fewer natural resources and materials.
- *R3 Reuse*. Reuse by another consumer of discarded product which is still in good condition and fulfils its original function.
- *R4 Repair*. Repair and maintenance of defective product so it can be used with its original function.
- *R5 Refurbish*. Restore an old product and bring it up to date.
- *R6 Remanufacture*. Use parts of discarded product in a new product with the same function.

---

[5] https://ellenmacarthurfoundation.org/.

- *R7 Repurpose.* Use discarded product or its parts in a new product with a different function.
- *R8 Recycle.* Process materials to obtain the same (high grade) or lower (low grade) quality.
- *R9 Recover.* Incineration of material with energy recovery.

The strategies are numbered according to the degree of circularity that can be achieved through them, inversely proportional: from R0, which coincides with maximum circularity (circular economy) to R9, which corresponds to low or no circularity (linear economy). The strategies are also divided into three main categories based on the circular economy goal to be achieved and elaborates possible courses of action to achieve them:

- Smarter product use and manufacture (R0 > R2).
- Extend life span of product and its parts (R3 > R7).
- Useful application of materials (R8 > R9).

The 9Rs framework is of particular interest because more than others it defines strategies for making systems truly circular. From 2008 to 2017, thanks to the work of several authors (Potting et al. 2017; Buren et al. 2016; Kirchherr et al. 2017), it has been possible to model and implement this framework, expanding on the so-called 3Rs—referring to the actions of Reduce, Reuse and Recycle—which dates to the first Earth Day in 1970.

2.2 Integrating Circularity Strategies with Responsible Advanced Design: An Extended Approach

According to Geissdoerfer et al. (2017) and Kirchherr et al. (2017), authors who have dealt with the Circular Economy have not yet equally considered the following factors: the commitment to strengthen and improve societal welfare at the level of social equity; long-term strategies in relation to future generations and thus the time dimension; and the role of consumers and the community within the process.

Based on this premise and what has been discussed so far, to develop a "'holistic view' of environmental quality, economic prosperity, and social equity as three dimensions of Sustainable Development" (WBCSD 2017; Elkington 1997, cited in Kirchherr et al. 2017, p. 227), as well as to respond to the need to include key concepts such as People, Planet, Prosperity, Peace, and Partnership, as encouraged by the SDGs, circularity and its strategies (9Rs) were considered to be integrated into the R-AD approach.[6] This led to the definition of an extended approach of

---

[6] This theoretical investigation is currently undergoing and is also being applied in research projects funded such as "Made in Italy Circolare e Sostenibile" (https://www.mics.tech), particularly within the thematic line "Spoke 1—Digital Advanced Design: technologies, processes, and tools". This project was developed under the Extended Partnership, line 11 of the PNRR (Italy's National Recovery and Resilience Plan).

which the dimensions, as described above, are Anticipation, Reflexivity, Inclusion, Responsiveness, Care, Transparency, Time, and Circularity.

This extended approach can thus be useful for analyzing both existing realities and for guiding more collaborative, ethical, and ecosystem-based practices for regenerative processes, which are essential for addressing current crises from a local and global perspective. Thus, it allows us to analyze case studies or guide design projects through a framework that verifies not only whether they have been applied e.g. recycling, reuse or repair systems from an environmental perspective (material aspects), but also how they have been implemented within a context at the social and cultural level (immaterial aspects), e.g., whether they have adhered to principles of equity, whether they have involved through transparent and inclusive processes the actors in the system, etc.

The extended approach was thus applied to the analysis of a selection of case studies, both from the Mediterranean region (Italy) and Latin America (Chile), choosing as an exemplary field of study the so-called "Packaging System" (Ciravegna 2010, 2017), which is a particularly complex, multidisciplinary, and cross-sectoral product-service system involving several different actors and interchange relationships.

In addition to its peculiar complexity, the choice of this field of application is due to the authors' participation in the activities of the "Packaging Innovation Observatory",[7] which is the result of a research project carried out by the Advanced Design Unit at the University of Bologna. The Observatory was conceived in 2019, then implemented and formalized in 2021, with the aim of monitoring, interpreting, narrating but also acting on innovation processes related to packaging, not only as an object but as a cultural phenomenon, an expression of contemporary complexity. Design, understood as a mediator between the knowledge (Celaschi 2008) of different scientific and technological fields, was the starting point of a process of analysis and design to create a system capable of spreading knowledge and multiplying collaborative projects in the packaging sector in its different declinations, as well as to bring out the trends of responsible innovation in this context. The role of the Observatory is therefore not only research and analysis around packaging to monitor the state of the art, but also storytelling and networking, with the aim of triggering innovation processes.

---

[7] https://adu.unibo.it/osservatoriopack/.

3 Packaging System as an Exemplary Scope in a Trans-Territorial Perspective

3.1 Packaging System as a Complex Sector Coping with Current Crises

Packaging has considerably expanded its functions over time. It is an object of use, which allows to contain the product, protect it, store it, transport it, and at the same time, facilitate its physical interaction with the user. Also, it is a communication device (communication prosthesis), characterized by functions, among others, of appellative, persuasive, informative and prescriptive kind (Bucchetti 2005; Ciravegna 2010, 2017).

Packaging is a complex artifact that, beyond its nature as an object of use, is in effect a driver and accelerator of the social, economic, and environmental phenomena in which it is immersed and can convey socially or environmentally relevant messages and information, affect quality of life and climate, monitor safety, and connect people and businesses.

For its multifaceted nature, the project of a package requires different skills and the intervention of multiple disciplines: for that reason, it should be understood as the result of an integrated set of choices made by a plurality of actors, where each performs a specific role—direct or indirect—in its definition. These actors, together with the relationships that are established throughout the product life cycle, define the so-called "Packaging System" (Ciravegna 2010, 2017), in which design may play an important role in direction and mediation, potentially giving shape through the project synthesis to solutions which are an expression of the confluence of the needs of the different parties involved and the multiple functions of the artefact, linking the communicative dimension to the operational and instrumental aspects, along with the perspective of process and the anticipatory vision of experimenting with materials and production systems.

Within this system, all the actors of the supply chain are involved, from the designer to the producer, from the distribution system to the end consumer: a plurality of subjects operating in an interconnected manner, in a value chain that is referred to as the "Value Web" (Bouwman and Janssen 2010).

If we look at innovation processes—to be understood as the ability to adapt to change—within this system, the most effective ones have indeed proven to be those where quintuple-helix forms of collaboration are activated (Carayannis et al. 2012), where all stakeholders can collaboratively intervene in packaging development from the earliest stages of design, and where the results of experimentation and design practices with a regenerative approach can have an immediate impact on the market, not only in terms of the product or service, but also of a means of communication that can convey responsible and ethical messages, responding to current emergencies (social inequalities, climate change, etc.). Packaging, in fact, is considered by public opinion to be one of the causes of the various crises that plague the planet, crises of an economic, social, environmental and health nature (Ciravegna 2020); moreover, it is a ubiquitous artifact in our daily lives, significantly influencing our needs as a society, community and productive culture (Ciravegna 2010, 2017).

The community of people working in the Packaging System today must be resilient and sustainable, an indispensable but difficult transition for the manufacturing world. The need to embrace change and thus trigger innovation requires a new way of reading the design and production process, the relationships that underlie it and the impacts it generates and must rely on a renewed system of tools and practices to move through this constantly adapting reality. As Carmelo Di Bartolo explains, quoting Renzo Piano in turn, "when you make innovation you have to create the tools to make it" (Parilli 2020).

3.2 Advanced Responsible Design and Circularity as Keys to Changing the Packaging System

The stakeholders in the Packaging System operate in a sector that is highly V.U.C.A.—i.e., with a high degree of Volatility, Uncertainty, Complexity and Ambiguity (Bennis and Nanus 1987)—and highly globalized, but at the same time very specialized and tied to the territory. A significant example in the Mediterranean area is the concentration in Italy of packaging machinery manufacturers in the so-called "Packaging Valley" in Emilia-Romagna region; on the other hand, in Latin America, taking the example of Chile, in addition to the high concentration of packaging manufacturers and processors in the country's capital, manufacturing districts have sprung up in the northern and southern areas in recent years, linked to the food sector, both to foster economic growth in the more peripheral areas of Chile and because of the strong link with local production in those territories.

In a globalized and connected world, where packaged goods circulate nationally and internationally, this specific ecosystem of actors has similar characteristics in different and even geographically distant places, such as in Chile and Italy. Therefore, the differences between these two realities do not relate to the Packaging System itself, but to the specific socio-cultural, political, and economic factors of each country, as well as the degree of maturity specific to each geographic area.

The variables of complexity of the territorial and cultural context add to the already articulated network of supply chain relationships, forcing companies to innovate, to continuously adapt to changes induced by current and sudden emergencies, drawing on solutions that also come from other territories and in a logic of cooperation with the ecosystem, while implementing responsible and regenerative solutions. This ability to adapt to transition is challenging and complex, not only for small and medium-sized enterprises but also for planners, professionals, and trainers.

In this changed global scenario, Responsible Advanced Design intertwined with circularity strategies can thus take on key value in driving change actions in the Packaging System. In addition, precisely because of the kind of "regenerative" and circular vision that is required of us at several levels, packaging is also configured as a cultural path to be traversed, in which the consumer can benefit both in educating to new, more conscious and ethical behaviors, and as a tool for transmitting his or her thoughts, becoming an active player in the process, also thanks to enabling technologies, in a multidirectional dialogue that can also start from the user and reach the brand and any other player.

Consequently, the application of the extended approach—Responsible Advanced Design intertwined with Circularity Strategies (Kirchherr et al. 2017)—to the Packaging System can lead, on the one hand, to an understanding of which projects are already in line with a regenerative process and, on the other hand, to an understanding of where it is necessary to intervene with more responsible and inclusive actions to move more and more toward Regenerative Sustainability.

3.3 Case-Studies

A few case studies were selected from Italy and Chile, which are two countries that, as already anticipated, are geographically distant but not so different in terms of the complexity of the packaging sector in each area and its close relationship with the territory. Three projects were chosen for each country in relation to the three main categories into which the 9Rs of Circularity strategies are grouped.

3.3.1 Smarter Product Use and Manufacture (R0 > R2)

This category includes Refuse, Rethink and Reduce strategies, which take place when products are conceived, designed, and developed; such strategies are precursor, enabling, and transformative. Precursors, because they occur before other strategies. For this category, the projects selected for analysis are Eco Piping Bags (Italy) and Algramo (Chile).

Eco Piping Bags

Eco Piping Bags[8] is a system of innovative and sustainable flexible containers for skin care products that aims to change the way we use cosmetics, by introducing a new packaging solution as a replacement for the traditional rigid bottles (Fig. 1). It is a project developed by UNIFARCO in collaboration with UNIRED, spin-off of the University of Padua. The main features of this product are the following.

- Packaging is minimized, as well as being fully compostable and biodegradable.
- The technological transfer of the shape, reminiscent of a pastry *sac à poche*, affirms new expressive coordinates and proposes consumer experimentation with new usage patterns.

---

[8] https://unired.it/insights/eco-piping-bags-tra-i-vincitori-del-best-packaging-2020/.

Fig. 1 Eco Piping Bags by UNIFARCO in collaboration with UNIRED

- The primary container consists of a custom-made "pouch" for cosmetic use coated with a special paper that provides a better grip and a pleasant tactile experience; the format of the secondary packaging, a parallelepiped with a square base, is designed to easily place the primary container on the store shelf or in home environments.

The project implements:

- Circularity strategies R0, R2;
- the R-AD dimensions of Responsiveness, Anticipation, Inclusion, Time, and Care.

A transversal analysis shows that more could be done on the dimension of Reflexivity, thinking about forms of active consumer education to make the paradigm shift understood in a positive way, and on Transparency, highlighting the process and connections within the supply chain.

Algramo

Founded as a startup in 2012 in Chile, Algramo[9] is a product-service system based on refill stations for bulk products (Fig. 2). Its mission is to create a solution that responds ethically to the environmental and social needs of a cross-section of people who want to make conscious purchases and care for their own well-being and that of the planet, but with limited economic capability. In addition to moving towards social justice, this product is also concerned with the impacts that packaging has on the environment. The main features of this system are:

---

[9] https://algramo.com/.

- buying food, personal or household products in bulk to get away from the logic of large-scale distribution, reducing waste, optimizing the cost of purchasing by consumers;
- combining in the same system the dispensing of a precise quantity of a product and payment through a vending machine;
- supporting a local economy (at neighborhood scale).

Currently, Algramo is implemented in several countries around the world, and it is based on a reusable and consumer-desirable smart packaging that, through the integration of an RFID chip, allows the consumer to be recognized, monitoring the number of reuses, and awarding them for each reuse.

The project implements:

- Circularity strategies R0, R1, R2;
- the R-AD dimensions of Responsiveness, Anticipation, Inclusion, and Care.

A transversal analysis shows that more could be done on the dimension of Responsiveness, bringing the project also to an immaterial dimension that also pushes it to be a lever to educate the community to a more aware and sustainable type of consume. In addition, work could be done on the Transparency dimension by also making the origin of the product visible.

Fig. 2 Algramo refill stations for bulk products

3.3.2 Extend Lifespan of Product and Its Parts (R3 > R7)

This category groups the strategies of Reuse, Repair, Refurbish, Remanufacture and Repurpose, meant to retain finished goods and their parts in the economy for longer, while maintaining or improving their value. For this category, the projects selected for analysis are Brown Lift Lock (Italy) and Green Glass (Chile).

IFCO Brown Lift Lock 6416

Brown Lift Lock 6416[10] is the design of a range of Reusable Plastic Containers (RPCs) made by IFCO for bread and bakery products (Fig. 3). It is collapsible and suitable for low and high temperatures; it is designed to ensure maximum quality and freshness from bakery to point of sale, and several benefits in terms of environmental impact. The main features of this product are the following:

- significantly reduced CO_2 emissions and water and energy consumption compared to single-use packaging, as well as less food waste and the generation of less landfill waste;
- IFCO RPCs are reused up to 120 times, cleaned and disinfected and, at the end of their life cycle, granulated and used to produce new IFCO RPCs;
- ventilation holes that keep the bread at the right temperature while reducing waste, and a robust structure that prevents damage and crushing;
- a smooth base and inner walls that protect the contents, as well as an interlocking design that allows for safer palletization;
- a mechanism called Lift Lock, designed to close and open quickly and safely;
- ergonomic design ideal for automated systems and equipped with handles for safe manual picking;
- compared to standard disposable packaging, it has good cold storage characteristics in the freezer and is not affected by moisture;
- IFCO RPCs produce up to 60% less CO_2 and 86% less solid waste; they also save 64% energy and recover 80% water compared to single-use packaging.
- with its 28 mm height when closed, the Brown Lift Lock 6413 boxes makes maximum use of available space in trucks or storage compared to stackable or rigid bakery boxes;
- finally, it allows an economic advantage for retailers in their fresh produce supply chains: the adoption of this standardized packaging system enables a 25% reduction in handling, storage, and equipment costs.

The project implements:

- Circularity strategies R3, R4, R5, R6;
- the R-AD dimensions of Care, Anticipation and Responsiveness.

The analysis of the case study shows that to make the product more responsible, more work could be done on the Transparency dimension of the entire supply chain

---

[10] https://www.ifco.com/it/soluzioni-alimentari/pane/.

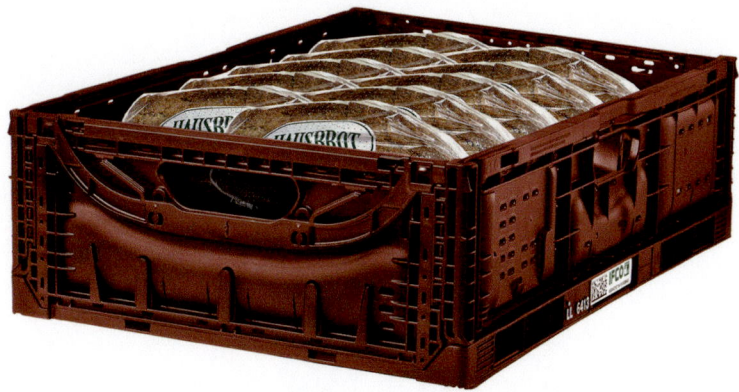

Fig. 3 IFCO Brown Lift Lock 6416

and the type of relations activated with suppliers and consumers, e.g., using enabling technologies already incorporated on the product.

Green Glass

Green Glass[11] is a company that recovers waste products, specifically glass bottles, which, thanks to a rethinking process, take on a new form and a new purpose within a different sector (Fig. 4). The main features of this project are:

- recovering containers that were so far seen as waste and transforming them into an everyday object: the glass, a product that no longer has the function of a "disposable" container but increases its life cycle because it retains a function of use within the context in which it is placed;
- generating employment for the most fragile sectors of the population such as the "cartoneros" or people working in the process of transforming the product from bottle to glass;
- maintaining the characteristic of a means of communication through the graphitization of the surface of the glasses. In fact, the glass also becomes a tool for telling stories;
- promoting environmental and social sensitivity and awareness. For example, for every purchase a tree is planted, or part of the proceeds are donated to charitable projects. This type of activity increases the social responsibility of the company.

The project implements:

- Circularity strategy R7;
- the R-AD dimensions of Time, Inclusion, Responsivity, and Care.

---

[11] https://www.greenglass.cl/.

Fig. 4 Recovered and repurposed glass by Green Glass

The analysis of the case study showed how the Inclusion dimension could be enhanced within the product itself, particularly in the graphic image, which could become a vehicle for narrating the concept of social responsibility.

3.3.3 Useful Application of Materials (R8 > R9)

This group of strategies (Recovery and Recycling) covers solid waste otherwise destined for landfill or burned without heat recovery. For this category, the projects selected for analysis are Rinascimento (Italy) and TheUpcyclingco (Chile).

Rinascimento

Rinascimento[12] is a circular economy project of ICMA Sartorial Paper that provides a high value-added service for the company-user, the end consumer, and the environment. It transforms pulp waste recovered from companies into recycled creative papers for new packaging projects by the same companies (Fig. 5). The project is a virtuous example of material reuse:

- at all production stages, waste materials are tracked to ensure their identification;
- the production chain is shortened: waste is directly used as raw material to produce new paper, eliminating the intermediate step of virgin or recycled pulp production, a process that uses chemicals, energy, and water;
- material handling is reduced due to both the shortening of the process and the design of the service: the company itself designs the circularity chain tailor-made

---

[12] https://www.icma.it/rinascimento/.

Fig. 5 Rinascimento circular papers by ICMA Sartorial Paper

for the customer and chooses partners based on the geographic location of the pulp mill and on the characteristics of the creative project;
- this process management results in less CO2 production.

The project implements:

- Circularity strategy R8;
- the R-AD dimensions of Transparency, Responsiveness, and Care.

Again, the project analysis shows that significant action could be taken on the dimensions of Anticipation and Reflexivity by having several actors of the territorial eco-system collaborating in the design (Fig. 5).

TheUpcyclingco

TheUpcyclingco[13] is a project born in 2019 that works on responsible plastic recovery and recycling. It transforms plastic waste, from different collaborations with Chilean communities, organizations, and universities, into new home and play products made of 100% recycled plastic (Fig. 6). The project is characterized by:

- a local supply chain from material recovery to its processing; the startup follows the complete cycle from the collection of the plastic material to its revaluation;
- a system of valorization of the partner companies by proposing the development of new products to be returned in a new version to the company; this also supports a form of collaborative environmental awareness that emphasizes how one's own

---

[13] https://theupcyclingco.cl/.

Fig. 6 Responsible plastic recovery and recycling by TheUpcyclingco

waste can also generate new life within one's own institutional or entrepreneurial reality;
- a policy of also raising consumer awareness by sending a recycling guide with the purchased products.

The project implements:

- Circularity strategy R8;
- the R-AD dimensions of Reflexivity, Inclusion, and Care.

Here it emerges that the dimensions of Anticipation and Responsiveness are minimally considered in the project development phases and could be enhanced.

4 Conclusions

From the research and the synthesis overview of the case studies (Table 1), it emerges how the R-AD model integrated with the circularity framework allows projects to be observed through a filter that not only investigates the stages of a product's production process, but also focuses on the relationships among the actors in the system, the social and cultural impacts of the product-service, and the intangible principles implemented to achieve sustainability.

Finally, the application of R-AD to the analysis of existent cases can contribute to promoting three actions:

Table 1 Overview of the analysis of selected case-studies

Selected packaging case studies	Circularity main categories (from 9Rs framework)	Circularity specific strategies (from 9Rs framework)	R-AD dimensions
Eco Piping Bags	Smarter product use and manufacture	R0, R2	Responsiveness Anticipation Inclusion Time Care
Algramo	Smarter product use and manufacture	R0, R2	Responsiveness Anticipation Inclusion Care
IFCO Brown Lift Lock 6416	Extend lifespan of product and its parts	R3, R4, R5, R6	Care Anticipation Responsiveness
Green Glass	Extend lifespan of product and its parts	R7	Time Inclusion Responsiveness Care
Rinascimento	Useful application of materials	R8	Transparency Responsiveness Care
TheUpcyclingco	Useful application of materials	R8	Reflexivity Inclusion Care

- mapping and systematizing good practices, design, production, and distribution systems that in some way already respond to these above-mentioned principles, useful for further implementations of the model;
- consolidating the idea that an interactive and integrated approach between Advanced Design, Responsible Innovation and Circularity can be a line of action to deal the needs of the complex systems that are around us today;
- bringing out critical issues and gaps in current product-service systems (e.g., lack of gender equality, inclusion, or other ethical aspects, etc.) by verifying through the conceptual model which R-AD tools can respond to process change and improvement.

To increase the practical and theoretical value of the R-AD model enhanced with the dimension of circularity, it is necessary to continue the ongoing process of investigation and mapping and to experiment with it in other contexts of intervention. It is hoped that these future actions can outline guidelines and a set of semi-processes to achieve, with a holistic approach, regenerative sustainability throughout a product's value chain.

Acknowledgements This chapter is the result of a joint work of the three authors; nevertheless, the authorship of "1 Responsible Advanced Design as a New Approach to Change and Regeneration" is to be attributed to L. Succini; "2 Responsible Advanced Design and Circularity for Regenerative Processes" to E. Ciravegna and L. Succini; "3 Packaging System as an Exemplary Scope in a Transterritorial Perspective" to E. Ciravegna and C. Giardina; "4 Conclusions" to L. Succini, E. Ciravegna, and C. Giardina. This study was carried out within the MICS (Made in Italy—Circular and Sustainable) Extended Partnership and received funding from the European Union Next-GenerationEU (Piano Nazionale di Ripresa e Resilienza (PNRR)—Missione 4 Componente 2, Investimento 1.3—D.D. 1551.11-10-2022, PE00000004). This manuscript reflects only the authors' views and opinions, neither the European Union nor the European Commission can be considered responsible for them.

References

Bennis W, Nanus, B (1987) Anatomia della leadership. Le 4 chiavi della leadership effettiva. Franco Angeli

Bezzi M (2021) Ricerca e innovazione aperta, partecipativa e responsabile: un glossario. Apre

Bouwman H, Janssen M (2010) Dealing with technology and actor views in designing ICT service systems. J Des Res 8(4):359–374. https://doi.org/10.1504/JDR.2010.035677

Bucchetti V (2005) Packaging design. Franco Angeli

Burget M, Bardone E, Pedaste M (2017) Definitions and conceptual dimensions of responsible research and innovation: a literature review. Sci Eng Ethics 23(1):1–19. https://doi.org/10.1007/s11948-016-9782-1

Carayannis EG, Barth TD, Campbell DF (2012) The quintuple helix innovation model: global warming as a challenge and driver for innovation. J Innov Entrep 1(1):2. https://doi.org/10.1186/2192-5372-1-2

Celaschi F (2008) Il design come mediatore tra bisogni. Il Design come mediatore tra saperi. In: Germak C (ed) L'uomo al centro del progetto. Allemandi, pp 40–52

Celaschi F (2015) Advance design points of view. In: Celi M (ed) Advanced design cultures. Springer International Publishing, pp 3–17

Celaschi F (2016) Non industrial design: Contributi al discorso progettuale. Sossella

Celaschi F, Celi M, Formia E (2014) Quando il design incontra il futuro. When design meets future. Plan Des Technol 3:20–29. Utopia. passato, presente, futuro–past, present, future

Celaschi F, Formia E, Franzato C (2018) Ritorno al futuro. Il fattore tempo e i futures studies nell'approccio design driven all'innovazione. diid — Disegno Industriale Ind Des XVIII(64):126–133

Celaschi F, Formia E, Iñiguez Flores R, León Morán R (2019) Design processes and anticipation. In: Poli R (ed) Handbook of anticipation. Springer International Publishing, pp 773–793. https://doi.org/10.1007/978-3-319-91554-8_48

Celi M (2010) Advanced design: Visioni, percorsi e strumenti per predisporsi all'innovazione continua. McGraw-Hill

Celi M (2015) Advanced design cultures. Springer

Celi M, Morrison A (2019) Anticipation and design inquiry. In: Poli R (ed) Handbook of anticipation: theoretical and applied aspects of the use of future in decision making. Springer International Publishing, pp 795–819

Ciravegna E (2010) La qualità del Packaging. Franco Angeli

Ciravegna E (2017) Diseño de packaging. Una aproximación sistémica a un artefacto complejo. RChD: Creación y Pensamiento 2(3) https://doi.org/10.5354/0719-837X.2017.47825

Ciravegna E (2020) Repensar los envases en tiempos de crisis: implicancias éticas y enfoque sistémico en el Diseño de Packaging. RChD: Creación y Pensamiento 5(9):1–6. https://doi.org/10.5354/0719-837X.2020.59536

Cole RJ (2012) Transitioning from green to regenerative design. Build Res & Inf 40(1):39–53. https://doi.org/10.1080/09613218.2011.610608

Deppeler J, Aikens K (2020) Responsible innovation in school design – a systematic review. J Respons Innov 7(3):573–597. https://doi.org/10.1080/23299460.2020.1809782

Elkington J (1997) Cannibals with forks: the triple bottom line of 21st century business. Capstone

Ellen MacArthur Foundation (n.d.) The circular economy in detail. How to build a circular economy. https://ellenmacarthurfoundation.org/the-circular-economy-in-detail-deep-dive

Escobar A (2018) Designs for the Pluriverse: radical interdependence, autonomy, and the making of worlds. Duke University Press

European Commission (2008) Directive 2008/98/EC of the European Parliament and of the Council of 19 November 2008 on waste and repealing certain directives. https://eur-lex.europa.eu/legal-content/EN/TXT/PDF/?uri=CELEX:32008L0098

European Commission (2013) Options for strengthening responsible research and innovation: report of the expert group on the state of art in Europe on responsible research and innovation. Publications Office. https://data.europa.eu/doi/10.2777/46253

European Commission, von Schomberg R (2011) Towards responsible research and innovation in the information and communication technologies and security technologies fields. Publications Office. https://data.europa.eu/doi/10.2777/58723

Formia E (2017) Storie di futuri e design: Anticipazione e sostenibilità nella cultura italiana del progetto. Maggioli Editore

Geissdoerfer M, Savaget P, Bocken NMP, Hultink EJ (2017) The circular economy – a new sustainability paradigm? J Clean Prod 143:757–768. https://doi.org/10.1016/j.jclepro.2016.12.048

Ghisellini P, Cialani C, Ulgiati S (2016) A review on circular economy: the expected transition to a balanced interplay of environmental and economic systems. J Clean Prod 114:11–32. https://doi.org/10.1016/j.jclepro.2015.09.007

Gibbons LV (2020) Regenerative—the new sustainable? Sustainability 12(13):5483. https://doi.org/10.3390/su12135483

Gibbons LV, Cloutier SA, Coseo PJ, Barakat A (2018) Regenerative development as an integrative paradigm and methodology for landscape sustainability. Sustainability 10:1910

Grimpe B, Hartswood M, Jirotka, M (2014) Towards a closer dialogue between policy and practice: responsible design in HCI. In CHI '14 proceedings of the SIGCHI conference on human factors in computing systems. ACM, pp 2965–2974. https://doi.org/10.1145/2556288.2557364

Iaione FC, De Nictolis E (2016) La quintupla elica come approccio alla governance dell'innovazione sociale. In: Montanari F, Mizzau L (eds) I luoghi dell'innovazione aperta. Modelli di sviluppo territoriale e inclusione sociale. Fondazione Giacomo Brodolini, pp 75–89

Ibrahim I, Ahmed N (2022) Investigating regenerative ideation within sustainable development goals. Sustainability 14(16):10137. https://doi.org/10.3390/su141610137

Iñiguez Flores R, Celaschi F, Formia E, León Morán RM (2019) The extended praxis of design: towards a characterization of the advanced design cultures. In: The design after. Cumulus conference proceedings 2019, pp 449–458

Iñiguez Flores R, Hernandis Ortuño B, Holliger C, Monterrubio Soto JC (2014) Advanced design as a process for knowledge creation. Delivering knowledge to stakeholders and fostering future skills for innovation. In: 5th international forum of design as a process. The shapes of the future as the front end of design driven innovation, pp 151–156

Jakobsen SE, Fløysand A, Overton J (2019) Expanding the field of responsible research and innovation (RRI) – from responsible research to responsible innovation. Eur Plan Stud 27(12):2329–2343. https://doi.org/10.1080/09654313.2019.1667617

Jenkin S, Pedersen Zari M (2009) Rethinking our built environments: towards a sustainable future: a research document. Ministry for the Environment, New Zealand

Kirchherr J, Reike D, Hekkert M (2017) Conceptualizing the circular economy: an analysis of 114 definitions. Resour Conserv Recycl 127:221–232. https://doi.org/10.1016/j.resconrec.2017.09.005

Long TB, Blok V, Dorrestijn S, Macnaghten P (2020) The design and testing of a tool for developing responsible innovation in start-up enterprises. J Respons Innov 7(1):45–75. https://doi.org/10.1080/23299460.2019.1608785

Lyle JT (1994) Regenerative design for sustainable development. Wiley

Maffei S (2021) Expanding the galaxy: designing more-than-human futures. DIID — Disegno Industriale Ind Des 75:12–19. https://doi.org/10.30682/diid7521a

Mang P, Haggard B (2016) Regenerative development and design: a framework for evolving sustainability. Wiley

Mang P, Reed, B (2013) Regenerative development and design. In: Loftness V, Haase D (eds) Sustainable built environments. Springer, New York, pp 478–501. https://doi.org/10.1007/978-1-4614-5828-9_303

Margolin V, Margolin S (2002) A "social model" of design: issues of practice and research. Des Issues 18(4):24–30. https://doi.org/10.1162/074793602320827406

Mozota BB (2006) The four powers of design: a value model in design management. Des Manag Rev 17(2):44–53. https://doi.org/10.1111/j.1948-7169.2006.tb00038.x

Papanek VJ (1971) Design for the real world: human ecology and social change, 1st American edn. Pantheon Books

Parilli M (2020) Il design bionico, quando l'uomo si ispira alla natura (e la migliora) Corriere della Sera, 19 March 2020. https://www.corriere.it/bello-italia/notizie/design-bionico-quando-l-uomo-si-ispira-natura-migliora-8efa3302-6888-11ea-9725-c592292e4a85.shtml

Poli R (2010) An introduction to the ontology of anticipation. Futures 42(7):769–776. https://doi.org/10.1016/j.futures.2010.04.028

Poli R (ed) (2019) Handbook of anticipation: theoretical and applied aspects of the use of future in decision making. Springer

Potting J, Hekkert M, Worrell E, Hanemaaijer A (2017) Circular economy: measuring innovation in the product chain. PBL Publishers

Reh L (2013) Process engineering in circular economy. Particuology 11(2):119–133. https://doi.org/10.1016/j.partic.2012.11.001

Schon DA (1983) Il professionista riflessivo: Per una nuova epistemologia della pratica professionale. Dedalo

Sihvonen S, Ritola T (2015) Conceptualizing ReX for aggregating end-of-life strategies in product development. Procedia CIRP 29:639–644. https://doi.org/10.1016/j.procir.2015.01.026

Stilgoe J, Owen R, Macnaghten P (2013) Developing a framework for responsible innovation. Res Policy 42(9):1568–1580

Succini L (2022) Processi e pratiche di innovazione responsabile: il ruolo dell'Advanced Design nella trasformazione collaborativa dei territori. Dissertation thesis, Alma Mater Studiorum - Università di Bologna. Doctoral Program in Architecture and Design Cultures, Cycle 34

Succini L (2023) Leggere l'Innovazione Responsabile attraverso l'Advanced Design. In: Formia E, Gianfrate V, Succini L (eds) Design per l'Innovazione Responsabile. Guida per processi formativi in trasformazione. Franco Angeli, pp 21–36

Succini L, Ciravegna E (2022) Design and responsible innovation. Ethics and caring as keys to addressing contemporary crises. DIID—Disegno Industriale Ind Des XX(77):24–36. https://doi.org/10.30682/diid7722b

Tronto JC, Fisher B (1990) Toward a feminist theory of caring. In: Abel E, Nelson M (eds) Circles of care. SUNY Press, pp 36–54

van Buren N, Demmers M, van der Heijden R, Witlox F (2016) Towards a circular economy: the role of Dutch logistics industries and governments. Sustainability 8(7):647. https://doi.org/10.3390/su8070647

Van der Ryn S, Cowan S (2007) Ecological design, 10th anniversary edn. Island Press

Von Schomberg R (2013) A vision of responsible research and innovation. In: Owen R, Bessant J, Heintz M (eds) Responsible innovation. John Wiley & Sons, Ltd., pp 51–74. https://doi.org/10.1002/9781118551424.ch3

Wahl D (2016) Designing regenerative cultures. Triarchy Press

WBCSD (2017) Our approach. http://www.wbcsd.org/Overview/Our-approach

Wynne B (2011) Lab work goes social, and vice versa: strategising public engagement processes: commentary on: "what happens in the lab does not stay in the lab: applying midstream modulation to enhance critical reflection in the laboratory." Sci Eng Ethics 17(4):791–800. https://doi.org/10.1007/s11948-011-9316-9

Zamenopoulos T, Alexiou K (2007) Towards an anticipatory view of design. Des Stud 28(4):411–436. https://doi.org/10.1016/j.destud.2007.04.001

Zhang X, Skitmore M, De Jong M, Huisingh D, Gray M (2015) Regenerative sustainability for the built environment – from vision to reality: an introductory chapter. J Clean Prod 109:1–10. https://doi.org/10.1016/j.jclepro.2015.10.001

Zhu L, Zhou J, Cui Z, Liu L (2010) A method for controlling enterprises access to an eco-industrial park. Sci Total Environ 408(20):4817–4825. https://doi.org/10.1016/j.scitotenv.2010.06.035

The Emergence of Practice

Delfina Terrado and Martin Bunge

Abstract In the last few years Regeneration and all possible ways of making it visible (agriculture, forestry, economy, social field) has become a trending topic for all those who are sensitive to the context we are living in. There is scientific agreement (as you can find in the IPCC report), on the urgency for a shift in our way of intervening, of treating land and air, and water, and people. Some people see regeneration as the natural movement from sustainability to a more committed, more fully embodied way of participating, some consider that we are dealing with a larger step that is needed in order to respond appropriately. The question then is: how do we build a more committed and conscious way of participating? What does the bubbling fountain of a regenerative practice look like? In this chapter we portray three experiential learning moments of different projects, that allow us to observe together how the inner movement that unfolds regeneration in our practice. We dive into the learning process that not only reveals the honing of a conscious participating action, but also the plunge into the human soul and the rooting of our quest to reunite with mother earth. The reader will walk with us into these questions, and perhaps on its way find his/her own questions and movements towards a regenerative practice.

Keywords Regeneration · Practice · Observation · Reflective social practice · Participation · Delicate activism

---

You are what you do, not what you say you'll do.
  Carl Gustav Jung

D. Terrado
Iniciativa Murmullo, Buenos Aires, Argentina
e-mail: terradodelfina@gmail.com
URL: https://iniciativamurmullo.com/

M. Bunge (✉)
Experiencia Raiz, Buenos Aires, Argentina
e-mail: martin.bunge@gmail.com
URL: http://experienciaraiz.com/

1 Introduction

Imagine yourself living in a very small countryside community, a beautiful environment surrounded by hills and forests, small streams of fresh water shimmering with the reflections of the sun. Your hands are full of moist earth, as they come out of the vegetable garden you are taking care of, a bee visits you, flying around just as they fly around every flower. A small breeze makes the wings of a butterfly move, slowly and gently sitting on the top of a wildflower. Your senses feel open, feeling the humming of insects all around you, the growing of plants and flowers that you have been taking care of since seed, the sun on your skin and the freshness of earth in your hands. Standing in the midst of that garden you are not a visitor, you are an active part, reading the needs and potential, observing how each species moves into new phases, witnessing time through your own engagement with living processes all around you. The bell rings, it is the signal for all to meet, so you take your tools and basket and start walking towards the meeting room, as you see others walking from different places all over the land. Smiles and greetings, and also reminders of things that will need attention afterwards.

The meeting is about to begin, there are some simultaneous conversations happening in the room: people commenting about the weather (everyone is expecting an afternoon rain), also about some tools that need maintenance, and people asking about family. There are more or less twenty people gathering together in a circle, inside a beautiful room that almost everyone helped to build a couple of months ago. Sitting in that circle you can feel the expectation living in the air, almost like a sown field, coming out of some rainy days into the sun. This is not the first time the group meets, there is a regular practice of meeting and conversations around the weekly agenda and mid-term initiatives. There is already a "way of meeting" that the group shares, one that was built out of itself, that demanded many frustrating moments (but also celebrations), and the acknowledgment that like the land you walked just a couple of minutes ago, there is something very alive in the room that cannot be touched, but can be sensed. It is the potential of social phenomena living *under the surface*, touched by the sun and soil, receiving humidity and breeze; inside that room many things happened, the group learnt to deal with the emergence of a community, learning from the collective process but also bringing the individual fully into the room, just like the wildflowers do outside.

But the world we are living in does not always look like these two instances in a possible life, even more in these very particular and challenging years. The pandemic context (of COVID19) unleashed an accelerating process towards individuation, our ways of engaging and relating (between human and non human beings) has profoundly shifted, in some cases it eroded some of our wonder towards the living, towards the meeting of others. There seems to be less time than before, we might think we became more productive, and time…, time is the master. We don't need to do exhaustive research to get to trends and realities that speak about how the world (and our participation in it) is moving towards uncertain times. The earth is warming up, at the same time our hearts and courage become tamed, and cool down. It is almost like we are moving to opposite directions in many aspects of our way of inhabiting our shared home. These are challenging times because real issues speak more than ever directly into our faces. We see new erratic climate events, new migration causes, increasing inequalities, declining trust in the mainstream economic rationale, governments promising more than they are capable of, our fragile farmlands losing their vital attributes. We could continue naming elements to describe the world's pains, that are also our pains. This is perhaps one of the most important things, since we keep thinking about ourselves outside the story, we are even investing in ways to move to other planets. It is important to stay close to *our pains*, even more if we are willing to develop ways of doing that sustains all forms of life, seeking for regeneration. We can't consider ourselves outside these pains, we are interwoven in them, and they can become truths for ourselves, because we (as a particular species) have the capacity to feel and make sense out of them. This is where the gem of our participation lives (and that we want to explore in this chapter), through our own experience, and the intensity of our way of getting involved with it. Feeling can never be replaced by an algorithm, it demands our minds, hearts and bodies, it demands our perceptions, it demands presence. The way we experience the world depends on our inner world, and our inner world is shaped through the experiences we live. This chapter is an invitation to look into the emergence of that quality of presence, a way of participating that is shaped by our attentiveness towards life, and the situations we encounter. It becomes a challenge for the reader (and inevitably to writers) to engage with a piece like this, because we are directing our attention towards our inner process while we encounter the outer. Perhaps this kind of attentiveness (a quality of the meeting) is what really allows us to meet phenomena (or the stories shared) to become one, and to see them from within. Within a phenomenological context, method is never just an engine that will unerringly produce insightful outcomes (Van Manen 2014, p. 29). We must stay close to the observed, with our senses awake, with our inner eyes open and with our capacity to embrace the new, which as Goethe says: every new knowledge starts with wonder. This piece is an attempt to journey with you into some nooks of our understanding of what a Regenerative practice is.

We will embody Regeneration through observing some cases that will help us illuminate the relevance of the cultivation and recognition *practice*, more than the particular outcomes of our action. *How can our ways of acting become more conductive to life?* We reach the world through our professional practices, something we will also explore here. Perhaps the very notion of what a practice means, enables us to walk towards ways of experiencing and engaging that help us to deepen our understanding about the world and ourselves. *How can our doing help us to move deeper into understanding life?*

2 Practicing a Practice

Why are we choosing to explore practice? Because we need to find ways to understand our participation. If there is something paradigmatically different in Regeneration, it is the reunion of human species with the greater integrity we are living in. There are for sure different ways to do this, but as mentioned before, we are attempting to come closer to life (and the way we intervene in it), and we must become more conscious about our participation in order to stretch our understanding and ways of acting. If we are about to look into *practice* we find it relevant to consider Alasdair's McIntyres understanding in his book After Virtue:

> By a practice I am going to mean any coherent and complex form of socially established cooperative human activity through which goods internal to that form of activity are realized in the course of trying to achieve those standards of excellence which are appropriate to, and partially definitive of, that form of activity… (MacIntyre 1981, p. 186)

The first thing that stands out in his definition is the fact that a practice is a form of activity, so there is a doing involved in it, and as we can discover in the work of the craftsman and in any other (architecture, farming, medicine, chess), we have socially established what these practices look like; we have formed an understanding of 'what it is'. He also gives a high relevance to the *goods that are internal* to that activity, this means "…those which cannot be had in any way but by playing chess, or some other game of that specific kind", that shouldn't be confused with those *external* (prestige, status, money) that can be achieved in different ways, not necessarily to a particular practice.

There are standards of excellence, says McIntyre, that are particular to every human activity, they are even constitutive to that particular practice. There is a "way of doing things", we recognize other practitioners through the recognition of particular knowledge, approaches and methodologies (or tools) used. The way we do things becomes something relevant then, not only because it speaks into the standards we pursue, but also because the variation of our approaches and practices can push the borders of a practice into new fields of exploration. As an example, imagine when machinery was introduced into farming, and all its new (and some unexpected) different impacts. Practice is something dynamic, that shapes how things are being done, and is also being shaped by those who engage in such activities (that's why it becomes so important to move from a practice to a Practice). Over time there are new tools, new frameworks, new approaches and of course new experiences (owns and others). We are not alone, many others are dealing with similar situations and questions, stretching their own concepts and ways of dealing with challenges.

As we said before, Regeneration as a realm of human thinking and acting has found many expressions in different activity clusters (agriculture, business, economy). When we listen to the word Regeneration, a word that has now become visibly repeated and focused for some people as "the next step of sustainability", we usually tend to think about the green. We mean that sometimes it is easier to find examples of our actions in the so called "natural world" (conservation, ecosystem restoration, agroecology and organic farming), since we are capable of experiencing the movement of life through our direct participation, through our eyes and hands. Social organisms that are also alive and therefore demand a different way of understanding and intervening, like those of the natural world, ways that enable that community or impulse to thrive by their own capacity to guide themselves through their own challenges.

In every practice well recognized and shared there is a thriving, an attempt to reach out into the world, as McIntyre points out, looking for those standards of excellence. On the way there is also the coining of the inner goods that are particular to that kind of practice. For gardeners their inner life emerging through practice is completely different than the one experienced by a writer, a runner, or a craftsman. There are nuances in the activities, in the reflections and especially in the way they relate to the "outer world". A Practice (when we use a capital P it is because we are pointing to those who have been observed and recognized by its practitioners), offers a real possibility to dive into our very particular way of meeting the world. In the end it is through our expressions in the world that we build ourselves, hopefully doing what we love, hopefully being aware of how the world speaks back to us, inviting us to move on into ourselves at the same time we move more fully into the world.

It is not strange to find people who have already walked a long journey inside a Practice, speaking from another kind of grounding, with a sense of blurred division lines between knowledge acquired "outside" and the one built from within. It is a powerful thing to hear someone living the Practice they embody, they are not separated entities, practices live through human beings, they shape and are being shaped by them. Ask yourself about the way you touch the world, the kind of activities you carry on, how you express yourself through it, you might find yourself everywhere. Craftsmen do not only build something out of wood (that could be the outcome), they immerse themselves into a creative process where they give themselves fully, and in that surrender, they also shape their own muscle, their own tools and frameworks. *Practicing a Practice* is about our own life, and our own expression into the world, like the wildflower offers itself full to the mystery of which it is part of. It is about meaning and purpose, but it is also about action. *How can we engage with ourselves, others and the world in such a way that we bring forth a world conducive of life? How do we create spaces that promote the flourishing of our human capacities? How can we deepen and strengthen a practice in these indelicate times?*

> Regeneration is
> the essential self-organization
> of life
> Capra and Wahl (2023)

3 Regeneration as a Verb

We are alive, at least while we write we see letters appearing on the screen, feeling and experiencing the movement into new ideas, perhaps you too are aware of your experience of the reading. Outside the wind is blowing, and the freshness of the ground under our feet. We experience, we feel, we are awake and present here. We carry that gift with us since we were inside the womb of our mothers, until we return to the womb of our great mother. And it is a gift that needs to be discovered, that needs space and time to find itself inside each one of us. Like the gardener in the story at the beginning, we are participants, and that notion of participation depends on our capacity of considering ourselves part (or not) of what we encounter. In the previous section we explored the qualities of a practice as a way to become aware of our particular way of participation, by illuminating it, some of our inner world becomes also illuminated.

It is especially interesting when one observes the activities and practices related to Regeneration, where life and biodiversity seem to be held in the center as source and result of our initiatives. There is a great opportunity inside the whole Regenerative movement, and we mean by this in a broader sense that just the outcomes of our action in the natural and social world. It has to do with a more deeper, complex and engaged way of participation, one that seeks for the living, caring for the patterns and the threads that build the space for regeneration to happen. Regeneration (or perhaps a Regenerative Practice) must remain alive and flexible as life is. If we focus too much on the measurable outcome (CO_2 captured, hectares of crops converted, or any other way of possible measuring techniques) we will miss the process. Talbott (2021) expresses this in one of his lectures in quite a unique way "The more one wants to grasp the truth of Nature through mathematical precision, the more it slips through our fingers as if one wants to grasp a handful of water." We need to ask ourselves where our focus is, regeneration must not become a set of techniques or approaches, it must help us stretch our understandings, as we stretch our acting to the world, one that as we pointed out before seems to be screaming aloud for something different from us.

How can we strengthen a regenerative practice? What are the main threads that weave into a coherent way of practicing regeneration?

In the next section let us explore together three cases to illuminate some of the aspects presented before. With Iniciativa Murmullo we will be diving into what we call inner ecology. Which involves our different ways of knowing such as thinking, feeling, sensing and intuition and how they impact the world that we bring forth with others. In the case of Quinta Esencia we will be exploring outer ecology, how we can engage locally (and/or eco-systemically) with regenerative practices and build together a desirable future. Experiencia Raíz, as a space for the exploration and strengthening of a living practice, of our ways to engage the world as a conversation between who we are and who we are becoming.

Each one of them has encountered a unique entry point to a practice that engages with the world regeneratively.

4 Iniciativa Murmullo: The Awakening of Inner-Ecology

In 2018 the public educational system of the Province of Buenos Aires was going through intense teacher strikes that burst demanding restorative action from the government due to a gas leak explosion that originated in a small public school in Moreno. The government failed to comply with systemic regulations of the building state of quite a few public schools. As this incident occurred early in the morning before the arrival of children, many lives were spared except for the vice-principal and one of the school assistants. This event unleashed anger and frustration within the educational community.

At the time the largest School Climate and Emotional Education Program of Argentina (Terrado 2021) was taking place within the province working with neighboring public schools. The aim was to enable a holistic approach (seeking to address in teaching other ways of knowing the world that have been long forgotten such as our senses and emotions) to learning with the available resources that each vulnerable school already had within the constellation of people working in them. The plan was to influence the creation of communities of practice within different territories, to develop human potential and foster the conditions for contextual wellbeing to emerge. Untapped potential rested within a whole school approach ready to be awakened.

The social unrest that triggered the fateful event, awakened what was thought was the opposite effect of what the program was looking for. There was profound pain flowing into each community of practice throughout the territory. What was emerging was not only that unfortunate event but the sum of all negligent experiences that had not been emotionally digested over time by the educational community. Many of its repercussions came up rushing to be expressed within the territorial communities of the social-emotional educational program. The group of facilitators that lead the program had the challenging task of holding and caring for a safe space to process and engage in difficult conversations. To maintain and restore healing practices in such a disruptive moment in time. To stay open and attentive to the current needs of each individual and community.

5 The Quality of Experience

Bortoft (2016) repeated in his lectures at *Schumacher College* that the human experience is intrinsically holistic, we can intentionally make it deeper when we engage in our thinking, sensory, emotional and intuition processes of knowing. Engaging in a learning process in order to become educated by the world, means to open up to diverse ways of knowing, to truly observe and get involved in the moment to moment unraveling of the present. Education today has to do more with the concern of preparing students for the future, than the living of the present and in this urgency there is no time for deepening participation in experience. *What kind of education can enable us to experience the present more fully?* These indelicate times call forth a rather committed participation. As Dewey (1938, p. 49) expresses beautifully:

> What, then, is the true meaning of preparation in the educational scheme? In the first place, it means that a person, young or old, gets out of his present experience all that there is in it for him at the time in which he has it. When preparation is made the controlling end, then the potentialities of the present are sacrificed to a suppositious future. When this happens, the actual preparation for the future is missed or distorted. The ideal of using the present simply to get ready for the future contradicts itself. It omits, and even shuts out, the very conditions by which a person can be prepared for his future. We always live at the time we live and not at some other time, and only by extracting at each present time the full meaning of each present experience are we prepared for doing the same thing in the future. This is the only preparation which in the long run amounts to anything.

If we are able to understand it this way, experience and its unique quality can serve as a self-revealing learning process. When we open up as learners to the context (the outer phenomena) through the process of knowing (thinking, feeling, sensing and intuition), we in turn dive deeper into our *inner ecology* (inner phenomena) allowing new meaning to emerge from the current situation. It is much easier to participate in experience when the emotions and senses that emerge are pleasurable. On the contrary, when the emotion is unpleasant or threatening the human tendency is to close oneself to the experience and set up controlling defenses. The controlling tendencies that individuals, communities or systems have originated as a (necessary) way of avoiding more unpleasant situations.

As we said before, Practice is the intentional participation of this encounter between the inner and the outer, between world and Self. The program facilitators held a Practice of intentional exploration of the unraveling of human potential within the participants and the territorial communities. There was a consistent reading of the situations, the inner process of facilitators, and the emergence of that relation, and through repetition it became alive for the community. This experience became transformative in itself, staying open to meet the new together. In the end, the only way of exploring new journeys within the human range of possibilities, is by being open to the mystery of life. However, the situation became very unpleasant and with it questions about ourselves, the practice and the ecology of our places emerged.

How can we as facilitators stay present to all human experience in these uncertain times? How can we model and cultivate healing human capabilities within educational territorial communities?

6 Emergent Learning

The facilitators community gather once every month to engage and develop the content, design new activities and to reflect on the emergent learnings of each community spread out around the province. These meetings were crucial to the development and growth of the individual and collective practices, they felt as if we were incubating a new way of being in service to a higher purpose for education. Our gatherings became alive with emergent learnings that travel from the students' classrooms, to the teachers' relationships, from the different projects to our small group of facilitators. The first gathering after the gas leak explosion was unique and had another flavor to it. There was a lot of silence, as if we were trying to process how to move forward, what was the best way to proceed. We opened the conversation where each facilitator brought their concerns about the state of their communities.

The community of the facilitator's employer was the government, to the educational community we were their representatives. So the first decision that we took collectively that day was that although we weren't responsible for the public school explosion, we needed to see ourselves as part of the story that was unfolding. We decided to stay close to our pain and others' pain, we decided to stay open and aware to process the emotions that were coming up in ourselves and within the community. We decided to make sense out of them individually and collectively. Our main aim as facilitators was to be present, in the moment and to bring awareness to every interaction and within the community.

There were 18 facilitators cultivating the territorial communities of practice. The Moreno community was the most challenging to support and the one that needed it the most. One day the facilitator of this community received a message from a teacher asking for help on how to introduce the phenomena of mourning to the children at her school. This teacher was a participant of the Moreno community and had noticed that the daughter of the school assistant that lost his life in the explosion was part of her school. Accompanying her and other students in making sense of such a difficult situation became a priority to her. Through the work with the program she realized that strengthening and supporting her school community developed individual and collective resilience and was also an opportunity to put what she had learned about emotional education into practice as a contribution to the community. Collecting and responding to this need deepened the conscious awareness of what emotional education in real time meant to the facilitators. Furthermore, noticing these emergent learnings, addressing them as facilitators strengthen the relationships with the territorial communities infusing trust and humaning what was previously eroded through years of negligence. Being in service to the emergent needs of each community became the most pressing matter, the first step to regenerating the social fabric. Of course there was a curriculum to follow and content to address but the precedence here was to model in community the approach to such uncertain times. This was the beginning and the turning point.

7 Presence

The visit to the damaged school occurred on a rainy Wednesday. The principal and the teachers were preparing and decorating the building to open classes soon and received the children for the first time. The school structure was already fixed and painted as nothing happened, there was only one small metal sign that honored the event. The facilitator sent by the program to the school arrived at 9 am, she rang the bell and was received by the principal in her office. At the initial moments of that encounter, time felt like it slowed itself for some seconds, and all the fear of not being received or welcomed came rushing in and overwhelming the facilitator. In the mind of the facilitator a few judgments were playing on as a broken record, she did not look like from that neighborhood she was too tall, too white and too blond to fit in. She knew her physical appearance represented the spitting image of historical victimizers. She was at that moment representing the government that failed this public school hugely, and that inflicted pain by neglecting their requests. And as an addition she could not give any material promise of any kind. She breathed in and out deeply, she remembered herself to stay calm and open to whatever happened in that visit. She reminded herself that she also knew pain and suffering and that nothing human was foreign to her.

Sometimes really strong emotions can detach ourselves from the present moment and in that detachment lose our capacities to understand what is asked of us. This situation asked for a more humane response than what the government had historically offered. This situation asked for deep listening, challenging but calm conversations and profound empathy. This situation demanded the facilitator be present, open and vulnerable to embrace the pain and the love that was in every corner of the school. The quality by which we engage with experience is determined by our participation in it. The more we engage with curiosity and contemplation with ourselves and our surroundings the more information of a situation we can discover. If we fly by a situation or we want to control it most certainly we will have a shallow understanding of it or even relate to it from a representation that we have already stored in our mind. That day and during the whole visit the facilitator made herself remember every time she felt herself closing to the experience to open up to be at service to what the experience was asking of her. Through her discipline practice be present to what was emerging moment to moment.

After these two particular occurrences and many others in different territorial communities, the group of facilitators learned how to respond to the emergent needs. Their expertise and practice unfolded from experience to experience, to become more themselves and showing up humble for others. They embrace the task to reveal the power of human capabilities to respond to the current challenges, to realize that the true potential lies in the hands of the people and that as a nested community of networks they could tap in unlimited resources. That year the public school that suffered the incident became part of the school climate and emotional education program, they were interested in supporting their community emotionally and becoming part of a larger network. The facilitators community saw this as a positive response to their approach in such difficult times.

8 Psychological and Social Regeneration

In 2019 the School Climate and Emotional Education Program came to an end. The community of facilitators that learned and worked together for two years had to be dissolved. They were at the time experiencing something unique within the network of territorial communities of practice, a recovery, growth and strengthening of the social fabric within the educational community. The relationship between principals, teachers and students were infused with creativity and innovation. There was more participation of parents within schools. Communities were supporting each other and promoting emotional education within other public schools outside the network.

Facilitators understood at the time that experience is the starting point, the relational moment where learning occurs. That initial moment is all-encompassing, being awake to our *inner ecology* affects our capacities of meaning making. How we approach experience has an effect on learning, and has an effect on us as individuals. If we change the way we relate to experience we can draw new meaning from it individually and collectively and thus regeneration may occur. The coming into life with new meaning that arises from approaching experience holistically and collectively, can bring hidden potential to the surface and new connections to act accordingly.

The community of facilitators became a cooperative of professionals called Iniciativa Murmullo with the purpose to co-create the conditions that promote the development of conscious, reflective and collaborative practices, towards the evolution of people, communities and organizations. What those two years of practicing together taught them is that as we open ourselves with our most alive human capabilities individually and collectively we can regenerate and draw the path forward to create conditions conducive to life.

9 Quinta Esencia: Hands on Outer-Ecology

> Every great people, each culture, each form of life has its own world.
> In this way the Andean world has its own
> peculiar mode of being and therefore experiences,
> in its own way, the events of its life" (Fernandez 1998, p. 172).

Each place has its own essence, its own uniqueness of expression, its own particular environment. It is an alive, never-static, ever-changing field of entanglements that continuously expresses itself sometimes silently sometimes with a roar. In the Buenos Aires pampas near the Atlantic sea there is a 30 hectares place called Quinta Esencia, a kind of oasis placed in the midst of a rural area where large farming corporations live together with centenary traditional small farmers, and a place like this where new explorations around agroecology, permaculture and regeneration is happening Quinta Esencia is an eco- literacy center, located in a unique natural environment that has worked for the past eight years on the implementation of an inspiring educational model. They have engaged with the territory through working directly "on the ground" in areas such as regenerative food production, education, and the promotion of comprehensive sustainability policies in the local community. Quinta Esencia seeks for coherence in every activity they undertake, applying permaculture design at the heart of the school, its buildings are bioclimatic and the energy used is renewable, the workshops and processes they open for people are oriented to have a real immersion in a very practical way.

> Every time the sun rises, it rains or the wind blows, all the productive and energy systems are activated to the rhythm of nature.
> Christian Tiscornia (Quinta Esencia founding member)

In a conversation with Christian Tiscornia, one of the founding members, he shared with us that from the beginning they intended to be: "*a living classroom that starts from the premise that in order to learn to love nature, experiential and sensory experience are decisive*". The educational model (that is in itself an experimental project) promotes holistic education that encompasses head, hearts and hands in the learning process. Its mission is to inspire positive systemic change, contribute to the design of a new development model that is inclusive, regenerative and sustainable. They promote the search for a way of living in harmony with nature and community participation, reaching out to the world in a very particular environment, inside the municipality of Mar Chiquita, where they are able to experience the movements, the turning phases, the flourishing and decaying of their activities and projects. They embody a glocal (think global, act local) mindset, always holding close to their particular actions a way of thinking and responding to wider contexts and complexities.

What becomes interesting in their experience is that their learning is strongly tied to working in the field, touching the ground, being outside. The vegetable gardens, many of the buildings, and the agroforestry (that were built during different kinds of workshops) invite people to put their hands on the ground, offering an opportunity very quickly and in a direct way how our action creates new conditions and outcomes. Quinta Esencia has developed a way of intervening that particular space in a very particular time, finding their own way of reading context, intervening as a learning process, reflecting on their experiences and impacts, and finding new ways. As Berry (1987) writes "*What is here? What will nature permit us to do here? What will nature help us do here?*", their intention has been to grow a way of doing that is in resonance with life.

It hasn't always been easy for Quinta Esencia. From the very beginning, relating to the land took many years of observation (natural processes throughout the cycles of the year), which enabled the core funding team to relate differently and more deeply to the questions that Wendell Berry asks. Said by its members, they understood from those initial moments that rhythm and the quality of their presence would be central to the development of a way of intervening in that particular place. For us, who does not live in Quinta Esencia, and were not part of that living process of observing, intervening, learning, we must imagine (and perhaps even looking into our own places of action), how important it is to understand ourselves embedded in place, nothing lives un-related, everything moves into other things, and is moved and transformed by others. Building a practice around these ways of understanding intervention, lives at the heart of how the Quinta Esencia understands things should be designed and created in that place.

If we are attentive to the way things are being done at Quinta Esencia, Seamon's (2018, p. 88) understanding of a Sense of place might be very helpful. He proposed that place and place experience might be understood via the relationships among three components: environmental ensemble, people-in-place, and common presence. There is a physical environment that is being intervened and also shapes the way people participate in it. We cannot be without participating, even with the way we think and feel. They develop their interventions (people in place) being very attentive to the way things express in that particular place, and how it relates to their journey into a more coherent and strengthened way of holding Quinta Esencia's purpose alive inside all their struggles and challenges. Being in the midst of the pampas, surrounded by larger farms and more mechanistic and profit-oriented approaches, trying to be influential in the policy-making processes, must be a demanding task. The question perhaps is: what is holding them together? What unites them to engage in such a way? What would the land say about their way of living/acting in it?

> The relative togetherness of entities in space sustains an environmental common presence that emerges as a sensible quality shared by the entities that are a part of that space.
> Seamon (2018, p. 88)

10 Experiencia Raíz: About Practice

Experiencia Raíz was born out of the intention of creating a safe space for the exploration of our professional practices. What we mean with safe space: a place where we can find ways to observe and recognize how we are meeting the world, and how it speaks about who we are and the journey we are walking in our lives. It might sound quite romantic, and even abstract, but what has been at the center of their exploration is how we experience things. As we explored at the beginning, there is a quality in our presence and our awareness that enables us to connect more deeply with our experience, and through the engagement with it, finding threads of our own path. *How can we help our uniqueness to find ways of expression?*

One of the thinkers that influences thought and action at Experiencia Raíz is J. W. von Goethe, who besides of being a magnificent writer, through his scientific explorations he also worked intensely in discovering a way of approaching natural phenomena that carries a powerful seed for our current times. In his scientific work there are indications and references for the reader to understand what happened to him and his way of seeing nature. Through dedication to rigorous observation and reflection, and also integrating the perspectives of other contemporaries (who responded to his ideas and approaches), he discovered a particular way of engaging with nature that revealed more about the becoming of natural form, the qualities and nuances than only a physical-materialistic way of seeing.

> Every process in nature, rightly observed,
> awakens in us a new organ of perception.
> Johann Wolfgang von Goethe (Seamon and Zajonc 1998)

The rigor in this kind of observing has to do with strengthening our inner capacities to try and hold phenomena as "untouched" as possible by our limiting thoughts and conclusions. We meet the world through who we are, through our own life stories and experiences, and in doing so we have full participation in everything we see. The hard thing is that we usually don't see the way we see things, we somehow see through them, and it takes time and effort to recognize the qualities and consequences of our seeing. Goethe indicated (and can be explored through practice) that there is something revealed through us in this process, at the same time we discover more about the world, we discover more about ourselves. There is a conversation that is cultivated by our participation and our reflection, by our capacity to hold our inner world together with the world we meet through our perception. There is a healing process going on, inner and outer, when we accept things as they are, honoring their particular expressions.

At Experiencia Raíz, learning experiences are designed in order to facilitate this process to happen, some of them demand an extended time in nature, others might happen virtually. In all of them there is dedicated time in observation, in bringing together elements and situations that illuminate aspects of how nature expresses, how we have been expressing ourselves along our lives, looking for patterns that reveal an underlying movement, a thread that can be followed to deepening in our understandings of ourselves and the world. There is where the heart of a Practice beats, in the recognition that we are entangled with life, that we live in a context of action and sensemaking, and that we are fully participants.

Speaking to the funding team of Experiencia Raíz, they shared a story of one of their programs where they saw the expression of the kind of work they aim to do in the world. During a long walk in nature with a group, they finally reached a camping place that was just before the meeting of two rivers that became one. That camping place was situated a couple of meters away from the very place where one can see both flowing rivers melt into only one. One person was standing on a stone a meter away from water, just contemplating the situation. One of the facilitators of Experiencia Raíz approached and they entered into some kind of reflective-sensemaking conversation around what this place was offering to them. Questions about relationship, about polarities, about our inner and outer meeting ran into the conversation, almost as both rivers were inviting them to see more deeply into the phenomena. What is still separated in you? Asked Martin, one of the members of Experiencia Raíz, the conversation dived into places where both realized how our lives moved into paths and places where inevitably separation was demanded in order to understand and live things more fully. "Sometimes we need to split, in order to see" was said. A whole new range of possibilities for them was raised, nothing really shifted outside, but they felt different.

A Regenerative practice must be something close to this process, where our ways of understanding the world and ourselves travel together. They are dynamic, they move along our path through our intentionality, that must stay close to our own contexts. There are important things happening worldwide, huge issues to be addressed, and we have already distracted time in trying to find solutions that are scalable, that fit the needs of many, generating large impacts. Life doesn't work this way, like every tree that expresses its uniqueness in a very particular way in the particular context they inhabit, life moves trough intimacy, through closeness, through connection. There is also a conversation happening between the inner movement and intentionality living inside the seed, and the ever changing context it inhabits. Everything is always in relationship, so in order to address real situations and issues: *how are we approaching our contexts of action? What are our recognized ways of seeing? What remains still hidden to our consciousness?* Our consciousness moves together with our questions, specially with those still unanswered. Perhaps regeneration has more to do with the witnessing of those answers, and the new questions that emerge from our participation, as a vessel, creating space for the new, opening up to the world through wonder, just as a bee does when visiting a new flower. It demands our own will and rigor to stay awake, not falling into inherited findings, that limit our capacity of experiencing life as it is.

Goethe also spoke about the concept of *intensification* in his studies about optics and morphology. "Intensification is a change in quality which makes something new of the phenomenon, bringing forth an unexpected and higher form which Goethe would characterize as a clear manifestation of the underlying idea through the material nature of the phenomenon" (Miller 1995, p. 10). We are living beings participating in this one mother earth, so we also carry the capacity to experience intensification, to experience ourselves moving into new phases and horizons in our professional and personal lives. *How can we live intensely? What will move us into our own discovery processes?*

> However, the union may occur in a higher sense
> if what has been divided is first intensified;
> then in the union of the intensified halves
> it will produce a third thing,
> something new, higher, unexpected. (p. 10)

11 Conclusion: Returning Home

As a participatory practice, regeneration must start with wonder, opening ourselves to the situation or phenomena we are witnessing. It is a human attribute, to be aware of how we approach things, it is a moral act. To be aware demands presence, calls us to what is happening right here, right now. Like the conversations inside the community of facilitators, sharing situations that happened in each one's contexts, staying true to what emerged through recognizing emotions, and nuances of those experiences.

And also the particularities of intervention on the ground in the fields of Quinta Esencia, where knowledge and practice emerges through participating and acting in the visible world, finding alternative ways to those which brought us to this place. It does not mean to undervalue the past, but to integrate it to respond appropriately to the present.

Something is revealing out of those processes, something is becoming, we can feel it happening, we stretch ourselves into new questions and inquiries, we face our intimate challenges of what it means to stay active in these times. As we said, these are not easy times, we are faced with huge challenges for every living organism on this earth, we are not separate from: being perpetrators and victims at the same time.

> How can we be aware of those evolving processes?
> How can we develop the capacities to hold and enable emergence to happen?

Perhaps we are speaking to the heart of regeneration here, the underlying thread that can be found in many of these expressions of practice in so many different contexts. If they have to do with ways of acting that are conducive to life, we might be also speaking into the heart of life itself. Even if we are still finding our way into the fact that we are part of a larger living whole, trying to make sense of what that really means for us and our activities, and how it might shift our way of thinking and acting. There is still a long journey ahead for us, to trust in our own experience, and our own different ways of learning and knowing. It has been said to us that there are methods, tools and techniques that are going to help us in intervention, some of the hard work of understanding will be solved by others' inventions. And it is true, many of the tools and approaches are helping us in the deepening and questioning of our actions, they are helpful, but they can also lead us to the idea that they will read reality in our name, give meaning to it. Oversimplification, complexity reduction, risk management and other hidden intentions live below the surface of approaches and methods. *Where are we left then? Which human capacities are being tamed by imposing a particular way of seeing to the world?*

We can not heal or innovate standing outside of the present moment, this sounds like a cliche phrase but we keep on building plans and artifacts based on a different story that we tell ourselves to sleep. Waking up to the present disconnect is like walking through our worst fears, realizing we have abandoned ourselves, our children, our community and the earth that supports us. Just like Carl G. Jung (Beebe 1996) (9) said so many decades ago *"One does not become enlightened by imagining figures of light, but by making the darkness conscious."* Once we have accepted individually and collectively where we are at by honoring what we have lost, what has been eroded or damaged and who we are now, we are ready to own our story. This is the path of a living Regenerative practice, one that emerges from an individual (and collective) perception of the world that is not interfered with by elements that soften it. We are the most delicate and precise organ of perception, understanding lives at the heart of who we are as a species. We cannot give this gift to others, and certainly not to technology. We carry (we are) the algorithm of life, we know how to understand it, and intervene in such a way that enables it to express itself more fully.

Undelicate times come with unsettlement, a deep unknown of how to connect, repair, restore or heal ourselves and the world around us. Urgency and fear lead us to close ourselves, to look for what keeps us alive. It is not an easy task to remain open in these current times, but if we don't make that effort, nobody will. We can explore with curiosity, love and kindness our present pains and fears. *What are they trying to tell us?* What happens *when we are open to feeling them? What meaning comes through them?* Feelings remind us that we care, that we are alive and in communion with our environment. Life is in us and all around us. When we engage with it, our human creativity awakens.

A Regenerative Practice is a conscious doing that needs to be embraced with discipline and discernment from our inner ecology to care and transform our outer ecology. Practice is the continual involvement with the present experience that allows learning and meaning to emerge. We are inside of it, whether we like it or not. Through that continual movement (between the inner and outer) the path will unfold, because we are open to it, responsive and active. We will be able to make meaning of alive patterns within social situations and in relationship to the context, creating and intervening from within, helping the situation to reach a wider integrity. We will be able to understand potential and to nurture it with our participation.

References

Beebe J (1996) Jungian illumination of film. Psychoanal Rev 83(4):579

Berry W (1987) Home economics: fourteen essays

Bortoft, H (2016) Goethean science part 1, 31 August 2016. YouTube. https://www.youtube.com/watch?v=nsH6-n7BUtw&list=PLQzSL8qUPHUHipm7LWnBHm8lojKFT1SNo. Accessed 19 May 2023

Capra F, Wahl D (2023) "Regeneration is the essence of life's self-organisation" Fritjof Capra & Daniel Wahl in dialogue. https://www.youtube.com/watch?v=MiUGKfJV5E0. Accessed 19 May 2023

Dewey J (1938) Experience and education. Macmillan, New York

Fernandez EG (1998) Development or cultural affirmation in the Andes? In: Apffel-Marglin F (ed.) The spirit of regeneration: Andean culture confronting Western notions of development. Zed Books, London, pp 1247–145

MacIntyre AC (1981) After virtue: a study in moral theory. University of Notre Dame Press, Notre Dame, Indonesia

MIller D (1995) Goethe. Scientific studies, vol 12. Princeton University Press

Seamon D (2018) Life takes place: phenomenology, lifeworlds, and place making. Routledge, London

Seamon D, Zajonc A (eds) (1998) Goethe's way of science: a phenomenology of nature. State University of New York Press

Talbott S (2021) Gestures of life. https://www.youtube.com/watch?v=2mnTOFKyVkg&t=795s. Accessed 19 May 2023

Terrado D (2021) Regenerative education. Medium, 14 May 2021. https://delfinaterrado.medium.com/regenerative-education-536e6f2ec1fd. Accessed 19 May 2023

Van Manen M (2014) Phenomenology of practice. Routledge Press

The Importance of Nature

Introduction

De La Torre et al. propose a new approach to the relationship between nature and the city through the study of the urban ecotone, understood as the interface between the natural ecosystem and urbanisation, from a regenerative perspective, seeking alternatives to overcome the relationship based on extraction and degradation to achieve co-evolutionary cooperation, which should allow urban and socio-environmental regeneration through a relationship of care, cooperation and valuation. Regional development should prioritise the valuation of ecosystem services as a vital support for the future, based on the recognition of the dynamics of the living systems that interact in the urban ecotone. Regional development should incorporate the perspective of regenerative design in the urban ecotone and participatory processes to articulate the narratives of citizen groups and thus define the path towards strategies for a 'regenerative evolution' according to the potential of the place, from the logic of living systems and their health. A regenerative paradigm can and should contribute to the capacity of all natural, cultural and economic systems of a place to grow and develop their continued health and viability, highlighting the importance of a change in thinking about the relationship between residents and nature. The urban ecotone is seen as an opportunity to rethink the periphery and improve social and environmental health conditions. From the dynamic vision of the ecotonal situations, it is possible to analyse the ecosystem services provided by each situation and identify the conflicts and citizen initiatives present, generating a way to improve the relationship between urban communities and nature. Regenerative paradigm can and should contribute from the capacity of all natural, cultural and economic systems of a place to grow and develop their continued health and viability, highlighting the importance of a paradigm shift in the relationship between residents and nature. The urban ecotone is seen as an opportunity to rethink the periphery and improve social and environmental health conditions. From the dynamic vision of the ecotonal situations it is possible to analyse the ecosystem services provided by each situation and identify the conflicts and citizen initiatives present, generating the way to improve the relationship between urban communities and nature.

Lara and Crespo review the importance of vegetation and nature in cities with hot climates, where heat waves have become more intense in recent years, especially under the singularity of the urban morphology of historic cities and from a critical review that identifies their strengths and limitations, while helping to understand the resistance that these types of proposals encounter, It highlights the difficulties faced by historic and heritage city centres in increasing the vegetation in their urban areas to effectively provide them with ecosystem services and points out the opportunity that exists in the interior of housing blocks in the historic urban fabric, which represents potential green areas with the capacity to become climatic refugees. The regulations for the protection of architecture and building typology in these cities are an important obstacle to increasing and promoting the presence of plants integrated into buildings in the already scarce free public spaces. Traditional cultural practices linked to the use of vegetation, which are already part of the socio-cultural landscape of the city, are also perceived as opportunities to devise practices that favour denaturalization and the knowledge of how the initiative develops in the future will offer keys that can be of great use for similar initiatives committed to achieving the desired sustainable transition for our cities.

Martín and Durán delve into the importance of affiliation with nature through biophilic design for the process of developmental evolution and regenerative design. With a thorough systematic review of the concepts of Biophilia and Biophilic Design in the Latin-Mediterranean context, they highlight how the topic is currently of great interest. There is a large scientific production on the subject in countries such as the United States, with a long way to go in Latin America to achieve these production quotas. There has been a boom since the COVID-19 pandemic and it would be encouraging if this trend were to continue, given the diverse benefits of Biophilia and Biophilic Design, increasing funding available for research on these topics, and there is a significant area of opportunity for Latin researchers to participate and collaborate. There are links between themes associated with biophilic design and the terms Architecture, Analysis, Sustainability, Strategy or Art, aspects that can be considered to be proper to the profession of architect, but also linked to the theme of Emotion, the term Effect and model, where aspects such as Change, Use, Interaction, Order, Climate Change and Temperature, which have to do with exterior or atmospheric aspects, are included. The main research in Spanish in the Latin-Mediterranean Region comes from Mexico, Colombia, Chile, Italy and Brazil. It is interesting to note that, both in terms of the volume of production and the connections between co-authors, the need to establish more links and relationships between Latin-Mediterranean researchers.

Exploring Urban Ecotones and Citizen-Driven Regenerative Actions

María Elena de la Torre, Miguel Ángel Bartorila, and José Javier Alayón González

Abstract The study of the urban ecotone of the Santiago River Canyon presents a renewed perspective on the nature-city relationship. This chapter introduces an alternative approach to transcend the extractive and degrading relationship, shifting towards a coevolutionary collaboration that enables the regeneration of the dual marginalization along the northern periphery of the Guadalajara Metropolitan Area. This is achieved through fostering a relationship based on care, cooperation, and appreciation, where regional development prioritizes the assessment of ecosystem functions as a vital foundation for the future. Drawing from the dynamics of interacting living systems within the urban ecotone, and considering the current era marked by an increased awareness of nature's alternatives that allow for harmonious cohabitation with and respect for nature. The study proposes that the agendas of these citizen initiatives within ecotonal contexts offer an alternative path towards regeneration, involving forms of protest, mobilization, and creativity. However, advocating for the preservation of living systems and a renewed approach to inhabiting the urban ecotone is not exempt from encountering confrontation, conflict, or suppression from urban development stakeholders. The process of regeneration draws upon specific resources: natural remnants, the consciousness of local inhabitants, and regenerative actions promoted by citizen potential to preserve planetary conditions in the face of the climate crisis, we observe two conflicting dynamics. On one hand, there exists an assessment of the financial potential of the territory linked with intensive urban development, often at the expense of essential ecosystem functions. On the other hand, various citizen-driven initiatives are emerging, fueled by the exploration of groups. These agreements bolster spaces for exchange, aligned with the principles

M. E. de la Torre
School of Architecture, Art and Design, Tecnologico de Monterrey, Guadalajara, Mexico
e-mail: medelatorre@tec.mx

M. Á. Bartorila (✉)
Facultad de Ingeniería, Universidad Autónoma de Querétaro, Querétaro, México
e-mail: miguel.bartorila@uaq.mx

J. J. Alayón González
Departamento de Arquitectura, Pontificia Universidad Javeriana (PUJ), Bogotá, Colombia
e-mail: alayon.j@javeriana.edu.co

of living systems to enhance health, thereby fostering increased opportunities for meaningful dialogue.

Keywords Biohistory · Ecosystem services · Regenerative design · Community initiatives · Coevolutionary mutualism

1 Introduction

The study of the urban ecotone of the Río Grande de Santiago Canyon proposes a new approach to the relationship between nature and the city. From the regenerative perspective, alternatives are sought to overcome the relationship between extraction and degradation to reach coevolutionary cooperation. This perspective should allow the regeneration of the urban and socio environmental marginalization of the northern edge of the metropolitan region of Guadalajara through a relationship of care, cooperation, and appreciation. The development of the region should prioritize the assessment of ecosystem services, as vital support for the future.

From the recognition of the dynamics of living systems interacting in the urban ecotone, and from a greater awareness of the potential of nature as a way to preserve the state of the planet in the face of the climate crisis, two dynamics emerge: first, that of intensive urban development at the expense of ecosystem services and second, the emergence of citizen initiatives motivated by the search for alternatives to cohabit with respect and care for nature. This chapter incorporates the perspective of regenerative design in the urban ecotone and articulates the narratives of citizen groups to define the path towards possible strategies for a "regenerative evolution" according to the potential of place.

The study suggests that the agendas of citizen´s groups in ecotonal situations point to an alternative path to regeneration from protest, assembly power and creativity. However, the aspiration to care for living systems and create a new way of inhabiting the ecotone encounters confrontation, conflict, or oppression from those who promote real estate development. Regeneration starts from specific resources: natural relicts, inhabitants' awareness of the environment and regenerative actions promoted by social groups. The social agreements strengthen the spaces for exchange with the logic of living systems for health and provide more opportunities for dialog.

A paradigm shift regarding the relationship between inhabitant and nature requires a closer look at the territory in which this exchange takes place. Giacomini (1979) stated that the urban–rural interface is the area where the most authentic and essential ecological problems take place, noting that this interface is the culminating place of the ecological commitment to the city. The interface is then understood as a spatial reality capable of resolving certain ecological conflicts which in turn, is the area where solutions to urban complexity are proposed. The concept of urban ecotone (Bartorila 2001) refers specifically to the processes and spatio-temporal transition in the simultaneous study of the city and its natural environment. The term has its etymological root in the Greek "eco", meaning home or habitat, and "tone", meaning tension. Urban ecotone represents both the intended equality of ecological and urban approaches and the hinge that allows the convergence between two realities with different logics.

Cities are constantly reinventing themselves due to their complex and particularly incomplete state (Sassen 2015). Thus, regeneration of the periphery through urban ecotones should therefore contribute to the definition of a new participatory approach. It is paradoxical that at the edges of urban sprawl, where the transition to natural or rural areas begins, pollution or environmental degradation affects the quality of life. However, this interface between natural ecosystem and city, defined as urban ecotone, offers a new opportunity to rethink the periphery. This encounters a threefold condition: the spatial, which is obvious, since these are areas far from consolidated and well-equipped urban centers; that of knowledge of the territory, since specific information is scarce and superficial; and finally, the few environmental regeneration measures, without systematically solving complex problems.

The work presented here was carried out in three stages. First, a new approach to the knowledge of the territory in the urban ecotone from a social and ecological point of view is presented. In a second stage, a diagnosis based on cartographies, workshops and interviews is described: the intersection of services, conflicts, and initiatives. Finally, considering the prospective interpretation of the practices of the pioneer collectives, some regenerative strategies and their actions are highlighted.

2 Urban Ecotone: Interface for Coevolutionary Mutualism.

The relationship between the natural ecosystem and urbanization offers opportunities to improve social and environmental health conditions. In this sense, it is known that the important potential of community initiatives focused on the care of the environment can influence public policies towards a better protection of ecosystems. Therefore, it is believed that a more harmonious and reconciled relationship between the city and ecosystems can be achieved at this interface.

The evolution of the natural environment and the urban environment have important similarities. On the one hand, ecosystem succession shows the dynamics of living systems and their spatial progress. On the other hand, urbanization and the cumulative development of infrastructures consider natural elements as obstacles. The relationship between ecosystems succession and urbanization processes, understood as evolutionary simultaneity, can be better explained by the concept of coevolution as mutual adaptation. The term coevolution was coined by Ehrlich and Raven in 1964 (Gual and Norgaard 2010). From a new perspective of the city and territory, this concept corresponds to both the historical adaptation process of ecosystems and that of the urban artifact. The coevolution of ecological networks and infrastructures shows parallelism and develops from a cumulative evolution.

Janzen (1980) points out that coevolution is related to specificity, reciprocity, and simultaneity. Coevolution in urban ecotones shows mutual evolutionary change due to this interaction: (a) specificity in the evolution of both the ecosystem and the urban artifact; (b) reciprocity, where both forms must co-evolve; and (c) simultaneity, where both manifestations evolve simultaneously.

Coevolution is the joint evolution of species that interact with each other. Mutualism and symbiosis are considered a form of coevolution of a positive, shared, and reciprocal nature between partner species that benefit from each other in terms of survival, growth, or ecological fitness. An example of coevolutionary mutualism is pollinating animals that feed on the nectar of flowers and disperse the pollen that helps them reproduce. In this interaction, both species benefit (Fig. 1). Similarly, this approach aims for both the urban ecosystem and the natural ecosystem interacting in the ecotone to develop a symbiotic relationship where both receive benefits and evolve together.

Based on a regenerative approach, the ability to build mutually beneficial relationships between communities and their ecosystems is believed to be achieved when the healthy qualities of an environment are present, and their benefits are understood and appreciated over time. That is, the reciprocal relationship among caring for the environment, economic entrepreneurship and social development occurs through a coevolutionary relationship.

From this perspective, regenerative development and design mean the reconnection of human aspirations and activities with the evolution of natural systems, essentially their coevolution. This means bringing human communities and economic

Fig. 1 Hummingbird Thalassinus (Mexican Violetear). *Source* Juan Jorge Hermosillo. Zapopan, Jalisco

Fig. 2 Río Grande de Santiago Canyon. *Source* authors file

activities into harmony with life processes. It means, then, that every human settlement is organized around the evolution of the watershed´s capacity to sustain life. Creative and economic activities can be directed towards the development of human potential by harmonizing the dynamic energies of nature (Mang and Reed 2012). In this sense, the starting point in urban ecotones means recognizing the latent potential of this biogeographical environment, from the essence of place (Fig. 2).

Understanding of place through local narratives, the innate and unique qualities are revealed and what it can evolve into. It is also essential to understand the system of nested relationships to which the place belongs in order to recognize its great potential, which goes beyond the specific local conditions that characterize it.

2.1 The Urban Ecotone in Its Bioregion

The Río Grande de Santiago Canyon is in the Central American realm (Neotropical) and is part of the Bajío Dry Forest ecoregion located in west central Mexico. It is part of the central trans volcanic belt. Its strategic location between two biogeographic areas (Nearctic and Neotropics) and in two physiographic provinces (Sierra Madre Occidental and Neo volcanic Axis) gives this region biological richness (Morrone 2005). In addition, the abrupt change in elevation allows the formation of microclimates and greater biodiversity (Fig. 3).

Fig. 3 The bioregion of dry and coniferous forests of Mexico. *Source* Modification by the authors from original image taken from https://www.oneearth.org/

The Santiago River Canyon is considered an important ecosystem in the region due to its biological richness (Valero et al. 2019; Schipper 2023). The Santiago River connects Mexico´s larger lake with the Pacific Ocean and plays an important role in the hydrologic cycle of the region (Fig. 4). However, it is in one of the most developed areas of Mexico, where agricultural, livestock, industrial, and urbanization pressures pose the greatest threats for ecosystems. Due to excessive deforestation and fragmentation by humans, a few relict habitats are found on isolated peaks and ridges, and on steep slopes in the case of the canyon.

The Mexican government´s environmental agency, CONABIO, has identified priority conservation actions at the national level: (1) increasing the number and connectivity of protected areas; (2) promoting outreach activities and educating local communities about the benefits of biodiversity and sustainable resource use; and (3) implementing restoration processes in degraded forest areas. These strategies must be implemented at the local level to be effective. This approach addresses the methodology of ecotonal situations.

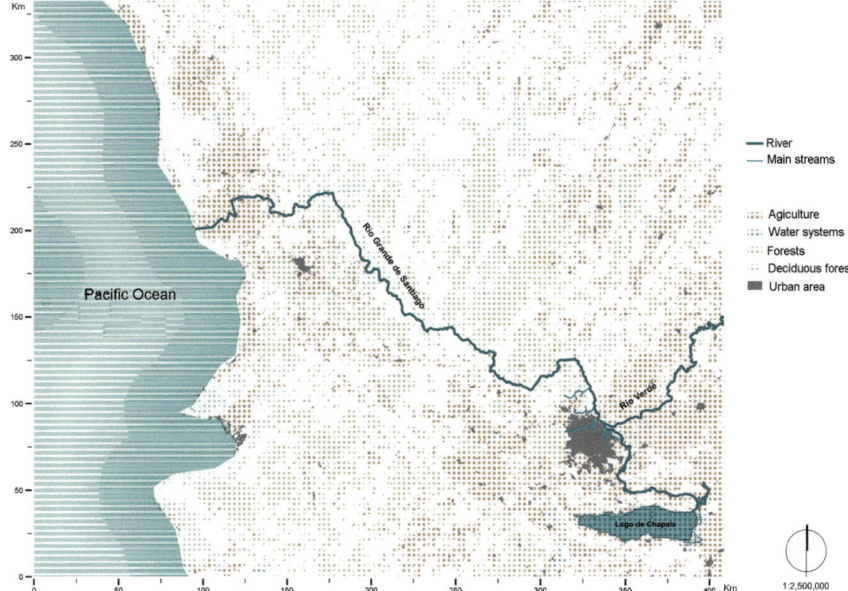

Fig. 4 Location diagram of the Santiago River in western Mexico. *Source* prepared by the authors

3 Ecotonal Situations: Ecosystem Services, Conflicts, and Initiatives

Ecotonal situations represent a dynamic vision between the previous situation, the current situation, and the foreseeable situation. In short, the intra-urban, peri-urban, and rural ecotonal situations are perceived (Fig. 5). These situations have a didactic character: they are the result of the interpretation of a previous reading about the coevolution of Guadalajara and its bioregion. This perspective interprets the reciprocal process of the urban artifact and the natural ecosystems as areas of encounter or barrier. In each ecotonal situation, the different ecosystem services, conflicts and citizen initiatives were identified through documentation, ethnographic studies, and community workshops. (de la Torre Escoto et al. 2022).

Ecosystem services are useful for analyzing the relationship between the ecosystem functioning and human well-being. This is key to linking the maintenance of the ecological functions of the urban ecotone with the quality of life of the community. Based on social perception and a multidimensional context diagnosis, the critical conditions in each ecotonal situation have been identified (de la Torre Escoto et al. 2022). According to the study "Ecosystems and Human Well-being: Synthesis" (Millennium Ecosystem Assessment 2005), ecosystem services are classified into the following categories: (1) provisioning services (timber, fossil fuels, medicinal plants, freshwater, food, fiber, energy); (2) regulating services (water purification, pollination, pest control, flooding, disease control, climate regulation);

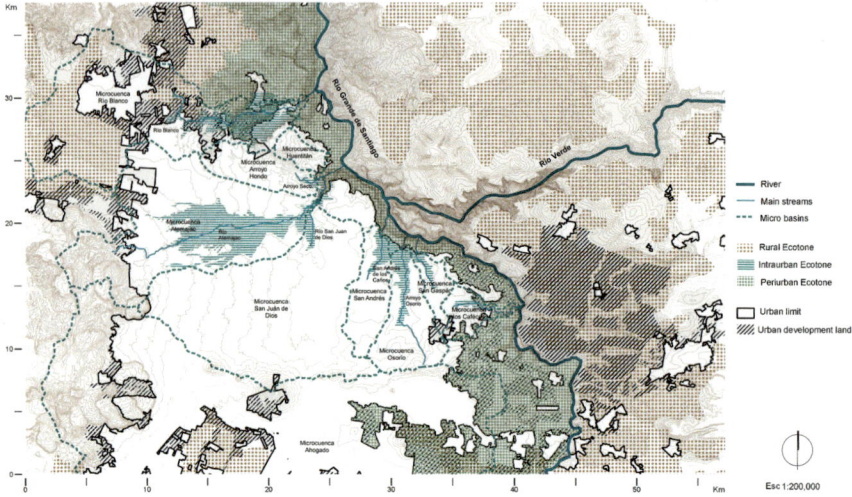

Fig. 5 Ecotonal situations and ethnographic studies location. *Source* prepared by the author

(3) cultural services (spirituality, education, recreation); and (4) supporting services that sustain all others (plants and wild life habitats, nutrient cycling, seed dispersal, soil cycling, primary production) (Fig. 6).

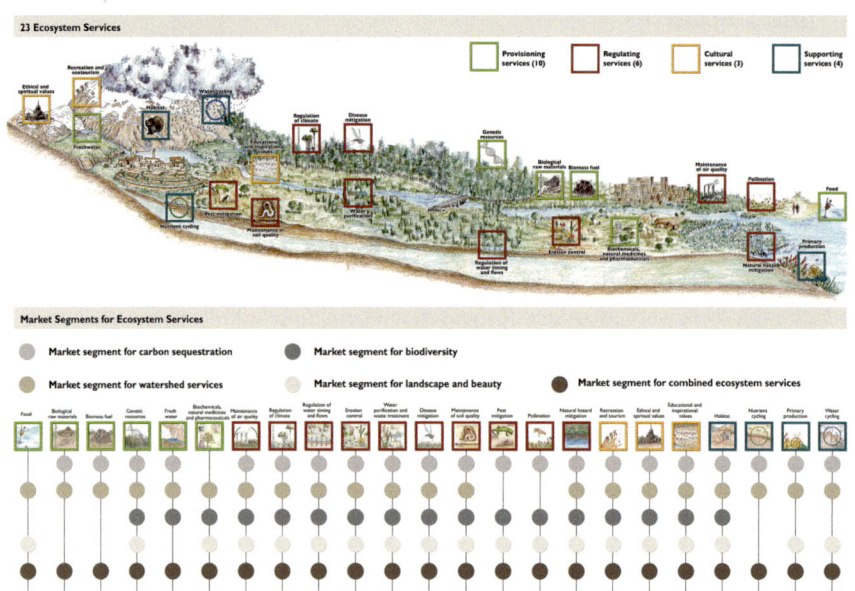

Fig. 6 Ecosystem services. *Source* Creative commons, creator Grid Arendal https://www.grida.no/resources/8141

To identify the specific ecosystem services perceived in each ecotonal situation, 30 interviews were conducted with citizens and organizations living or working in these areas (April 2021). Asking about the natural history of the place and the perceived environmental values allowed classification by ecosystem service categories. Likewise, specific technical information for the peri-urban ecotonal situation was obtained from the decree of protected natural area under the category of "Natural Formation of State Interest of the Barrancas of the Santiago and Verde Rivers." (Gobierno del Estado de Jalisco 2015).

From the citizens point of view, and despite the socio-environmental conflicts in each ecotonal situation, it is still possible to find positive aspects related to landscape assessment, environmental awareness and community organization. Research in this area has made it possible to identify various citizen initiatives that solve problems through their own regenerative practices, achieving both social and environmental benefits.

3.1 Intra-Urban Ecotonal Situation

This ecotonal situation is related to the strip formed by the tributaries of the Santiago River whose edges have been precariously urbanized with informal settlements. The socio-spatial structure of intra-urban ecotonal situations shows overpopulation with different degrees of marginalization.

The ecosystem services identified in the intra-urban ecotonal situation are:

<u>Provisioning services</u>: Family gardens, flower planting, clean water springs.

<u>Supporting services</u>: Relicts of biodiversity.

<u>Cultural services</u>: Parks, viewpoints, bathing places, floral landscape, ecotourism and canyon culture, legends and stories, memory of positive ecosystem values, good air quality in some areas (Fig. 7).

Here are the most densely populated areas of the metropolis (Fig. 8). The provision of land without urbanization or services, known as informal urbanization, was the only alternative for many families to obtain land to build a home. The infrastructure, services, and facilities that settlements require implies complex land regularization processes and the economic capacity of poorly funded local governments. Encroachment into the natural edge results in gradual degradation of ecosystem values. Rivers and streams are treated like sewage by domestic and industrial uses. These conditions affect the health and quality of life of residents (Pasarín and Diez 2013). In addition, both the regenerative capacity of ecosystems and the provision of ecosystem services are reduced.

In terms of socio-environmental conflicts in informal settlements, water and odor pollution, garbage, dengue fever, respiratory diseases, gastrointestinal diseases, kidney diseases, vandalism, insecurity, drug use and distribution, teenage pregnancies. Among the neighborhoods on the edge of the Atemajac, San Juan de Dios and San Gaspar streams there lies a landscape of risk and abandonment caused by treating riverbeds as waste sinks and exterminating flora and fauna.

Fig. 7 **a** Atemajac River in Guadalajara municipality **b** San Gaspar stream in Tonala municipality. *Source* authors file

3.2 Peri-Urban Ecotonal Situation

This is the space between the urban limit and the south bank of the Santiago River at an altitude between 1,540 and 800 m a.s.l. This strip is located on a steep slope and forms a riparian ecosystem with great biodiversity and exceptional geological formations. With few points of contact or integration with the urban artefact, this ecosystem is ignored by much of the metropolis on the plateau. The ecosystem services are more evident, and due to the steep slope, that makes up the canyon, natural conditions are preserved despite latent environmental degradation. The few accesses and visibilities of the city to the ecosystem show the indifference or lack of vision to reconcile urbanization at the edge with this extraordinary environment. The unclear relationship between urbanization and conservation makes this edge a diffuse boundary and shows that urban planning and management tools are insufficient to

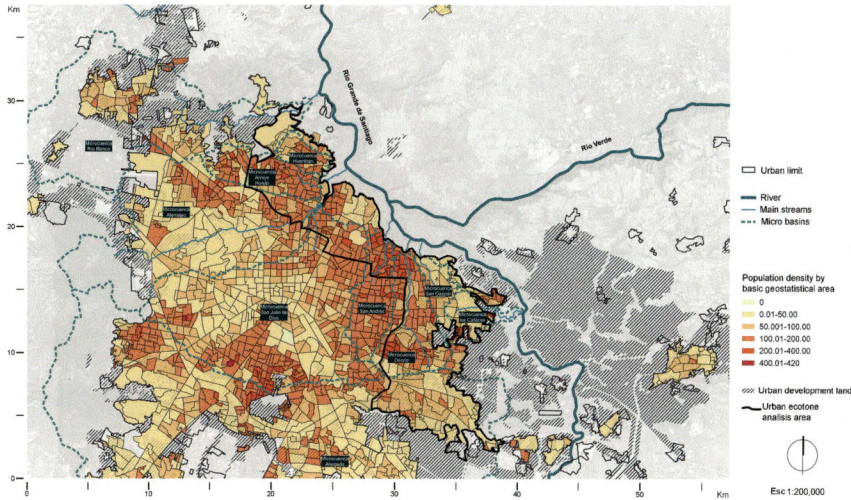

Fig. 8 Population density by AGEB 2020. *Source* prepared by the author based on IMEPLAN (2016) and INEGI (2020)

effectively regulate and protect the edge, even though 21,422.98 hectares in nine municipalities are protected by decree (Government of the State of Jalisco 2016).

One of the main challenges for the urban ecotone is the interface between two types of land regulation tools: conservation area regulation and urban development plans. On one hand, the natural reserve recognizes the area of influence as the contiguous territory that influences the environmental processes and events that affect the ecosystem. Management programs for natural areas only make recommendations, but they do not have the authority to regulate what happens beyond their boundaries. On the other hand, the urban development plans of the area of influence mention that urban actions may only be carried out if they are compatible with the management program of the adjacent natural area (Ayuntamiento de Guadalajara 2022). Neither of these land regulation tools, however, is specifically designed to control the impacts of development on the natural urban fringe.

On the other hand, the Santiago River system has been exploited for hydroelectric power since the late 19th century (Ramos-Gutiérrez and Montenegro Fragoso 2012). Since then, pressure on the canyon and river has increased due to population growth, land use change, water and solid waste discharge, and construction of water infrastructure. The extraction, storage, transfer and massive detour of water from rivers for various purposes have altered the environmental characteristics of connectivity and habitat (Gómez-Balandra, Díaz-Pardo and Gutiérrez-Hernández 2012), resulting in a decline in organisms, a loss of species, and a change in ecosystem balance.

In October 2021, a workshop was convened with the groups that defend the environmental values of the Barranca del Río Santiago to identify the common agendas among them and the possible collaboration for the conservation of the ecosystem.

Fig. 9 Canyon landscape. *Source* Authors archive

In this peri-urban ecotonal situation, the communities have identified the following ecosystem services:

Provisioning services: nopal, corn, Barranqueño mango (endemic mango), plum, chia, family gardens, flower cultivation, clean water sources, mining.

Regulating services: Climate thermoregulation, water collection and filtration, soil formation and erosion control, air quality regulation, good air quality in some areas.

Supporting services: refuge and biological corridor for flora and fauna, relics of biodiversity.

Cultural services: Local culture related to the characteristics of the canyon, stories and legends, research, parks, viewpoints, bathing areas, floral landscape, scenic value, recreation and ecotourism, rappelling, hiking (Fig. 9).

According to the Estudio Técnico Justificativo y Programa de Manejo de las Barrancas del río Santiago y Verde (Secretaría de Medio Ambiente y Desarrollo Territorial 2015), the main socio-environmental conflicts are: urbanizations in risk areas such as canyons and riverbeds, poor water quality of inner-city streams in waterfalls toward visitor areas, flooding in urban canals, illegal dumping, pollution of the Santiago River. According to citizen´s groups, the problems of the canyon are related to the lack of management and coordination capacity of the different levels of government to protect natural areas, the unregulated impact on marginal settlements, government megaprojects in the area, perceived insecurity due to drug and alcohol abuse, and the presence of graffiti. In addition, there are conflicts such as illegal entry into the protected natural area, ignorance of risky trails and areas, extinction of flora and fauna, lack of maintenance of trails and roads, visual, olfactory and acoustic pollution, garbage, abandoned animals, and vandalism.

3.3 Rural Ecotonal Situation

These are non-urbanized areas with agriculture rural communities. The ecosystem services identified in the rural ecotonal situation are:

Provisioning services: Cultivation of nopal, corn, barranqueño mango, plum, chia, vegetables.

Regulating services: Climate thermoregulation, water collection and filtration, soil formation and erosion control, air quality regulation.

Supporting services: Refuge and biological corridor for flora and fauna.

Cultural services: Scenic and landscape value, hiking, rappelling, research.

The following environmental conflicts are identified: Pollution of the Santiago, Blanco, Juchipila, and Milpillas rivers with the presence of leachate affecting water and soil, pollution of food sources: Fishing, orchards, vegetable production, lack of water supply, flood; pests in plantations, respiratory and gastrointestinal diseases, cancer; poisoned livestock, migration and displacement, poor public services, lack of environmental awareness, crime and drugs.

In the rural area of Zapopan there is a Municipal Hydrological Protection decree (Congress of the State of Jalisco 2004). However, there is a serious socio-environmental problem in two specific places. In the first case, residents of the Ex-Hacienda del Lazo lived from fishing for many generations because of its proximity to the Santiago River. Four decades ago, dead fish were found on the banks of the river. As a result of this environmental pollution, residents began to grow fruit trees and vegetables. However, about twenty years ago, most of the trees in the historic orchards were affected by contaminated water. As a result, the remaining residents of the town now only grow nopales (Opuntia ficus-indica), the only edible species relatively far from water and soil pollution (Fig. 10). In the second case, the water consumed by the indigenous population of San Francisco de Ixcatán comes from ancient springs. Private settlements on slopes not only affect soil quality and affect the aquifer, but also have drilling permits which are denied to the indigenous organizations. As mentioned earlier, landfills contaminate the aquifer and allow leachate to filter through the earth into nearby streams such as La Soledad or Milpillas.

In the three ecotonal situations, citizen initiatives are seen as seeds of change to protect ecosystem services. Civil society has been organized to work on solutions and develop a better relationship between nature and society. Ochoa García (2012) recognized community organizations in Jalisco that expose conflicts related to water, mining, biodiversity, forests, or conservation of natural areas. They have in common that they speak out against environmental degradation and the violation of collective rights. They form networks with the aim of fully disclosing the damage and injustices they suffer. The people affected by the environment, especially the most vulnerable, show a material interest in the environment and their territory, since this is the source and condition of their livelihood (Martínez González and Hernández González 2010).

Fig. 10 a Nopales field **b** local authority next to a polluted water canal. Ex Hacienda de El Lazo, **c** Río Santiago. *Source* authors file

4 Canyon Pioneer Citizen Groups: Practices

Canyon Pioneer citizen groups are non-governmental organizations whose activities focus on protecting, valuing, and promoting the "Barranqueña culture" (local culture associated with the characteristics of the canyon), a mix of natural and cultural aspects. Despite their different agendas, they have in common the promotion of a harmonious relationship between the ecosystem and the communities that inhabit and visit that place.

Social networks and press releases were used to identify the citizen groups that would later be interviewed at their activation sites. They were also invited to participate in a workshop to identify their common agendas and possible alliances for working together to conserve this natural heritage. This workshop was held at IMDEC (Instituto Mexicano de Desarrollo Comunitario). All previously identified groups defending the environmental values of the barranca were convened. Neither CEIBA nor Huerto Tlacuache attended this workshop due to their agendas. Through their testimonies, it was possible to verify the diversity of their goals and circumstances, having in common the search for alternatives in their relationship with nature as a source of regenerative health. The differences between the groups are characterized by the nature of their relationship with the territory and with nature.

In the rural ecotonal situation, the effects of the current urban development model are evident, where water and soil pollution affect their production and livelihood. In the peri urban ecotonal situation, as stated by the community guardians, the love for the Canyon is the reason for their activity to promote environmental values among the hundreds of athletes who descend the sloping trails every day. Finally, in the case of the intraurban ecotonal situation, there are two groups. On the one hand, the proponents of urban agroecology, and on the other hand, the network of neighboring groups in the Huentitán area that demand the protection of the environment against real estate development accepted by the municipality in different administrative periods (Fig. 11).

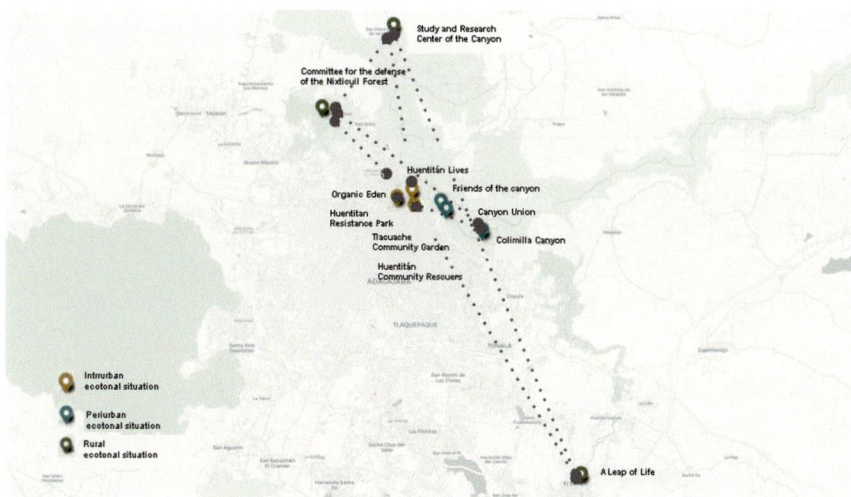

Fig. 11 Map of social organizations on the urban ecotone. *Source* Prepared by the authors http://umap.openstreetmap.fr/es/map/defensores-medio-ambiente-ecotono-urbano_681197#11/20.6594/-103.3474

The following are the environmental reasons of some of the groups interviewed in the study area:

In the rural ecotonal situation there is the Center for Studies and Research of La Barranca (CEIBA). It is composed of a group of inhabitants of the villages of the canyon (Jalisco). It was founded in 2021 by the General Assembly of community members of the indigenous community of San Francisco de Ixcatan. They carry out regenerative community activities (orchard, vermicompost, nutrient cycle). They recognize rivers as living beings and the regenerative dynamics of water (Fig. 12). Popular knowledge, indigenous roots, ethnobotany are valued, and archaeological remains are protected. They promote rainwater harvesting in the domestic settings and recognize the peaceful environment as a value. Their main activities include: La Ceiba community library, courses in traditional medicine, reforestation, hiking, research, workshops, and festivals. https://ceibarranca.com.mx/

The Community Rescuers Network was identified in the peri urban ecotonal situation. This group was founded 42 years ago in the Huentitán Canyon. It carries out various activities such as care and cleaning the canyon, although it is not always supported by the authorities. The rescuers are inspired by the love for the beauty of this space and are aware of the ecosystem services for the preservation of life in the region. They work actively to preserve the flora and fauna, as well as to provide environmental education on the bio history of this territory.

Key activities include assisting visitors and support with rescue operations, marking trails and pathways, creating the general and universal map, identifying sites, maintaining, and protecting flora and fauna. They participate in updating the risk atlas, knowing in detail the risk points. They analyze content on social networks

Fig. 12 Barranca in Ixcatán. *Source* authors archive

to avoid misinforming visitors, they promote the language code to ask for help, and reforestation with personal control, among other activities (Fig. 13).

In the intra-urban ecotonal situation, there is Edén Orgánico, a group of women who promote and work urban gardens on the edge of the canyon in Lomas del Paraíso neighborhood. Founded in 2005, the women have become a reference point in the agroecological agenda in the Metropolitan area of Guadalajara and its surroundings. They have achieved this by producing compost, growing organic vegetables, and selling food made from the vegetables they produce. They also promote environmental education and training for other women. Key activities that support ecosystem services include maintaining the earth´s cycle through cultivation, reducing erosion by planting native and edible vegetation, and promoting neighborhood interaction. They offer alternative prehispanic food courses (Fig. 14).

The main concerns of Eden Orgánico are the constant plagues, drought, hail, and the interruption of the cycles of nature due to human factors. Another problem is the lack of environmental education and community participation. The group aims to increase community by promoting active listening and creating a support network.

4.1 United Citizens Groups of Huentitán

One area that requires special attention in this study is Huentitán, located between the Atemajac River and the San Andres Stream. This section of the urban ecotone manifests a historical wealth significant for the metropolis. Settled since pre-Columbian times, its name in Nahuatl means "place where offerings are made." In this point, roads led to the north of the country, giving rise to various cultural and commercial

Fig. 13 **a**, **b** Hikers, **c** Huentitan community rescue group, **d** barranca view, **e** path and signs. *Source* authors file

Fig. 14 Urban Garden Edén Orgánico. *Source Estefanía Hernández Cerpa.* https://www.zonadocs.mx/2019/09/04/el-eden-organico-guardianas-de-la-tierra-las-semillas-y-la-vida/

exchanges. The place was the scene of fights between the Huentitán Indians and the Spanish, as well as battles during the Mexican Revolution and the Cristero Rebellion.

In the early 20th century, the Rio Grande de Santiago or Santiago River became the most important resource for the hydroelectric industry. Dams, roads, and buildings that housed hydroelectric plants were built in various locations along the river. The energy supply enabled the development of industries and transportation that influenced the city expansion. The canyon also became a place for hiking and for the increasingly frequent use of the hot springs on the outskirts of the city. In the Huentitán area there is the Las Juntas hydroelectric plant, built in 1901 and whose tracks, known as Malacate, connected the city to the worker's camp. Today it is a landmark for the practice of sports and ecotourism.

Currently, Huentitán stands for the defense of the natural values of the city. Over several decades, different groups have formed, differing in the way they interact and mediate with the local authorities. Although they have been torn apart by different conflicts, they all have in common the defense of the environment and the recovery of public spaces. The different groups demand that this area become the buffer area of the Barrancas del Río Santiago and Río Verde Natural Protection Area because they are strips of vegetation incorporated into the landscape that influence ecological processes and provide a variety of goods and services. They want this site to be declared Municipal Ecological Park.

The social groups recognize the urban edge as an Environmental Management Unit-209 (UGA). This area is the sum of relevant public and natural spaces such as the Huentitán natural park, the Guadalajara Zoo, the El Disparate, an area of more than 50 hectares, and the Mirador Independencia Park. The UGA is defined in the "Modelo de Ordenamiento Ecológico Territorial del Estado de Jalisco. Municipio de Guadalajara", according to the decree Published on July 28, 2001: Publication of the Agreement on the Territorial Ecological Planning of the State of Jalisco, in the Official Newspaper "El Estado de Jalisco" and its modification on July 27, 2006.

The group called Tlacuache Community Garden was interviewed. It is an autonomous social group founded in 2017 and based on agroecology, environmental education, and social occupation of common spaces, specifically the Huentitán Natural Park. They support community development and food sovereignty. Among the activities they carry out are solidarity work of tequio for milpa planting, organic farming, compost, and backyard gardening workshops (Fig. 15). The group also organizes forums and festivals such as the so-called #SinMaízNoHayPaís (Without corn there is no country). They were key actors in the activities of the so-called "Parque Resistencia Huentitán", a 13-hectare plot of land in Huentitán. It was occupied by different groups for four months to demonstrate against real estate development on land designated as a green space in 1980 (through decree N° 10,421 of the Congress of the State of Jalisco). In 2008, this land was ceded by the municipality to a foreign company to develop a real estate project, and in 2012 this company ceded its rights to another similar company.

In a manifesto, they call for new ways of inhabiting cities and interacting with people and nature, "in the face of a common crisis, alternatives must be built in common". They promote resilience processes of degraded ecosystems, and protection

Fig. 15 a Huentitan Park and **b, c** Huerto comunitario Tlacuache. *Source* Authors archive

of those that are preserved. Barter o Trueque as exchange of popular knowledge and research. Action as a proposal to overcome struggle and resistance, to promote community and ecological life.

Located near the Barranca de Huentitán, Distrito Iconia is developed on just over 13 hectares, offering various apartment alternatives. The company's advertisement says: "Our apartments for sale in Guadalajara have the best amenities for a future life project, is your best real estate investment thanks to the added value offered by its location and proximity to various points of entertainment and growth." As an alternative to this real estate project the group proposes a space to promote regenerative culture:

- An educational, organizational, and communicative space to transform reality. "A school of resistance, to raise awareness and learn respect and care for nature."
- A place to restore collective memory. "Milpa resistance, bioconstruction, herbal medicine. The place to claim our cultural identity, our deepest identity".
- A place to connect with the land and a space for diversity, community, and action.
- An intergenerational place.
- A school to transfer the experience to other spaces.
- A lung to regulate the microclimate in the city.

After four months of resistance and activation, and despite the violent eviction, the Parque Resistencia Huentitán (Fig. 16) represented a space for collective action that

Fig. 16 Huentitan Resistance Park. The photograph reads: "*We are recovering what belongs to the people. Huentitan Resistance Park. Come join us, we need you*". *Source* Authors archive

made visible the struggle for public spaces, green areas and a healthy environment. In addition, solidarity and support grew among various resistance groups which defended the territory inside and outside the metropolitan area of Guadalajara. The group Tlacuache community garden signs its Manifesto with the following sentence:

From the north-northwestern metropolitan barrancas, never again a territory without us, never again us without territory. The defense of the Barranca is the defense of life.

5 Towards a Regenerative Perspective in Urban Ecotones

Based on these findings, it can be stated that a possible scenario of harmonious integration between the inhabitants of the peripheral areas and nature must be approached from a cooperative perspective. Considering the ecosystem services in the described ecotonal situations, the following strategies are proposed to improve the encounter of communities with nature:

1. Promote environmental education as a change in consciousness towards systemic assessment to be able to visualize a possible regeneration scenario.

2. Defining a public policy with an ecosystemic vision, that articulates and coordinates environmental protection and development, which requires agreements among all actors through planning and management tools at the different levels of government.
3. Suppressing actions that damage the natural environment and minimizing threats that reduce biodiversity and endanger health.
4. Elaboration of strategies for ecological and social regeneration in a system of citizen initiatives, working harmoniously with nature over time, overcoming extractivism.

From this last strategy, and based on the pioneering and regenerative role of citizen´s initiatives, the following actions are derived:

5.1 Promote Environmental Awareness and Deep Experience of Nature

(a) Defending the "barranqueña" culture, protecting its biohistory and valuing its cultural and environmental heritage. (b) Creating educational nodes on "deep ecology" about the Santiago River Canyon ecosystem and its bioregion. (c) Communicating natural cycles, ecosystem services and their relationship with holistic health. (d) Organizing hiking and activities to appreciate nature and its cycles, promoting mindfulness.

5.2 Strengthening Community and Nature Civic Groups that Promote the Care for Community and Nature

(a) Organizing nature activities that strengthen community cohesion and the protection of natural areas. (b) Promoting ecotourism and environmental education. (c) Opening spaces for scientific, academic and community research on canyon biodiversity and environmental services. (d) Supporting and providing knowledge and tools to community guardians to detect illegal impacts, as well as assisting in visitor's care.

5.3 Harmonize the Urban and Environmental Regulatory Framework

(a) Integrate all open spaces into urban and environmental planning tools into an articulated system of protected natural areas, public spaces and unused areas and promote their regeneration and protection. (b) Strengthen management to integrate, enhance and increase nature in urban sprawl. (c) Set limits to urban use of natural areas and strengthen sanctions against pollution and destruction of ecosystems.

5.4 Reconfigure Urban Ecotones Through a System of Inner-City, Peri-Urban and Rural Green Spaces

(a) Increase and articulate the area of neighborhood natural spaces and their connection to the natural edge. (b) Reforest with native species from the canyon. (c) Promote urban gardening programs at neighborhood level to strengthen community links and access to healthy food. (d) Renew parks as part of a system of community and environmental use.

5.5 Integrated Water Management with an Approach that Considers Watersheds, Ecosystems, and Human Needs

(a) Immediate rehabilitation of riverbeds. (b) Mapping of discharge points. (c) Application of nature-based solutions in areas of opportunity at the micro-catchment level. (d) Urban design sensitive to the water cycle, rain gardens along streets to allow rainfall infiltration. (e) Involvement of all stakeholders to achieve effective action.

5.6 Comprehensive Improvement of Urban Neighborhoods with a Biophilic Approach

(a) Protect the current use of agricultural land by indicating it with this category in urban planning, thus balancing residential use with productive use. That is, residential use requires low density and agricultural use requires a collective and agroecological vocation. (b) Provision of facilities that qualifies the life of the residents of the periphery, with programs that link integral health and the culture of care between the community and nature. (c) Renovation and design of public spaces with a regenerative approach.

6 Conclusion

This chapter addresses the relationship between the city and nature through the study of the urban ecotone, which is understood as the interface between the natural ecosystem and urbanization. A co-evolutionary, cooperative relationship at the northern edge of the Guadalajara metropolitan area is proposed to regenerate the urban and ecological boundary conditions. Based on the recognition of the dynamics of living systems interacting in the urban ecotone, we identify two conditions in tension: intense urban development at the expense of ecosystem services and the emergence of citizen´s initiatives motivated by the search for alternatives for living together in respect and care of nature.

As Mang and Reed (2012) recall, the regenerative paradigm can and should contribute to development based on the capacity of all natural, cultural, and economic systems found in a place to grow and develop their health and continued viability. In line with the narrative of place, a methodology "for understanding the uniqueness and significance of these places as a basis for engaging in regenerative development and planning" (Mang et al. 2016), we have focused in this study on understanding and conceptualizing the right relationship with place. The purpose of this approach is to understand the unique dynamics and potential of a site, project, and community in relation to their place of life, and to conceptualize how the project can be a regenerative force through the right relationship.

In this order, our study suggests that the perspective of regenerative design in the urban ecotone and the narratives of the collectives enable possible strategies for "regenerative evolution" according to the potential of the place. Furthermore, it highlights the importance of a paradigm shift in the relationship between residents and nature by looking at the area where this exchange is most evident. In this case, it is the interface between the city and its natural environment, defined as areas where the most authentic and essential problems of ecology emerge. Therefore, the urban ecotone is considered in this study as an opportunity to rethink the periphery and improve social and environmental health conditions, recognizing the important potential that community initiatives focused on caring for the environment represent to influence policy.

Based on the dynamic vision of the ecotonal situations in Guadalajara, which includes intra-urban, peri-urban, and rural situations, it is possible to analyze the ecosystem services provided by each situation and identify the conflicts and citizen initiatives present. This chapter summarizes the actions of pioneer collectives working to protect, enhance, and promote "barranqueña culture," a mix of natural and cultural aspects in the Barranca region of Jalisco, Mexico. Despite their different goals, all these groups strive for a harmonious relationship between the ecosystem and the communities that inhabit and visit this space.

Finally, strategies aimed at improving the relationship between urban communities and nature in urban ecotones are proposed theoretically. Mentioned are the promotion of environmental education, the introduction of public policies with an ecosystem approach, the prohibition of environmentally harmful acts, and the linking of social and environmental regeneration strategies in a system of citizen´s initiatives working in harmony with nature. Finally, specific measures are proposed, such as promoting environmental awareness, strengthening the agendas of community and nature advocacy groups, harmonizing urban and ecological regulatory frameworks, and reshaping urban ecotones through a system of intra-urban, peri-urban, and rural green spaces. In short, the goal is to create a system of spatial and functional continuity that enables the coexistence of intersecting social and ecological networks as specific measures for the coevolution of living systems.

References

Ayuntamiento de Guadalajara (2022) Planes Parciales de Desarrollo Urbano. https://transparencia.guadalajara.gob.mx/planesparciales

Bartorila MÁ (2001) Ecotonos urbanos, interfaces ambientales y proyectación, Colonia Caroya (Arg.) In: I Congrès Ecologia i Ciutat. Universitat Politècnica de Catalunya, Barcelona

Decreto # 20611. Área Municipal de Protección Hidrológica la Barranca del Río Santiago en el Municipio de Zapopan, Jalisco. Periódico Oficial. Congreso del Estado de Jalisco, 7 de octubre de 2004

De la Torre Escoto ME, Bartorila MA, Alayón González JJ (2022) Situaciones ecotonales y servicios ecosistémicos: salud ambiental en la Barranca del Río Grande de Santiago. Área Metropolitana de Guadalajara, México. urbe. Revista Brasileira de Gestão Urbana 14. https://doi.org/10.1590/2175-3369.014.e20210357

Decreto por el que se establece Área Natural Protegida bajo la categoría de "Formación natural de interés estatal de las Barrancas de los ríos Santiago y Verde". DIGELAG Dec.003/2016. Periódico Oficial. El Estado de Jalisco, 7 de octubre de 2004

Giacomini V (1979) Urban ecology applied to the city of Rome; MAB Italia Project 11. Roma. https://unesdoc.unesco.org/ark:/48223/pf0000042720

Gual MA, Norgaard RB (2010) Bridging ecological and social systems coevolution: a review and proposal. Ecol Econ 69(4):707–717. https://doi.org/10.1016/j.ecolecon.2008.07.020

Janzen D (1980) When is it coevolution. Evolution 34(3):611–612

Mang P, Reed B (2012) Designing from place: a regenerative framework and methodology. Build. Res. & Inf. 40(1):23–38. https://doi.org/10.1080/09613218.2012.621341

Mang P, Haggard B, Regenesis (2016) Regenerative development and design. A framework for evolving sustainability. Wiley, New Jersey

Martínez González P, Hernández González E (2010) La lucha por la justicia ambiental en Jalisco: un Salto de Vida, por la defensa del Santiago. Universidad de Guadalajara, CUCI Ediciones

Millennium Ecosystem Assessment (2005) Ecosystems and human well-being: synthesis. Island Press, Washington, DC. https://www.millenniumassessment.org/en/Reports.html

Morrone JJ (2005) Hacia una síntesis biogeográfica de México. Rev. Mexicana De Biodiversidad 76(2):207–252

Ochoa García H (2012) Mapeo de conflictos ambientales y alternativas en Jalisco, aportes para una metodología. In: Tetreault DV (coord) Conflictos socioambientales y alternativas de la sociedad civil. Guadalajara. ITESO, México, pp 67–92

Pasarín MI, Diez E (2013) Salud comunitaria: una actuación necesaria. Gac Sanit 27(6):477–478. https://doi.org/10.1016/j.gaceta.2013.10.001

Ramos-Gutiérrez L, Montegro Fragoso M (2012) Las centrales hidroeléctricas en México: pasado, presente y futuro. Tecnología y Ciencias Del Agua III(2):103–121

Sassen S (2015) La ciudad compleja pero incompleta. [Entrevista por Antonio Martínez Velázquez] Arquine, 5 October 2015. https://www.arquine.com/compleja-pero-incompleta-conversacion-con-saskia-sassen/

Schipper J (2023) Bajío dry forests. One earth, 4 May 2023. https://www.oneearth.org/ecoregions/bajio-dry-forests/

Secretaría de Medio Ambiente y Desarrollo Territorial. Gobierno del Estado de Jalisco (2015) Formación Natural Estatal Barranca de los ríos Santiago y Verde. Estudio Técnico Justificativo y Programa de Manejo. Guadalajara

Valero A, Schipper J, Allnutt T (2019) Southern North America: Southern Mexico. https://www.worldwildlife.org/ecoregions/nt0204

Gómez-Balandra MA, Díaz-Pardo E, Gutiérrez-Hernández A (2012) Composición de la comunidad íctica de la Cuenca del Río Santiago, México, durante su desarrollo hidráulico. Hidrobiol 22(1):62–78

Citizen Action to Renaturing the Historic City of Cordoba (Spain)

Francisco Crespo-García and Ángela Lara-García

Abstract Vegetation is able to cool the environment through evapotranspiration. This ability is useful for cities in hot climates, where heatwaves have become more intense in recent years due to Climate Change and the urban heat island phenomenon. The singularity of the urban morphology of historic cities causes special difficulties in the introduction of tall-growing vegetation. Furthermore, taking into account the heritage values and tourist interests involved. A case in point is the city of Cordoba and one of the main areas in its historical centre: Axerquia. Drawing on the theory of sustainable transition, a critical review of the *Axerquía Verde* (Green Axerquia) process allows us to identify its strengths and limitations as a *niche* experience, while also helping us to understand the resistance that this type of proposal faces from the institutional regime. Lastly, the case study identifies traditional cultural practices related to the use of vegetation as opportunities to promote the renaturalisation of historical cities. This is the case of the *Fiesta de los Patios de Córdoba*, which was inscribed to UNESCO'S Representative List of the Intangible Cultural Heritage of Humanity in 2012.

Keywords Sustainable transition · Renaturing · Historical city · Córdoba · Citizen action

1 Introduction

The aim of this chapter is to analyse the difficulties faced by citizens' initiatives promoting the renaturation of historic city centres in Mediterranean cities in the framework of sustainable transition process. A case study methodology is adopted through the critical review of the experience of *Axerquía Verde* (Green Axerquia).

F. Crespo-García (✉)
AMASCE SCA, Cordoba, Spain
e-mail: curro@amasce.coop

Á. Lara-García
Universidad de Sevilla, Sevilla, Spain
e-mail: anglargar@us.es

© The Author(s), under exclusive license to Springer Nature Switzerland AG 2025
C. Cobreros et al. (eds.), *Regenerative Design*, Cities and Nature,
https://doi.org/10.1007/978-3-031-76890-3_7

Axerquía Verde (AxV) is a citizen initiative that addresses the challenge of renaturing a sector of the historical city centre of Cordoba (Spain) known as Axerquia, as an adaptation strategy to face the heatwaves that affect Mediterranean cities, in the context of their intensification as a result of Climate Change.

The process started with the identification of spaces in the Axerquia where the presence of vegetation could be increased. The end goal of the initiative was to bring down the temperature by renaturing the urban grid, for which several strategies were put forward to increase vegetation density. The proposal was shaped by superimposing a series of actions designed to create a green infrastructure (GI) that enhanced the bioclimatic conditions in the area.

This work presents the systematisation and review of the AxV process based on the first author's participatory observation experience as a member of the team that developed the proposal. The group was formed in May 2020 in the heat of the debates that arose during the Covid-19 lockdown on the prevailing urban model and its effects on health. The team of 8 residents of the Axerquia neighbourhood, all of them with a technical background, counted on the shared prior experience of taking part in participatory processes and defending Cordoba's urban woodland. The AxV proposal began with an analysis of Axerquia's spatial and environmental context based on an examination of reports and the existing bibliography, fieldwork, and synthesising cartography. After identifying and delimiting the areas where it might be possible to take action, several strategies were defined and datasheets were prepared for each. The proposal to renature AxV was especially focused on the tree as the most appropriate plant species due to its impact on the environment through evapotranspiration and its ability to provide shade. The primary strategy in the proposal was the plan to plant 24 new urban woods.

A technical proposal was drafted collectively with descriptive cartography, and even the cost of executing the project was estimated. The work counted on advice from external experts. A round of public presentations was begun in January 2021 to present the proposal to social and institutional agents. The first author of this review was involved in 10 of these presentations, which served as instrument to evaluate the social perception of the proposal and its acceptance.

As an innovative proposal that arises from citizen action to have a real effect on urban sustainable transition, AxV faced several difficulties and limitations that can be explained through the Multi-Level Perspective Theory of sustainable transition (Kemp et al. 1998; Loorbach et al. 2017) both in respect of the sociocultural framework (*landscape*) and the institutional and regulatory framework (*regime*). The socio-technical transition toward a paradigm of greater sustainability and resilience in which nature-based solutions (NbS) play a relevant role is a process driven by global physical and discursive mechanisms influenced by local social constellations and cultures. There are abundant theoretical and methodological contributions on the topic that are continually being discussed and enriched. Nonetheless, it is imperative to dig deeper into specific highly-significant processes to confirm and refine the results in a contextualized way.

This is the goal of this paper. The proposal of AxV to extend an urban green network over public and private spaces in a densely populated historical sector of an average-sized Mediterranean city has been seen as an opportunity to investigate the still underexplored limitations and barriers to considering renaturalisation as a Climate Change adaptation strategy in historical city centres.

2 Theoretical Framework

2.1 Green Infrastructures as Urban Climate Adaptation Strategies

City and nature are two different entities that only appear when a line is drawn to differentiate and separate them. However, humans still need to include trees, vegetation, flora and fauna in their cities despite their withdrawal from nature to build a haven for themselves. The current biophilic city movement seeks to raise awareness about the function of internally rebalancing human beings with all things living, which includes vegetable species (Wilson 2021). A greater presence of vegetation around a dwelling is associated with a 20% lower self-perception of poor health in general and also improves mental health indicators (Triguero-Mas 2019).

Vegetation has been a common component of the urban system used to cool the environment since ancient times and civilisations (Guerrero et al. 2016). Mediterranean cities have traditionally included garden areas that combine elements of vegetation with flowing water and these have had an obvious social and climatic function and shaped their cultural identity. Discourses supporting the greening of urban areas are gaining traction in urban resilience policies nowadays, and nature-based solutions (NbS) are currently international institutions' preferred option to improve cities' resilience to climate risks (European Commission 2013, 2016; Faivre et al. 2017; Gutiérrez et al. 2017; IUCN 2012; UNESCO 2018). The idea of urban greening has evolved from gardens to enhance public space to Green Infrastructure (GI) capable of providing cities with ecosystem services in the current quest for the benefits that nature brings (Elderbrock et al. 2020). GI connects and gives consistency to the resources of the natural heritage: from the trees that line streets and avenues, neighbourhood gardens and city-scale parks, to sub-regional agricultural, forestry and natural spaces (Boada and Sánchez 2012; Rueda 2009).

As mentioned above, vegetation in the city has a major impact on the regulation of the microclimate and people's physical and mental health, and also contributes to improve biodiversity and to reduce the risk of ecosystem collapse. Due to their cooling effect[1] and the shade that they afford, trees tend to be at the forefront among the wide range of vegetable species that can be considered in renaturing initiatives, although some critic voices suggest that the prominence given to them is excessive.

---

[1] In ideal conditions, with a dry temperature of around 20°C, a tree with a 50 m^2 top generates a cooling power of approximately 11.3 kW per day.

Well aware of trees' bioclimatic function and the ecosystem service that they provide, many grassroots protest movements have emerged to denounced the trees problems in the urban public spaces and the different types of diseases that they suffer: illnesses, infestations, incorrect pruning… In heritage and tourist cities, trees are reduced to decorative elements, with no concern for the minimum conditions that they required for their viable development. And this is why these movements call for a new way of designing urban spaces with a greater attention to the proper integration of vegetation.

2.2 *The Challenge of Naturalizing Historic Cities*

Historical cities are known to accumulate extremely dense urban areas due to the overlapping of building and renovation processes in highly delimited spatial areas. The result is a combination of narrow streets and squares with disconnected and changing pathways where fullness prevails over emptiness. When we attempt to introduce contemporary space usage practices for which historical cities are not prepared, then issues arise that are inherent in the cities' own morphology.

In the case of the bioclimatic adaptation of the consolidated city, Higueras (Higueras 2009) states the need to consider a series of constraints:

- Regarding the exposure to sunlight due to tall buildings in narrow streets and the direction that they face, which leads to a lack of sunlight in winter and inadequate heat dissipation in summer.
- Reduced ventilation conditions due to filling of open spaces and the natural ground in gardens and courtyards in blocks of multi-family dwellings, or 'patio houses' being lost.
- Air and noise pollution as a result of the private mobility model and the constrictions imposed by the narrow roads.
- Not enough vegetation to provide ecosystem services.

Other problematic constraints that should not be overlooked when addressing the pressures on historical city centres (Calle 2019):

- The presence of historic-artistic monuments and the 'touristification' processes that escalate the specialisation of public space in their most immediate surroundings.
- Loss of residents due to rising housing costs, the lack of retail outlets nearby and the difficulty of finding a solution to parking problems.
- Excessive tertiarisation and loss of complexity in some specific areas.

In addition, a number of factors concur in historical city centres that cause the regional climate in the surroundings to change due to a phenomenon known as the *urban heat island*: urban morphology, the heat released from buildings, the lack of green areas and impermeable soils that prevent the ground from retaining moisture and the air pollution (Domínguez 1999).

How should we set about mitigating Climate Change in the area where the most intense heat island effect and the heritage and urban value of the city coincide? Considerations around the harmonised workings of materials, vegetation, shade and moisture as researchers are now pointing out were not unknown to traditional architecture and town planning and they offer valuable teachings that can contribute to understanding the use of vegetation to produce microclimates adapted to the rigours of summer (Gartland 2012). Horizontal energy flows are especially important for this as they are what make the area's general climate prevail over vegetation's ability to eliminate energy through evapotranspiration. This gives two possibilities to have a real impact on the microclimate in these spaces with vegetation: either the planted area has to be large in size, as in the case of urban parks, or the environment created by evapotranspiration has to be enclosed, so the horizontal transport of energy can be physically neutralised (Guerrero et al. 2016). It is this second strategy that can be recognised in the majority of traditional Hispano-Muslim gardens and courtyards, which everyone agrees, can be considered the most sophisticated achievements as far as interventions in environmental comfort are concerned, given the rigorous summer conditions.

In the particular case of Cordoba, the tradition of naturing the inner spaces of multi-family dwellings was the origin of the so-called "Patios of Cordoba". This custom consists of adorning traditional houses' *patios* or courtyards with flowers, and its heritage value lies in its ability to create social links such as in the knowledge and use of nature that it represents (UNESCO 2012). This tradition evolved into a popular celebration, the "Fiesta of the Patios", which UNESCO also recognises as Intangible Cultural Heritage.

2.3 Urban Renaturing Processes in the Framework of the Sustainable Transition.

In today's city, urban renaturing process are clear examples of sustainable transition initiatives that responds to the NbS conception: living solutions underpinned by natural processes and structures designed to address various environmental challenges while simultaneously providing economic, social, and environmental benefits (European Commission 2015). In this sense, urban renaturing has the capacity to provide ecosystem services that improve urban resilience (Bowler et al. 2010; Frantzeskaki and McPhearson 2022; Gill et al. 2007; Guerrero et al. 2016; Gutiérrez et al. 2017; Juvillà 2019; Kabisch et al. 2016; Yang and Lee 2021), and, thus, heighten the importance of 'using and living with' natural and semi-natural ecosystems in cities to improve human well-being (Escobedo et al. 2019).

In addition to their direct benefits, NbS have a transformative social impact since they mediate new social relations and new social configurations that contribute to social innovation in cities and change the perception of nature and human-nature relations in urban contexts. In this sense, NbS promote community- and policy-based initiatives to improve sustainability and livability and the aspiration to foster inclusivity and social justice, thus contributing to accelerating sustainable transitions in cities (Frantzeskaki et al. 2017; Gutiérrez et al. 2017; Sarabi et al. 2019). Social and community movements have been increasingly demanding urban green spaces due to progressive social recognition of their socio-ecological values (Campbell et al. 2016; García et al. 2022). In the context of the climate emergency, these demands are currently being addressed in discussions on urban resilience.

The implementation of NbS to enable the urban sustainability transition requires the collective stimulus of transition initiatives and the participation of urban change agents to mediate and catalyse transformation processes (Sarabi et al. 2019). Thus, civil society can advocate for more radical and progressive ideas. These characteristics of rapid experimentation adapted to the local context cause civil society to function as a driver of sustainability transitions (Frantzeskaki et al. 2017; Forrest and Wiek 2015; García et al. 2022; Seyfang and Longhurst 2013; Seyfang and Smith 2007).

Even though the advantages of NbS are generally accepted, their implementation is hindered by obstacles framed in the difficulties that confront the wider sustainable transition process (García et al. 2022; Frantzeskaki et al. 2017; Kabisch et al. 2016; Lara et al. 2022; Sarabi et al. 2019). Grey infrastructural solutions continue to dominate the delivery and management of services for historical and cultural reasons, although in many cases NbS may present more efficient and cost-effective solutions than more traditional technical approaches (Davies and Lafortezza 2019; European Commission 2015; Liquete et al. 2016).

NbS, understood as "soft engineering approaches" (Short et al. 2019), conform the basis of the GI that provides essential ecosystem services to address urban environmental challenges in a context of sustainable transition. Based on transitions research literature (Loorbach et al. 2017), one of the theories that we can call on regarding the difficulties faced by the sustainable transition process is the Multi-Level Perspective (Geels and Schot 2007; Kemp et al. 1998; Loorbach et al. 2017; Smith et al. 2010), which describes socio-technical transitions through the inter-relationships of three analytical levels: *regime, niche,* and *landscape.* The *regime* is the complex structure of scientific knowledge, engineering practices, production processes, procedures, norms and institutions. It gives stability to social and technical relations and displays resistance to changes that, should they occur, would be more oriented toward optimising the existing regime than changing it. The *landscape* is composed of socio-political and macro-economic factors, cultural patterns, the global and environmental setting, etc. Changes to the *landscape* come about very slowly (over decades) and have to be perceived and interpreted by stakeholders to have any influence. *Niches* are the spaces in which individual actors, alternative technologies and local practices operate and are expressed as new ideas, initiatives or innovative techniques (Lafuente et al. 2020).

According to this perspective, the sustainable transition occurs when the *landscape* evolves (the global and environmental setting; political, international and cultural conditions, etc.) and puts pressure on the *regime*, where internal tensions are intensified. This situation can give rise to an opportunity for solutions that emerge from the *niches* to replace the *regime's* technologies and structures (Brown and Clarke 2007). In contrast, resistance to transition resides in the socio-technical *regime*: institutional logic could influence actors' perceptions and the dissemination of new practices (Hughes et al. 2013; Lafuente et al. 2020). Applying this perspective has led to other elements being included in the analysis: interactions between social actors, social movements, power relationships, etc. (Markard et al. 2012; Seyfang and Smith 2007).

Adopting this sustainable transition framework, the AxV initiative is understood as a *niche space*. A local-community proposal led by a citizen collective that seeks to generate innovative dynamics to integrate GI into an urban area, recognised as world heritage, as a climate change and heatwave adaptation strategy. In this process, AxV will face resistance from the *regime* in the form of institutional, regulatory, sociocultural and budgetary limitations and even the physical and spatial difficulties that the inclusion of vegetation represents in the complex morphology of Cordoba's old town. These limitations mostly coincide with those identified in the scientific literature as common barriers to the implementation of NbS in urban areas (Frantzeskaki et al. 2017).

The core ambition of transitions research is to better understand such transitions, to anticipate and adapt to undesirable transitions (e.g., high-impact climate change), and to explore possibilities to advance and accelerate desired transitions (Loorbach et al. 2017). In order to contribute to this goal, the study presented is located in the field of *actionable science*, understood as a type of research designed with the stakeholder in mind that aims to inform decisions, improve the design or implementation of public policies, and (or) influence the strategies, planning, and behavior that affect the environment (Gallardo et al. 2020).

Improving the knowledge of sustainable transition processes against the backdrop of climate change continues to be a fundamental scientific challenge. There are abundant theoretical and methodological contributions on the topic that are continually being discussed and enriched. Nonetheless, it is imperative to dig deeper into specific highly-significant processes to confirm and refine the results in a contextualized way. As Sarabi et al. conclude, "In-depth analysis of barriers and enablers in each documented case of NbS uptake is required to build a reference based to identify and predict barriers to and enablers of NbS uptake and implementation" (Sarabi et al. 2019, p. 16).

3 Case Study: The Green Axerquia Initiative

3.1 The Axerquia District of Cordoba

Axerquía, Axarquía or Ajarquía (Arabic Al-Sarquiyya, 'the east') is the name given to the eastern part of Cordoba's historical quarter (Fig. 1). Along with the western sector, it is the primaeval nucleus of the Roman colony founded in the year 171 B.C.E. that represents the part of the city that has remained unchanged from the 10th century to the present day. 'La Axerquía' occupies an area of 106.17 Has, 19% of which is roadways and public space. The rest of the area is privately owned and has a sponge-like quality and is able to breathe thanks to the characteristic system of courtyards and free spaces inside blocks of multi-family dwellings (patio houses).

Until the beginning of the 20th century, Cordoba was enclosed behind a wall. The city was characterised by a mediaeval-style urban layout with alleys, raised walkways and houses with public right of passage, and no large constructions or empty streets (Torres 2021). The current appearance of the historical quarter and, within it, of Axerquia, mainly conforms to transformations that occurred from the Christian military occupation (1236) onwards. This caused a universal crisis in the city and triggered a process of population loss. The existing Muslim urban shapes and forms were lost, enabling large building complexes such as mansions, stately homes and convents to take over and gradually occupy the empty and interstitial spaces, and the replacement of the existing housing configuration with an evidently mediaeval layout.

Fig. 1 The historical quarter of Cordoba

With the start of urban growth at the beginning of the 10th century, there was a large rise in the new city's social and economic power. As Federico Abad states, municipal governments' historical 'lack of interest in the city enabled Ajerquía to retain the features that it had inherited over 10 centuries up to the present today' (Abad 2016, p. 73). Its picturesque nature, therefore, remained frozen in time and thus preserved its future interest for the nascent tourist sector (Abad 2016).

A diverse range of factors come together in the geographical area of Axerquia and explain the multiple problems that it currently experiences. In 1994, UNESCO expanded the World Heritage listing of the Mosque (1984) to include a large tract of Cordoba's historical quarter. This led to a series of urban protection instruments being developed to protect buildings and free spaces, limiting any actions that could be taken and making management more difficult. In this sense, operations to pedestrianise and refurbish public spaces to improve the habitability of the city, the urban image and economic dynamisation in the wake of World Heritage recognition inevitably involved a gentrifying effect due to the associated increase in income and social displacement (Anguelovski et al. 2016). Parallel to this, public urban renewal programmes were developed in the area (URBAN Ribera; North Axerquia Preferred Regeneration Area) with mixed results, and without being able to halt the loss of the resident population that had been ongoing since the beginning of the 21st century (Torres 2021). Lastly, to the population crisis was added the recent and explosive phenomenon of the transformation of residential housing into tourist accommodation (Díaz and Barrero 2022; Estivill et al. 2019), which further drove tertiarisation in the district and caused the potential resident population to lose interest in living in the area.

The habitability of cities is strongly conditioned by the climate. From this point of view, Cordoba is characterised by a subcontinental Mediterranean regional climate that alternates cold winters with occasional frosts and long hot summers that intensify the urban heat island effect. Rainfall is usually concentrated in spring and autumn. Taking the daily mean temperature as a reference, Cordoba's climate can be seen to combine very long summers (157 days, 43% of the year) and normal springs (88 days, 24.1% of the year) with shorter than usual autumns and winters (50 days, 13.7% of the year and 70 days, 19.2% of the year, respectively) (Domínguez 1999).

The climate scenarios contemplated in Cordoba's Municipal Climate Change Plan (2023) for the 2015–2100 period foresee a gradual worsening of the local climate in summer. A large part of Axerquia is socially and environmentally highly vulnerable due to the combination of high population density, population mean age and unemployment; housing age and condition; the proliferation of sub-standard housing; accessibility issues and concentrations of people over 85 years of age living alone in blocks with no lifts.

3.2 Motivation, Agents and Process of AxV

The outbreak of Covid-19 in 2020 and subsequent enforced lockdown triggered an enormous number of conversations and debates in the world around city models and their direct impact on health, urban resilience to risks and the adaptive ability to Climate Change. In this generalised framework of debate, a group of residents in the Axerquia area contacted each other and agreed on the need to formulate a renaturing process in the area where they live as an urgent strategy to adapt to the various climate horizons.

A driving group was gradually formed at the beginning of the initiative. The people who set it in motion had previously taken part in activities organised by the Platform for the Defence of Trees in Cordoba (an informal citizen collective set up to denounce the deficiencies in tree maintenance in the city), so they were able to count on prior experience in collective participatory processes and relationships with the local administration. They also had the tools and sufficient instrumental knowledge to undertake the purpose of the project. A few weeks after it was set up, there were 8 members on the group, 3 women and 5 men between the ages of 40 and 76. The main profile of these people was architects with training in urban planning (4), but the fields of agricultural engineering (1), ecologism (1) and citizen participation (2) were also represented.

This initial group formed from the first contacts called itself the "Guerrilla Verde Axerquía" or Axerquia Green Guerrilla and presented itself, in the words of one of the members, as 'an exercise in collective intelligence to generate knowledge and action' (Axerquia Verde [AxV] et al. 2021).

The process developed spontaneously to tackle the initial objective of preparing a technical proposal to study and size up the renaturing actions required in the Axerquia area. The phases of the process are summarised below following the initiative landmarks:

- Work team formed (May 2020)
  - Identification of participants; driving group set up.
  - Ideas shared and first actions defined.
- Analysis of physical context (May–June 2020)
  - Walks to recognise empty plots, abandoned buildings, ruins and existing public spaces and evaluation of their condition.
  - Preparation of cartography to synthesise information compiled during walks.
- Definition of the proposal (June–July 2020)
  - Identification and delimitation of public and private spaces that could accommodate trees.
  - Definition of strategies to quantify the number of trees required and preparation of data sheets on each of the spaces classified into:

    Empty plots, empty spaces, empty spaces inside blocks of dwellings.

Public spaces.

- Contact with experts to seek specific advice on the proposed species to be planted.
- Calculation of the budgeted cost of the material implementation of the proposal.

• Drafting of the project (September–December 2020)

- Preparation of graphic and cartographic (GIS) materials, texts and tables.
- Final layout.

• Public presentation of the proposal to various agents in order to seek alliances to materialise all or part of the project (from January 2021)

- Citizen participation organisations and neighbourhood associations such as the Citizen Committee of the Central District and the Rehabilita Cordoba Association (Regenerate Cordoba Association) (meetings #1, #2).
- Cordoba City Council departments, including the Mayor's Office, the Area for the Environment, Parks and Gardens, Heritage and Finance, Historical Old Town branch office, Cordoba Royal Botanic Garden (meetings #3, #4, #5, #6, #7, #8).
- Research groups from the University of Cordoba such as the IN-HABIT group (meeting #9)
- Individuals from the academic and scientific worlds such as technical personnel from the Institute of Sustainable Agriculture (meeting #10).

• An association was set up to continue promoting and championing the project (February 2022).
• Cordoba City Council submits an application to the Biodiversity Foundation in order to implement a pilot project in the public space inspired by *Axerquía Verde* (November 2022).
• *Axerquía Verde* is mentioned as a reference within the lines of action proposed by both the Municipal Strategic plan and the Municipal Climate Change plan (February 2023).

3.3 Description of the Green Axerquia Proposal

The AxV initiative is one of the current urban renaturing movements in the city of Cordoba. In this sense, the proposal complements other public initiatives presented in Cordoba such as the Green Belt project and the parks and gardens system envisaged in the municipal plans, and citizen initiatives to raise awareness such as the Ecologists in Action initiative and its Neighbourhoods for the Climate programme.

The aim of AxV initiative is to reduce the temperature by 1.5–3 °C by renaturing this sector of the city. The first task carried out was the identification of existing spaces in Axerquia where vegetation can already be found and which would be able to accommodate more, and other spaces which can also be used for this purpose

(Fig. 2). The tool used was an examination of the research on empty plots, buildings in ruins and the interiors of abandoned blocks of dwellings in the whole of Cordoba's historical quarter carried out by the architect Esteban (2018), which generated some cartography that geolocated the proposed actions.

Then, a variety of strategies were defined to increase the presence of vegetation on these spaces:

1. Creation of woods or forestry masses in the large non-built-up empty spaces and the inside of blocks of dwellings in ruins.
2. Improvement and optimisation of public gardens, tree-lined squares and streets.
3. Improvement of both public and private gardens.

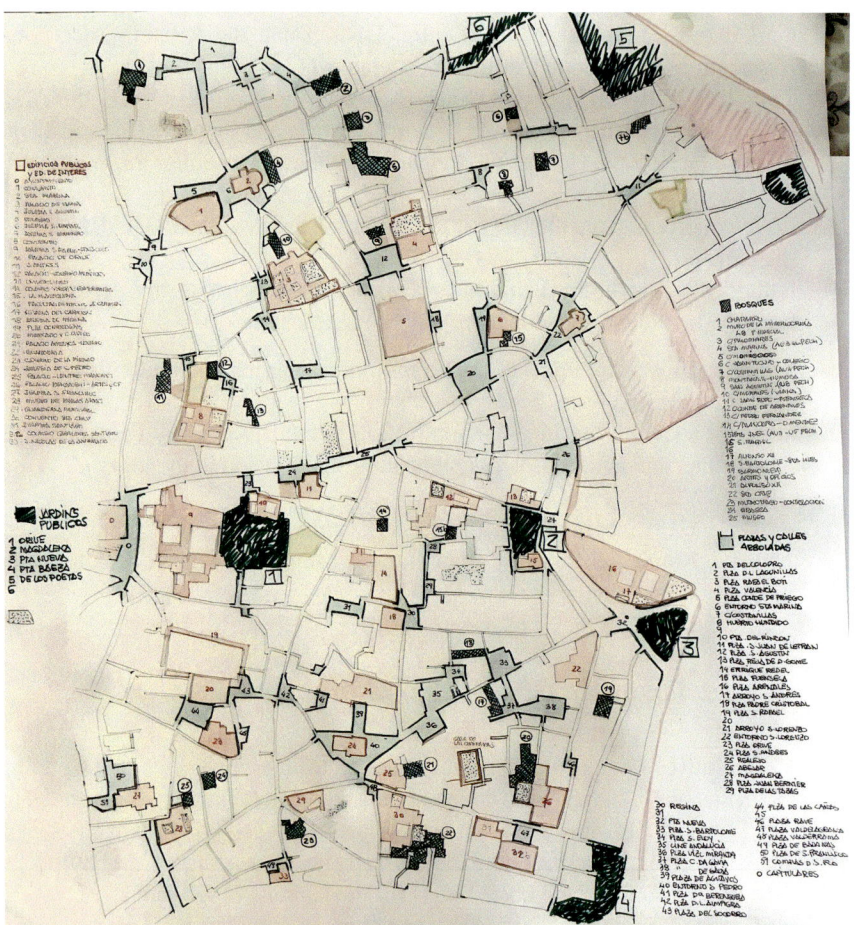

Fig. 2 The information compiled during walks by Axerquia Green Guerrilla was synthesised into a cartography (Axerquía Verde [AxV] et al. 2021). *Axerquía Verde. Arborización del espacio público y áreas degradadas en ciudades históricas mediterráneas.*)

4. Detection of areas of opportunity able to be included in the renaturing actions.
5. Other minor measures to resolve the continuity of the green sector are generated by carrying out the preceding four strategies.

Finally, these strategies took the form of the following actions:

Urban woods

AxV defines urban woods as the 'type of habitat with high tree density, land covered by treetops and a surface area of over 0.5 Ha' (Axerquía Verde [AxV] et al. 2021, p. 35). Based on this premise, AxV proposes that degraded spaces and areas in Axerquia be turned into urban woods, in some cases, micro woods given their size. The special feature of Axerquia's mediaeval urban layout means that there are some small cul-de-sacs that are difficult for vehicles and machinery to access but which are the only entrances into previously built and inhabited blocks of dwellings that are currently abandoned and in ruins. Of the combination of plots of this type, housing in ruins and even some foreclosed properties determined by the city's urban development plans, 24 spaces were identified as suitable for being turned into woods, involving 49 different cadastral parcels and a total surface area of 20,007 m^2.

Tree-lined streets and squares

Axerquia's streets and squares are an essential part of the make-up of the sought-after green network to cover the urban layout. The greatest difficulty in converting them into nature corridors is their geometry, given their narrow width and the fact that they are spaces that play a role in integrating urban mobility. From the point of view of pedestrian mobility, one of the main things lacking in these spaces is, precisely, appropriate shade to prompt journeys on foot when there are high levels of solar radiation. To make public space in Axerquia properly habitable, AxV proposes increasing the existing number of trees and renewing part of the urban pavement.

Actions on the paving are envisaged to complement this and to guarantee that the vegetation has optimal bioclimate effects: 'we need to ensure that there is a proportion of permeable surface that provides the soil with a minimum level of health and complements moisture from the water tables, which are getting lower and lower, and the very interesting underground currents in Axerquia. There is a further aspect which is as important as these considerations and which stresses the importance of the soil's permeability; the surface's porosity, or its ability to retain sufficient surface moisture and prevent rapid water drainage' (Axerquía Verde [AxV] et al. 2021, p. 50). The AxV proposal in this regard can be summarised as increasing the size of tree wells, the creation of continuous flower beds and other flower beds on top of the existing pavements, and or that paved areas should be completely replaced with earth.

Public, convent and private building gardens

AxV proposes the creation of some new gardens alongside those that already exist by the city delivering on the spaces that its plans have already set aside for green zones and the addition of some new spaces using private plots in an evident state of abandonment.

AxV did not consider whether the land suitable for contributing to Axerquia's Green Web is public or private. This was considered to be an urban development matter that could be postponed until the project's fulfilment stage. On the environmental level, all the parcels of land in Axerquia were considered to be part of a continuum and suitable for inclusion in the proposed GI web. AxV proposes the inclusion in the network of convent gardens, and gardens that form part of monuments or are in private properties but, in principle, does not envisage any tree densification actions. It only analyses the aspects of accessibility and the opening arrangements.

Areas of opportunity

Areas of opportunity are areas where, despite having a specific use, the planting of vegetation or some other renaturing measure would help to substantially improve their condition and provide greater benefits to the surroundings. The detected areas of opportunity include outdoor 'summer cinemas', which is a characteristic summer practice in population centres in the south of Spain with films shown in the open air in courtyards and blocks of dwellings with earthen floors and a large amount of vegetation around the edges. Some summer cinemas still exist in Cordoba, exclusively in the Axerquia area.

Some other areas of opportunity are surfaces used as car parks, school playgrounds and the odd empty public urban site that the urban development plan has set aside to be fitted out.

As has been seen thus far, AxV's proposed renaturing process is mainly tree-based. Among the available species, AxV advocates abandoning the mass use of purely ornamental species and recommended a preponderance of native plants and occasional exotic species that has previously shown capable of adapting to the environment and are characterised by its strength, rapid growth and environmental services.

As mentioned above, overlaying the area with woods, streets, gardens, courtyards and areas of opportunity gives an image of Axerquia crisscrossed by a system of hubs and green corridors and an increased number of trees where the area's biodiversity is concentrated (Fig. 3). This proposal supposed an increase in the free surface area set aside to accommodate 25,837 m^2 of vegetation and the planting of 15,246 individual plants. Investment in labour and plants in the whole of the area was calculated at 5,500,000 €. Finally, the opportunity to create 24 urban woods within a very dense historical quarter has been seen as an opportunity to research the strategy of a green network of public and private spaces in the city centre of Mediterranean cities, on which very little academic literature currently exists.

Fig. 3 Green Axerquia as a green network of public and private spaces (Axerquía Verde [AxV] et al. 2021). *Axerquía Verde. Arborización del espacio público y áreas degradadas en ciudades históricas mediterráneas.*)

4 Discussion

4.1 Limitations to the Renaturalisation of Historical City Centres

The idea of enhancing urban greening as a strategy to address Climate Change and its consequences is clearly established in the scientific literature (Faivre et al. 2017; Gutiérrez et al. 2017; IUCN 2012) and politics recommendations (European Commission 2013, 2016; Yang and Lee 2021). Also, the presence of different kind of barriers that hinder these processes of sustainable transition (Kabisch et al. 2016; Sarabi et al. 2019). However, the various constraints that renaturing process facing on their insertion into the dense urban areas of Mediterranean historic centres have not been studied as much, where morphological, heritage and socio-institutional issues overlap.

The first aspect that should be highlighted in this review of the AxV initiative is the ability to identify the potential of a series of parcels and private properties that, due to their state of ruin and abandon, would be suitable for transforming into land able to accommodate large quantities of vegetation in the historic centre. The proposal to transform these empty spaces into small urban woods opens up the possibility of considering these interstitial spaces as areas of opportunity for improving the naturing of mediaeval old towns. A more exhaustive study of this type of regreening strategy involving empty areas inside blocks of dwellings in other historical old towns in Spain or even Europe, usually as a citizens demand, is one of the future lines of research that comes out of the present work on the AxV experience.

Another key question to be analysed in the AxV initiative is the role of trees as an element of urban renaturalisation. From the point of view of the microclimate function, the presence of trees is sufficient to generate a beneficial effect on the surrounding area due to their ability to provide protection from solar radiation (generation of shade) and the cooling effect of their previously mentioned evapotranspiration. Evapotranspiration is intensified by the area surrounding the tree being moist and composed of earth (Hernández 2013), which, in cities, is something that is generally limited to parks and gardens. Historical city centres by definition lack a significant surface area of this type. Over time, the areas devoted to vegetable gardens and landed estates within the walls of historical cities were gradually encroached upon and turned into part of the built area.

Notwithstanding, we know that when walls or vertical constructions physically delimit garden spaces, this contributes to the vegetation's cooling effect by preventing the dissipation that is produced by horizontal energy chains. In the study of the historical gardens in Cordoba and its province by Domínguez (2002) based on measurements taken in situ, the microclimate effect of these concentrations of vegetation on their immediate surroundings is qualified as 'inappreciable' when there are no vertical elements to confine the achieved coolness. Only the non-built areas inside large blocks of dwellings and the private gardens of stately homes or convents provide the right conditions for trees' microclimate function to work optimally.

In these sense, at the meetings held during the AxV proposal's public presentation phase, some limitations on the proposed spaces renaturalisation emerged, that we can classify as:

- Spatial limitations (meetings #4, #5, #6, #7). Trees are the plant with the greatest volume at the end of their growth phase. During the development phase, factors such as buildings nearby, compacted ground and water in the ground determine the final form of the tree. This is why trees require adequate space to develop healthily and sufficient permeable ground around them to capture oxygen and water. These requirements are indispensable for a tree's good health and are difficult to meet in the streets of historical old towns in cities such as Cordoba, where the narrow streets and squares (only 20% of the total surface area is public space) prioritise the transit of pedestrians and vehicles and call for any obstacle to mobility such as trees to be avoided.
- Socio-cultural limitations (meetings #4 and #6). Like every living being, plants have the ability to affect their surroundings. They are magnets of biodiversity and supports for nests. They generate pollen. They affect the ground. And so they directly impact the lives of the people who live in their vicinity, both positively (biophilia, improvements to the environment and landscape) (Triguero-Mas 2019; Wilson 2021) and negatively (inconvenience, impacts). The first thing that the occasional concentrations of vegetation in abandoned plots face is resistance from the owners of neighbouring plots.
- Institutional limitations (meetings #3, #5 and #7). On the one hand, there is a prior resistance of the public administration to any intervention on private land, even though it has legal tools available to do so on behalf of the public interest. On the other hand, regulation in the area of urban development and urban vegetation in public spaces is a municipal responsibility in the Spanish regulatory framework. Cordoba currently lacks any bylaws that regulate public space or urban greening. Municipal actions closely comply with what is laid down in the General Urban Development Plan (2001) and the Special Plan for the Protection of the Historical Quarter (2003), which were both drafted at a time when the design criteria did not take into account the Climate Change paradigm or its consequences for the physical definition of urban space. Thus, the regulatory framework does not lay down any criteria for incorporating vegetation into the public space with this decision left to the will of the designer. What is more, integrating vegetation into constructions as coverings or green façades is banned in order to preserve the traditional cityscape, which is in contradiction with the landscaping of the traditional "*Patios of Cordoba*", which can be considered as a green façade typology in courtyards.
- Budget limitations (meeting #5). Managing and maintaining urban vegetation is the direct responsibility of Local Councils, which traditionally have multiple budgetary constraints and difficulties in handling public expenses. An increase in the number of garden spaces and trees in the city without a proportional increase in the budget is, in the final instance, perceived as a barrier to the implementation of urban renaturalisation processes.

One last barrier can be added to the above-stated difficulties and is related to the uniqueness of historical and heritage city centres:

- Aesthetic limitations. The urban image is one of the main resources for promoting tourism in historical towns and cities. Such places seek a broad uncluttered view of monuments with the removal of any element that interferes with their being viewed or photographed, and urban developments that contribute to monumentalising urban spaces in line with heritage settings (García et al. 2022).

Most of these limitations clearly respond to the factors that the scientific literature identifies as common barriers to the incorporation of NbS in urban space (Sarabi et al. 2019): *inadequate financial resources* (Frantzeskaki et al. 2017; Raymond et al. 2017), *path dependency* (Kabisch et al. 2016; Kronenberg 2015), *institutional fragmentation* (Davis and Naumann 2017; Frantzeskaki et al. 2017; Kabisch et al. 2016; Wamsler 2015), *inadequate regulations* (Davis and Naumann 2017; Kronenberg 2015), or *limited land and time availability* (Albert et al. 2019; Davis and Naumann 2017; Frantzeskaki et al. 2017). The *path dependence* factor can be related to other frequently identified factors: *lack of information, knowledge, and understanding in applying integrated, adaptive forms of management*, and *fear of the unknown (operational performance)* (Kabisch et al. 2016; Kirkpatrick et al. 2013; Kronenberg 2015; Lohr et al. 2004).

On the other hand, the above difficulties require to consider densely occupied and compact city centres such as Cordoba's as restricted spaces for exclusively tree-based renaturalilsation. In that sense, a critical review of the process and content of the AxV proposal could point the excessively prominent role given to trees as a major weakness due to spatial limitations also require other types of smaller-scale operations to be considered. Other NbS such as green building coverings and façades and shrub and bush species in the public space would contribute to expanding the area of the GI beyond free space. The combined presence of vegetation in public spaces, in courtyards and inside dwelling blocks and buildings could contribute to more effectively achieving the sought-after green corridors. Their future inclusion would serve to broaden the range of issues for discussion presented in the AxV proposal.

However, Cordoba possesses knowledge, experience, and local, technical, and community culture, which is a strong component of the local identity. This characteristic of the local culture appears to be a possible trigger for *valorizing and exploiting the existing tacit and expert knowledge* (Krasny et al. 2014; Moseley et al. 2013) but it requires *community engagement, empowerment, and participation* through *collaborative governance approaches* (Brown and Farrelly 2009). For this, the transition initiatives such as Axerquia Verde require sufficient resources to be made available to them, such as time, budget, space, or a political mandate to enable changes toward sustainability (Frantzeskaki et al. 2017). In our case, the extant will and knowledge of the implementation of NbS promotion policies are not being sufficiently exploited. No stable frameworks exist for collaboration with citizens or the local scientific community.

4.2 Green Axerquia as a Sustainability Transition Process

The theory of the Mult-level Perspective of sustainable transition processes identifies AxV initiative as a *niche* space (Kemp et al. 1998; Lafuente et al. 2020; Lara and Moral 2022; Loorbach et al. 2017) that is one that operates and expresses new ideas based on an innovative proposal: the identification of the latent opportunity presented by the inside of blocks of dwellings, undeveloped plots and abandoned buildings in historical cities that can be turned into GI.

In this framework, the AxV initiative itself can be read as an emanation of one part of Cordoba's sociocultural *landscape* open to the urban transformations made essential by Climate Change. More specifically, AxV has emerged out of the prior experience of some of its members in the Platform for the Defence of the Tree in Cordoba. The emergence of citizen movements worldwide to raise awareness of the ecological crisis (Extinction Rebellion, Fridays for Future, Green Guerrillas…) has meant that the demands of this Platform have become increasingly important and have had a political impact. However, the need to go beyond action confined to denunciation to proactive and creative actions focused, in this instance, on finding a solution to the particular case of Axerquia, led to the launch of the initiative under study.

AxV's status as a *niche* space partially explains the lack of greater social support and the initiative's marginal position with respect to the socio-political and cultural framework of Cordovan society, which is completely in step with other cities of a similar size and socio-demographic profile in its surroundings. The Spanish Centre for Sociological Research's barometer (November 2022) shows that the economic crisis and families' economic problems are the Spanish people's greatest concern (18% of people surveyed), well in front of Climate Change (1.7%) in a distant ninth position in absolute terms (CIS 2022). To what point do citizens feel the climate threat to be a major problem, despite being able to very clearly see the consequences of its effect in their daily lives? The lukewarm reception given to the AxV proposal by political representatives can be read as a symptom of the still insufficient pressure of the *landscape* on the *regime*. The meetings held with the Parks and Gardens Municipal Area only confirmed that a problematic and controversial view of trees persists in Cordovan society.

AxV focused its efforts on impacting very select areas of political representation, citizen representation, the sphere of municipal technicians and certain individuals in the academic world, with the belief that the proposal would permeate through these sufficiently to come to fruition. The Multi-level Perspective does offer an indication of the slow speed with which changes in societies' cultural *landscape* operate and also of the resistance inherent in the assimilation of innovations by the knowledge, procedure and practice system that is the *regime*, whose raison d'être is to provide stability in social and technical relations (Hughes et al. 2013; Lafuente et al. 2020). In this sense, the limitations identified in the AxV process are no different from all the *niche* initiatives with similar goals whose difficulties are rooted in this lopsided distribution of strength (Markard et al. 2012; Seyfang and Longhurst 2013).

Paradoxically, Cordoba is a city with major international projection based to a certain extent on its university's academic tradition in agriculture and the existence of a number of popular festivals based around vegetation, the greatest expression of which UNESCO's recognition of its 'Fiesta of the Patios', which takes place in May, on its List of Intangible Cultural Heritage. The traditional ways of adorning small-scale private spaces in private properties is a part of popular know-how passed down from generation to generation that precisely and sophisticatedly addresses the holistic integration of plants into an architectural space. This has been inherited from the Hispano-Muslim tradition linked to the typology of the 'patio house' that is so characteristic of the Axerquia district and the city's historical quarter. The heritage value of Cordoba's 'Fiesta of the Patios' lies in the patio's function as a place for intercultural exchange and encouraging a sustainable collective way of life based on the establishment of solid social links and solidarity networks and exchanges between neighbours, with the simultaneous stimulus to acquire knowledge and respect for nature (UNESCO 2012), as recognised in the UNESCO Convention's area of 'Knowledge and Practices concerning Nature and the Universe'. In this sense, the Cordoba's patios are a representative example of the socio-ecological and transformational values of NbS (Campbell et al. 2016) that foster community initiatives to improve urban sustainability and livability, thus contributing to accelerating ecological transitions in the cities (Frantzeskaki et al. 2017; Gutiérrez et al. 2017; Sarabi et al. 2019) (Fig. 4).

The AxV initiative has the opportunity to complete its proposal to renaturalize the Axerquia by integrating a practice deeply rooted in the heart of the culture of the Cordovan society. Its inclusion would allow to update its social and festive dimension to the keys and challenges of our time, which contributes to release it from the trivialisation that the exacerbated tourist promotion condemns it. Considered in this way, the Fiesta de los Patios is perceived as a relevant starting point to develop an integrated strategy of renaturalization of the public space of the historic center of the city, thus becoming a "courtyard of courtyards".

5 Conclusions

The experience of the process followed by AxV initiative shines a light on the difficulties that historical and heritage old towns of cities such as Cordoba face for increasing vegetation in their urban areas. AxV's proposal for renaturing the Axerquia district of Cordoba shows us the difficulties that exist in cities of this type to equip themselves with GI overlying the built city that would effectively provide it with ecosystem services.

From a Multi-Level Perspective point of view, the AxV initiative is understood as a *niche space* and as experience AxV helps to identify the main resistances that this type of proposal encounters in the institutional *regime*. As a *niche* space, AxV proposal is right to point to the opportunity that exists inside the blocks of dwellings in the historical urban fabric. They represent potential green areas with the ability to

Fig. 4 The "Fiesta de los Patios", part of the sociocultural landscape and a contextualized strategy to renature the historical city centre. Left, Festival de los Patios 1960 (Archivo Municipal de Córdoba; this file is licensed for research and non-profit purposes); right, Córdoba, Festival de los patios (author: Graham Stanley; this file is licensed under the Creative Commons Attribution 2.0 Generic International license; https://www.flickr.com/photos/grahamstanley/5701128583/in/album-721576 26673236062/)

be turned into climate shelters for people in cities. The three conditions that GI must meet to bring down the temperature in the environment are all met in these places: the presence of vegetation, natural and permeable ground and spatial delimitation by vertically constructed elements.

On the other hand, the unbending protection of the building architecture and typology in these areas of the city is a major deterrent to NbS increasing and boosting the presence of plants integrated into buildings in the already scarce free public spaces. This situation is especially conspicuous in a city like Cordoba, where a large part of the historical quarter is on the World Heritage List.

Lastly, the traditional cultural practices linked to the use of vegetation that are already part of the city's sociocultural *landscape* are also perceived as opportunities to devise practices that favour renaturing. Such is the case of Cordoba's Fiesta of the Patios, which is known and valued by all the citizens and is sufficiently important to be able to put pressure on and bring about changes in the regulatory and bureaucratic regime.

Finally, knowledge of how the initiative develops in the future will offer keys that can be of great use to similar initiatives committed to achieving the sustainable transition desired for our cities.

References

Abad F (2016) La barriada de Cañero. Una pieza singular en el desarrollo urbano de Córdoba. Utopía Libros

Albert C, Schröter B, Haase D, Brillinger M, Henze J, Herrmann S, Gottwald S, Guerrero P, Nicolas C, Matzdorf B (2019) Addressing societal challenges through nature-based solutions: how can landscape planning and governance research contribute? Landsc Urban Plan 182:12–21

Anguelovski I, Shi L, Chu E, Gallagher D, Goh K, Lamb Z, Reeve K, Teicher H (2016) Equity impacts of urban land use planning for climate adaptation: critical perspectives from the global north and south. J Plan Educ Res 36:333–348. https://doi.org/10.1177/0739456X16645166

Axerquía Verde [AxV], Bendala C, Esteban A, Dugo L, Pareja R, Donnier I, Larios J, AMASCE SCA (2021) Axerquía Verde. Arborización del espacio público and áreas degradadas en ciudades históricas mediterráneas. https://es.scribd.com/document/504169529/Axerquia-Verde

Boada M, Sánchez S (2012) Naturaleza y cultura, biodiversidad urbana. Ecoinnovación para la Mejoría Ambiental de Productos y Servicios. Ed. Diagrama

Bowler DE, Buyung-Ali L, Knight TM, Pullin AS (2010) Urban greening to cool towns and cities: a systematic review of the empirical evidence. Landsc Urban Plan 97(3):147–155. https://doi.org/10.1016/J.LANDURBPLAN.2010.05.006

Brown R, Clarke J (2007) Transition to water sensitive urban design: the story of Melbourne. Monash University, School of Geography and Environmental Science, Australia

Brown RR, Farrelly MA (2009) Delivering sustainable urban water management: a review of the hurdles we face. Water Sci Technol 59:839–846. https://doi.org/10.2166/WST.2009.028

Campbell LK, Svendsen ES, Sonti NF, Johnson ML (2016) A social assessment of urban parkland: analyzing park use and meaning to inform management and resilience planning. Environ Sci Policy 62:34–44. https://doi.org/10.1016/j.envsci.2016.01.014

Centro de Investigaciones Sociológicas [CIS] (2022) Estudio nº3384. Barómetro de noviembre 2022. https://www.cis.es/cis/export/sites/default/-Archivos/Marginales/3380_3399/3384/es3384mar.pdf

Davies C, Lafortezza R (2019) Transitional path to the adoption of nature-based solutions. Land Use Policy 80:406–409. https://doi.org/10.1016/J.LANDUSEPOL.2018.09.020

Davis M, Naumann S (2017) Making the case for sustainable urban drainage systems as a nature-based solution to urban flooding. In: Kabisch N, Korn H, Stadler J, Bonn A (eds) Nature-based solutions to climate change adaptation in urban areas: linkages between science, policy and practice. Springer

De la Calle M (2019) Turistificación de centros urbanos: clarificando el debate. Boletín de la Asociación de Geógrafos Españoles 83. https://doi.org/10.21138/bage.2829

Díaz I, Barrero M (eds) (2022) Turismo, desarrollo urbano y crisis. Las grandes ciudades andaluzas en su encrucijada. Comares

Domínguez P (1999) Clima, medio ambiente and urbanismo en Córdoba. Diputación de Córdoba

Domínguez P (2002) Clima regional and microclimas urbanos en la provincia de Córdoba. Universidad de Córdoba

Elderbrock E, Enright C, Lynch KA, Rempel AR (2020) A guide to public green space planning for urban ecosystem services. Land 9(10):1–23. https://doi.org/10.3390/LAND9100391

Escobedo FJ, Giannico V, Jim CY, Sanesi G, Lafortezza R (2019) Urban forests, ecosystem services, green infrastructure and nature-based solutions: nexus or evolving metaphors? Urban for Urban Green 37:3–12. https://doi.org/10.1016/J.UFUG.2018.02.011

Esteban A (2018) Córdoba vaciada. https://maphub.net/AmadorEsteban/cordoba-vaciada

Estivill M, Ibáñez A, Estivill L (2019) Estudio sobre la vivienda con fines turísticos en la ciudad de Córdoba. Viviendas Municipales de Córdoba S.A.

European Commission (2013) Green Infrastructure (GI) – enhancing Europe's natural capital. Communication from the commission to the European Parliament. The Council. The European Economic and Social Committee and The Committee of the Regions. COM/2013/0249 final

European Commission (2016) Next steps for a sustainable European future – European action for sustainability. Communication from the Commission to the European Parliament, the Council, the European Economic and Social Committee and the Committee of the Regions. COM/2016/0739

European Commission. Directorate-General for Research and Innovation (2015) Towards an EU research and innovation policy agenda for nature-based solutions and re-naturing cities. Final report of the Horizon 2020 expert group on "Nature-based solutions and re-naturing cities." Publications Office of the European Union. https://data.europa.eu/doi/10.2777/479582

Faivre N, Fritz M, Freitas T, de Boissezon B, Vandewoestijne S (2017) Nature-Based Solutions in the EU: innovating with nature to address social, economic and environmental challenges. Environ Res 159:509–518. https://doi.org/10.1016/J.ENVRES.2017.08.032

Frantzeskaki N, Borgström S, Gorissen L, Egermann M, Ehnert F, Frantzeskaki N, Borgström S, Gorissen L, Egermann M, Ehnert F (2017) Nature-based solutions accelerating urban sustainability transitions in cities: lessons from Dresden, Genk and Stockholm Cities. In: Kabisch N, Korn H, Stadler J, Bonn A (eds) Theory and practice of urban sustainability transitions. Springer Nature. https://doi.org/10.1007/978-3-319-56091-5_5

Frantzeskaki N, McPhearson T (2022) Mainstream nature-based solutions for urban climate resilience. Bioscience 72(2):113–115. https://doi.org/10.1093/BIOSCI/BIAB105

Forrest N, Wiek A (2015) Success factors and strategies for sustainability transitions of small-scale communities – evidence from a cross-case analysis. Environ Innov Soc Trans 17:22–40. https://doi.org/10.1016/J.EIST.2015.05.005

Gallardo G, Saunders F, Sokolova T (2020) Co-creating actionable science: reflections from the Global North and South. Cambridge Scholars Publishing

García A, Lara A, López D (2022) Espacios públicos desamparados, ciudades insípidas. In: Díaz I, Barrero M (eds) Turismo, desarrollo urbano y crisis. Las grandes ciudades andaluzas en su encrucijada. Comares

Gartland LM (2012) Heat Islands: understanding and mitigating heat in urban areas. Routledge, London. https://doi.org/10.4324/9781849771559

Geels FW, Schot J (2007) Typology of sociotechnical transition pathways. Res Policy 36(3):399–417. https://doi.org/10.1016/j.respol.2007.01.003

Gill S, Handley JF, Ennos R, Pauleit S (2007) Adapting cities for climate change: the role of the green infrastructure. Built Environ 33(1):115–133

Guerrero JJ, Caceres F, Giménez de Azcarate F, Moreira JM (2016) Servicios de regulación climática aportados por la vegetación urbana a la Ciudad de Córdoba. 1a Parte: Fundamentos and Metodología. REDIAM. Consejería de Medio Ambiente and Ordenación del Territorio, Junta de Andalucía

Gutiérrez L, García G, García I (2017) Soluciones Naturales para la adaptación al cambio climático en el ámbito local de la Comunidad Autónoma del País Vasco. S. P. de G. Ambiental. G. Vasco. Ihobe

Hernández A (coord) (2013) Manual de diseño bioclimático urbano. Recomendaciones para la elaboración de normativas urbanísticas. https://oa.upm.es/15813/1/2013-BIOURB-Manual_de_diseno_bioclimatico_b.pdf

Higueras E (2009) El reto de la ciudad habitable y sostenible. Distribución y Asesoramiento de Publicaciones Jurídicas (DAPP)

Hughes S, Pincetl S, Boone C (2013) Triple exposure: regulatory, climatic, and political drivers of water management changes in the city of Los Angeles. Cities 32:51–59. https://doi.org/10.1016/J.CITIES.2013.02.007

IUCN (2012) The IUCN Programme 2013–16. https://portals.iucn.org/librarand/node/10320

Juvillà E (coord) (2019) Renaturalización de la ciudad. Diputación de Barcelona

Kabisch N, Stadler J, Korn H, Bonn A, Frantzeskaki N, Pauleit S, Naumann S, Davis M, Artmann M, Haase D et al (2016) Nature-based solutions to climate change mitigation and adaptation in urban areas. Ecol Soc 21. ES-08373-210239

Kemp R, Schot J, Hoogma R (1998) Regime shifts to sustainability through processes of niche formation: the approach of strategic niche management. Technol Anal & Strat Manag 10:175–198. https://doi.org/10.1080/09537329808524310

Kirkpatrick JB, Davison A, Harwood A (2013) How tree professionals perceive trees and conflicts about trees in Australia's urban forest. Landsc Urban Plan 119:124–130. https://doi.org/10.1016/J.LANDURBPLAN.2013.07.009

Krasny ME, Russ A, Tidball KG, Elmqvist T (2014) Civic ecology practices: participatory approaches to generating and measuring ecosystem services in cities. Ecosyst Serv 7:177–186. https://doi.org/10.1016/J.ECOSER.2013.11.002

Kronenberg J (2015) Why not to green a city? Institutional barriers to preserving urban ecosystem services. Ecosyst Serv 12:218–227. https://doi.org/10.1016/J.ECOSER.2014.07.002

Lafuente R, Ganuza E, Paneque P (2020) Las resistencias sociales a la transición hidrológica en el sur de España: el apoyo de los ciudadanos a la construcción de nuevos embalses. XI Congreso Ibérico de Gestión y Planificación del Agua. Fundación Nueva Cultura del Agua

Lara A, Berraquero L, Del Moral L (2022) Contested spaces for negotiated urban resilience in Seville. In: Ruiz-Mallén I, March H, Satorras M (eds) Urban resilience to the climate emergency. The urban book series. Springer, pp 197–223. https://doi.org/10.1007/978-3-031-07301-4_9

Lara A, Del Moral L (2022) Nature-based solutions to hydro-climatic risks: barriers and triggers for their implementation in Seville (Spain). Land 11(6):868. https://doi.org/10.3390/land11060868

Liquete C, Udias A, Conte G, Grizzetti B, Masi F (2016) Integrated valuation of a nature-based solution for water pollution control. Highlighting hidden benefits. Ecosyst Serv 22:392–401. https://doi.org/10.1016/J.ECOSER.2016.09.011

Lohr VI, Pearson-Mims CH, Tarnai J, Dillman DA (2004) How urban residents rate and rank the benefits and problems associated with trees in cities. J Arboric 30:28–35. https://doi.org/10.48044/JAUF.2004.004

Loorbach D, Frantzeskaki N, Avelino F (2017) Sustainability transitions research: transforming science and practice for societal change. Annu Rev Environ Resour 42:599–626. https://doi.org/10.1146/annurev-environ-102014-021340

Markard J, Raven R, Truffer B (2012) Sustainability transitions: an emerging field of research and its prospects. Res Policy 41(6):955–967. https://doi.org/10.1016/J.RESPOL.2012.02.013

Moseley D, Marzano M, Chetcuti J, Watts K (2013) Green networks for people: application of a functional approach to support the planning and management of greenspace. Landsc Urban Plan 116:1–12. https://doi.org/10.1016/J.LANDURBPLAN.2013.04.004

Raymond CM, Frantzeskaki N, Kabisch N, Berry P, Breil M, Nita MR, Geneletti D, Calfapietra C (2017) A framework for assessing and implementing the co-benefits of nature-based solutions in urban areas. Environ Sci Policy 77:15–24

Rueda S (Coord) (2009) Plan de indicadores de Biodiversidad urbana de Vitoria-Gasteiz. Agencia de Ecología urbana de Barcelona y Dpto. Medio Ambiente del Ayuntamiento de Vitoria-Gasteiz

Sarabi SE, Han Q, Romme AGL, de Vries B, Wendling L (2019) Key enablers of and barriers to the uptake and implementation of nature-based solutions in urban settings: a review. Resources 8(3). https://doi.org/10.3390/resources8030121

Seyfang G, Longhurst N (2013) Desperately seeking niches: grassroots innovations and niche development in the community currency field. Glob Environ Chang 23(5):881–891. https://doi.org/10.1016/J.GLOENVCHA.2013.02.007

Seyfang G, Smith A (2007) Grassroots innovations for sustainable development: towards a new research and policy agenda. Environ Polit 16(4):584–603. https://doi.org/10.1080/09644010701419121

Short C, Clarke L, Carnelli F, Uttley C, Smith B (2019) Capturing the multiple benefits associated with nature-based solutions: lessons from a natural flood management project in the Cotswolds, UK. Land Degrad Dev 30:241–252. https://doi.org/10.1002/ldr.3205

Smith A, Voß JP, Grin J (2010) Innovation studies and sustainability transitions: the allure of the multi-level perspective and its challenges. Res Policy 39(4):435–448. https://doi.org/10.1016/J.RESPOL.2010.01.023

Triguero-Mas M (2019) Planificación urbana y salud: espacios verdes y azules. In: Juvillà E (coord) Renaturalización de la ciudad. Área de Territorio and Sostenibilidad de la Diputación de Barcelona. Diputación de Barcelona

Torres A (2021) Datos estadísticos del catastro de viviendas del municipio de Córdoba al 31/12/2020. https://rmbco.odilotk.es/info/datos-estadisticos-del-catastro-de-viviendas-del-municipio-de-cordoba-al-31-12-2020-00610398

UNESCO (2012) La fiesta de los patios de Córdoba. Convention for the Safeguarding of the Intangible Cultural Heritage. https://ich.unesco.org/es/RL/la-fiesta-de-los-patios-de-crdoba-00846

UNESCO (2018) Informe mundial de las Naciones Unidas sobre el desarrollo de los recursos hídricos 2018: soluciones basadas en la naturaleza para la gestión del agua - UNESCO Biblioteca Digital. https://unesdoc.unesco.org/ark:/48223/pf0000261494

Wamsler, C (2015) Mainstreaming ecosystem-based adaptation: transformation toward sustainability in urban governance and planning. Ecol Soc 20

Wilson EO (2021) Biofilia. Errata Naturae

Yang B, Lee D (2021) Urban green space arrangement for an optimal landscape planning strategy for runoff reduction. Land 10:897. https://doi.org/10.3390/LAND10090897

State of the Art of Biophilic Design Research in the Latin-Mediterranean Region

Lucía Martín López🄳 and Rodrigo Durán López🄳

Abstract Through a systematic review of the literature on the concepts of Biophilia and Biophilic Design, this article aims to identify the state of the art of research on these topics in the Latin-Mediterranean context. The Preferred Reporting Items for Systematic Reviews and Meta-Analyses (PRISMA) protocols were used for the selection of the literature to be reviewed. Four main research questions were formulated for the analysis: (1) What are the trends in scientific production in the Region over time based on the number of annual publications? (2) How is the co-occurrence network of important terms extracted from the specific literature, and how do they vary annually within the Region? (3) From where, within the Region, are the insights being produced? And (4) Which Funding Sponsors are providing the economic stimuli for research on Biophilic Design in the Region? The article includes the visualization of these aspects through diagrams made with VOSviewer that allow the reader to construct his or her own reflections to establish a debate on the state of the art of the subject.

Keyword Architecture · Bibliographic review · Biophilia · PRISMA

1 Introduction: *Biophilia* and *Biophilic Design* in the World

The research on *Biophilia* and *Biophilic Design* concerns diverse topics from multiple approaches. The first article registered on the subject of Biophilia in SCOPUS was produced by Landis (1975) in the United States within the field of psychoanalysis

---

L. Martín López
Escuela de Ingeniería de Fuenlabrada, Universidad Rey Juan Carlos (URJC), Madrid, Spain

Present Address:
L. Martín López (✉)
School of Architecture, Art and Design, Tecnologico de Monterrey, Querétaro, Mexico
e-mail: lucia.martin.lopez@urjc.es

R. Durán López
Faculty of Architecture, Universidad Anáhuac, Ciudad de México, Mexico
e-mail: rodrigodu@gmail.com

© The Author(s), under exclusive license to Springer Nature Switzerland AG 2025
C. Cobreros et al. (eds.), *Regenerative Design*, Cities and Nature,
https://doi.org/10.1007/978-3-031-76890-3_8

and around Eric Fromm's concept of Biophilia. While the first text that SCOPUS has registered with the term Biophilic Design is the chapter "Biophilia" by Stephen R. Kellert, published in 2008 in the United States, in the book "Encyclopedia of Ecology". In it, Kellert concludes by discussing the apparent decline in beneficial biophilic contact with nature in modern society and offers a new paradigm of design of the human-built environment (Kellert 2008).

In the Latin-Mediterranean Region, defined for this work by the territories of the countries listed in the Research Method, the first article on Biophilic is produced in 1991 in Italy by D. Vojnovic, G. Procida, and L. Gabrielli Favretto focused on the field of Chemistry and Food Additives (Vojnovic et al. 1991). Four years later, the concept of Biophilia appears for the first time in the SCOPUS records within the field of psychoanalysis in the text "Center-to-Center" Relatedness between "Analyst and Patient" (Biancoli 1995) by the Italian Romano Biancoli, where it is stated that "Biophilia makes psychoanalysis an art because it is applied to living things." The first article in the Region that deals with the importance of Biophilia in design is "Urban parks as opportunities for the interaction of children of animals," co-authored by Colombia and the United States. Although it does not explicitly use the term Biophilic Design, it discusses the use of public parks in urban environments as affordances for humans to interact with nature elements (Páramo and Mejía 2004) and deals specifically with Biophilia and urban planning. It is until 2013 that the use of the term "Biophilic Design" in the Latin-Mediterranean Region is registered in SCOPUS with the Egyptian article "Bio-regenerative rating technique: A critical review" by M. A. Hanafi and M. M. Naguib, which also presents a comprehensive analysis of the roles that bio-inspiration plays in developing the concept of sustainable, ecological design, and construction in contemporary architecture (Hanafi and Naguib 2013).

Worldwide, the thematic areas in which papers on Biophilia or Biophilic Design are found stand out for their diversity. According to SCOPUS bibliometric data, in a 1975–2021 span framed by Landis's article (1975), the two main ones are the Social Sciences (286 documents) and the Environmental Sciences (284 documents), together accounting for one-third of the publications. They are followed, in order of highest to lowest number of documents found, by Engineering (202 documents), Arts and Humanities (140 documents), Agricultural and Biological Sciences (123 documents), and Psychology (101 documents). Finally, with less than 5% of the documents obtained, are the thematic areas of Medicine (80 documents), Earth and Planetary Sciences (78 documents), Energy (76 documents), and Computer Science (65 documents).

Since 1975, the production of documents on these two topics has been distributed quite asymmetrically around the globe. According to data obtained from SCOPUS, the United States leads the production of texts on these topics with almost 280 products. It is followed far behind by Australia with almost 100 texts and the United Kingdom with almost 80 texts found in the database. Further down, still above 50 documents, stands Canada. Below the 50 reported documents are, listed in order of ranking from highest to lowest, Italy, Russia, Japan, Germany, Sweden, and South Korea.

Concerning the authors with the greatest production of texts on Biophilia and Biophilic Design, also with bibliometric data from SCOPUS at an international level, we find that in the 1975–2021 span the most prolific is the American author Timothy Beatley with 13 publications. With only one less, 12 in total, he is followed by Peter Newman from Australia. In third place, with seven publications each, are Syuhaida Ismail from Malaysia, Yannick Joye from Belgium, and André Potvin from Canada. With an output of 6 texts are the Italian Guiseppe Barbiero and the aforementioned Stephen R. Kellert. The authors Claude M. H. Demers, Thomas Panagopoulos, and Sung-Jun Park complete the list of the ten most productive authors.

In terms of affiliations, Curtin University and the University of Melbourne stand out, with 18 and 12 documents respectively, found in SCOPUS published since 1975. Directly proportional to the production of texts mentioned above, U.S. institutions occupy important places in the subject of documents by affiliation, being the universities of Virginia, Texas A&M, Cornell, Purdue, and Oregon, with a number ranging from 13 to 8 documents listed. Rounding out the top ten institutions are Yonsei University in South Korea with ten documents, and Sveriges lantbruks Universitet in Sweden and Université Laval in Canada with nine each.

Finally, reviewing the matter of funding sponsors at the international level is also important in the context of this work. The bibliometric data provided by SCOPUS in the same 1975–2021 span place the National Science Foundation, from the United States, as the organization that has supported the largest number of texts on the topics Biophilic Design and Biophilia, with 13 papers found in the database. The rest of the organizations that have supported texts on the topics in question do not exceed ten documents. The top ten are completed by the Japan Society for the Promotion of Science and the Russian Foundation for Basic Research, each with nine documents; the Australian Research Council with eight documents; the National Research Foundation of Korea with seven documents; the European Cooperation in Science and Technology (COST), the Natural Sciences and Engineering Research Council of Canada, and the Universiti Teknologi Malaysia with six documents each; and with one less (5 each), are the Italian Fondazione Cariplo and the Portuguese Fundação para a Ciência e a Tecnologia (FCT).

2 State of the Art of Bibliometric Analysis on Biophilia and Biophilic Design

It is considered essential to review works that, like this one, have the objective of analyzing the existing literature on the topics of Biophilia and Biophilic Design. For the purposes of this article, several databases were reviewed, and the 17 most significant bibliometric analyses of the last ten years were examined, considering works by researchers from all over the world. It was important to review the frequency of appearance of the different aspects and concepts addressed in the works and to be able to relate them to each other, mainly those dedicated to nature, health (physical and emotional), the built environment, and critical and theoretical aspects.

Of the above-mentioned selection of works, the bibliometric analysis most robust and closest to what is intended in this research is the one carried out by Tirri et al. (2021), both for its methodology and its presentation of results, although without addressing the Region proposed in this work. As for the rest of the works reviewed for this research, it is worth highlighting the specific relationship established by Camrass (2022) between Biophilic Design with Regenerative Design and Regenerative Thinking, which are core topics in this book. The relationship of these concepts with the 17 Sustainable Development Goals (the heart of the United Nations' 2030 Agenda for Sustainable Development) is also important, as pointed out in the works of Hung and Chang (2021) and Zhong et al. (2022). However, there is a direct relationship between these, and the challenges involved in their implementation, mainly in terms of governance and infrastructure, which is addressed by K. Camrass as well as Ferrerira et al. (2020). Criticism of the scope of Biophilic Design and the difficulty of its definition and implementation is also added to the work of Zare et al. (2021).

A factor frequently present in the works reviewed is the issue of scale in the built environment. While works such as Hu and Roberts (2020), Kahnehshenas et al. (2020), McGee and Park (2022), and Yassein and Ebrahiem (2018); focus their review on Biophilic Design bibliography to interior and furniture design. Other works such as Camrass (2022), Fisher (2019), Lee and Park (2022), and Zhong et al. (2022); comment on the importance of extending the application of Biophilic Design to bigger scales, such as architectural and urban.

When talking about sustainability and environmental regeneration, a common pair of themes tend to be energy efficiency and paradigm shifts, and interestingly, only the work of Erebor et al. (2021) addresses the case of energy efficiency; and among others, Heymans et al. (2019) deal with the case of revisiting old and new paradigms. In addition, most of the works reviewed point out both the importance of Biophilia evaluation and its difficulty, suggesting that insufficient tools have been developed. However, the authors Barbiero and Berto (2021), Gillis and Gatersleben (2015), Lee and Park (2022), Wijesooriya and Brambilla (2021), and Zhong et al. (2022); review existing certifications for this purpose.

Finally, it is interesting for the objectives of this research that out of the 17 selected, (Table 1), eight (almost half) restrict their study to works in English, ten (more than half) establish a timeframe for the body of work to be considered, and only two of them state explicitly that they do not receive funding of any kind, which could suggest that there is a significant amount of economic support for these topics.

3 Research Method

Through a systematic review of the literature on the concepts of Biophilic Design and Biophilia, this article aims to identify the state of the art of research on Biophilic Design in the Latin-Mediterranean context. The Preferred Reporting Items for

Table 1 Comparative table of the bibliometric analyses selected for the present work. Own elaboration, 2022

Bibliometric analyses	Barbiero and Berto (2021)	Camrass (2022)	Fisher (2019)	Erebor et al. (2021)	Ferrerira et al. (2020)	Gillis and Gatersleben (2015)	Heymans et al. (2019)	Hu and Roberts (2020)	Khanehshenas et al. (2020)	Lee and Park (2022)	McGee and Park (2022)	Hung and Chang (2021)	Tirri et al. (2021)	Wijesooriya and Brambilla (2021)	Yassein and Ebrahiem (2018)	Zare et al. (2021)	Zhong et al. (2022)
Location (**Latin-Mediterranean bold**)	**Italy**	Australia	Australia	Nigeria	**Portugal**	UK	Australia	USA	**Iran**	Korea	USA	Taiwan	USA	Australia	Egypt	**Iran**	Netherlands
Year	2021	2022	2019	2021	2020	2015	2019	2020	2020	2022	2022	2021	2021	2020	2018	2021	2021
Nature-based solutions and blue–green infrastructure					•		•						•				•
Public health and well being	•		•		•	•		•	•	•	•		•		•		•
Human emotions		•	•			•	•	•	•	•	•	•	•		•	•	•
English language								•	•		•	•	•		•		•
COVID-19		•		•									•	•			
Built environment	•									•	•	•	•		•		
Urban regenerative thinking and practice		•															
Timeframe		•			•	•		•	•	•	•		•		•	•	•

(continued)

Table 1 (continued)

	Barbiero and Berto (2021)	Camrass (2022)	Fisher (2019)	Erebor et al. (2021)	Ferrerira et al. (2020)	Gillis and Gatersleben (2015)	Heymans et al. (2019)	Hu and Roberts (2020)	Khanehshenas et al. (2020)	Lee and Park (2022)	McGee and Park (2022)	Hung and Chang (2021)	Tirri et al. (2021)	Wijesooriya and Brambilla (2021)	Yassein and Ebrahiem (2018)	Zare et al. (2021)	Zhong et al. (2022)
Bibliometric analyses	•																
Location (**Latin-Mediterranean bold**)	**Italy**	Australia	Australia	Nigeria	**Portugal**	UK	Australia	USA	**Iran**	Korea	USA	Taiwan	USA	Australia	**Egypt**	**Iran**	Netherlands
Year	2021	2022	2019	2021	2020	2015	2019	2020	2020	2022	2022	2021	2021	2020	2018	2021	2021
Scale of BD interventions		•	•					•	•	•	•				•	•	•
Challenges (governance and infrastructure)		•			•												
Paradigms (technology)		•					•			•							
Critique of current BD approaches		•												•	•		
BD Evaluation and certification		•			•	•				•				•			
Energy efficiency				•								•					
SDGs								•		•							•
Funding	•			•	•		•										

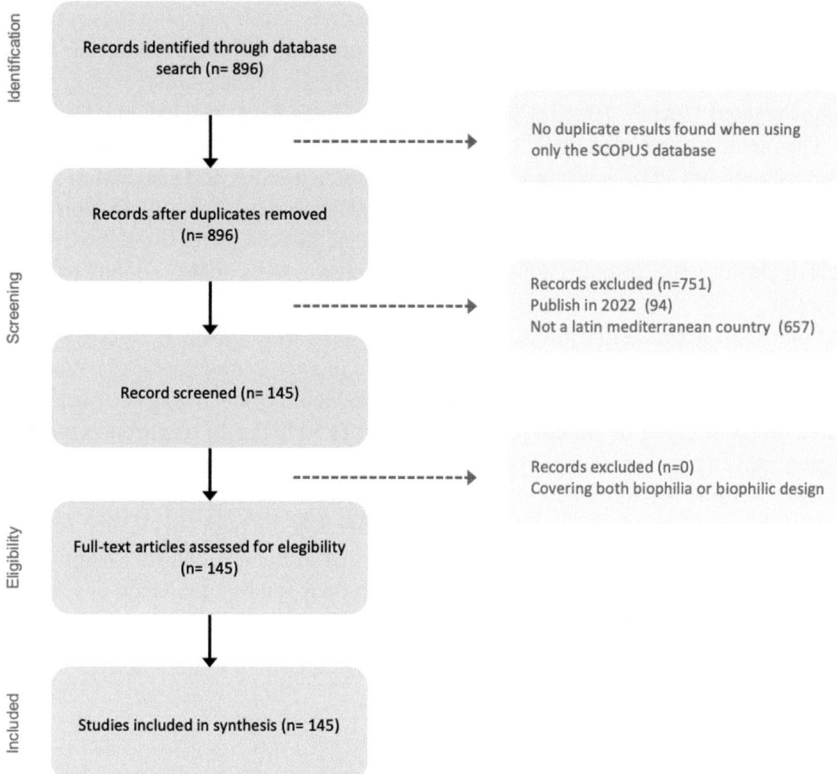

Fig. 1 Literature review diagram. Image by authors based on PRISMA (Moher et al. 2009), 2022

Systematic Reviews and Meta-Analyses (PRISMA) protocols were used to select the bibliography to be reviewed (Page et al. 2021) which improve the transparency, quality, and consistency of methodological information (Hutton et al. 2016) (Fig. 1).

Initially, all results were collected from the SCOPUS database focused on the topics "Biophilic Design" or "Biophilic" or "Biophilia" (896 records). Data were extracted on August 5, 2022, and results published in 2022 (94 records) were excluded to avoid inaccuracy of data for the current year. The main exclusion criteria applied was the restriction by geographic location, leaving out of the analysis all results not produced in the 43 countries of the Latin-Mediterranean Region. According to FAO (1998), 23 countries are part of the Mediterranean basin: Albania, Algeria, Cyprus, Egypt, Spain, France, Greece, Iran, Israel, Italy, Jordan, Lebanon, Libya, Malta, Morocco, Portugal, Syria, Tunisia, Turkey; and from the former Yugoslavia, Slovenia, Croatia, Bosnia and Herzegovina, and Serbia. And twenty are the American countries considered Latin because they have Spanish or Portuguese as their official language: Argentina, Bolivia, Brazil, Chile, Colombia, Costa Rica, Cuba, Dominican Republic, Ecuador, El Salvador, Guatemala, Honduras, Mexico, Nicaragua, Panama,

Paraguay, Peru, Puerto Rico, Uruguay, and Venezuela. After applying these criteria, 145 results were obtained, locating material from only 17 countries in the Latin-Mediterranean Region (Croatia, Egypt, Slovenia, Spain, France, Greece, Iran, Israel, Italy, Portugal, Turkey, Brazil, Chile, Colombia, Costa Rica, Mexico, and Peru).

The metadata provided by SCOPUS for these 145 documents were analyzed with spreadsheets and VOSviewer, a software tool for constructing and visualizing bibliometric networks, to visualize the scientific landscape around Biophilic Design in the Latin-Mediterranean context. In the results section, the criteria for the elaboration of each of the networks visualized through the VOSviewer software have been explained in detail.

Four main research questions, or set of questions, were formulated for the study from which several tangential issues were subsequently extrapolated: (1) What are the trends in scientific production in the Latin-Mediterranean Region over time, based on the number of annual publications; (2) How is the co-occurrence network of important terms extracted from the specific literature? And how do they vary annually within the Region? (3) From where, within the Region, are the insights being produced? And how is the network of linkages being built? (4) Which funding sponsors are providing the economic stimuli for research on Biophilic Design in the Region, where are they from, and what has been their funding tendency over time?

The article includes the visualization of all these aspects through diagrams that allow the reader to construct his or her own reflections in order to establish a debate on the state of the art of the subject.

4 Results and Discussion

The results extracted from the bibliometric analysis of the specific literature are shared below in an attempt to answer the four main research questions.

The five texts with the highest impact in the Region, considering impact through number of citations (Fig. 2), cover topics ranging from Social Ecology to Retail Marketing, including Landscape Design, Sustainable Ecosystems, Economics, Epidemiology, Public Health, Biomedicine, Mental Health, Environmental Health Sciences, Soil Science and Agricultural Chemistry, Business, Management, Agriculture, Food and Environment, Marine Physics and Chemistry, Human and Social Sciences, Sustainability, Mathematical Sciences, Ecology, and Science and Environmental Engineering, which shows the great complexity of the subject, as will be observed in point (b) of this section.

As for the productivity of researchers in the Region, the ten most prolific accumulated a total volume of 32 texts, some of them co-authored (Fig. 3). At the top of the list is Giuseppe Barbiero (Italy) with six papers, the one with the highest impact (cited by 51) being "An individual's connection to nature can affect perceived restorativeness of natural environments. some observations about biophilia" (Berto et al. 2018). In second place is Thomas Panagopoulos (Portugal) with five papers, the highest impact (cited by 28) being "Urban River recovery inspired by nature-based

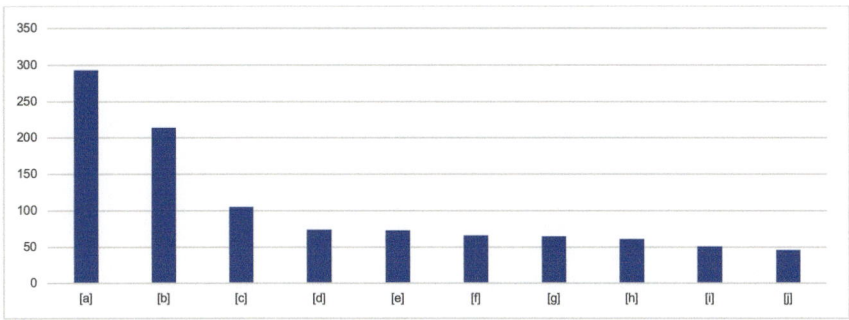

Fig. 2 Top 10 most cited texts from the Latin-Mediterranean Region focused on the topics "Biophilic Design" or "Biophilic" or "Biophilia": **a** "The non-economic motives behind the willingness to pay for biodiversity conservation" (Martín-López et al. 2007); **b** "Effects of moisture and temperature on net soil nitrogen mineralization: A laboratory study" (Guntiñas et al. 2012); **c** "Modern compact cities: How much greenery do we need?" (Russo and Cirella 2018); **d** "The association between lifelong greenspace exposure and 3-dimensional brain magnetic resonance imaging in Barcelona schoolchildren" (Dadvand et al. 2018); **e** "Experiences in nature and environmental attitudes and behaviors: Setting the ground for future research" (Rosa and Collado 2019); **f** "The restorative potential of shopping malls" (Rosenbaum et al. 2016); **g** "Wildflower green roofs for urban landscaping, ecological sustainability and biodiversity" (Benvenuti 2014); **h** "Iodine speciation in the northwestern Mediterranean Sea, method and vertical profile" (Tian and Nicolas 1995); **i** "An individual's connection to nature can affect perceived restorativeness of natural environments. some observations about biophilia" (Berto et al. 2018); **j** "The INDICARE-model—Measuring and caring about participation in higher education's sustainability assessment" (Disterheft et al. 2016). Own elaboration based on bibliometric data from SCOPUS, 2022

solutions and biophilic design in Albufeira, Portugal" (Blau et al. 2018). In third place is Rita Berto (Italy) with four papers. Her most cited text is the aforementioned in co-authorship with Barbiero (Berto et al. 2018), while her second most impactful text (cited by 10) is "Biophilia as Evolutionary Adaptation: An Onto- and Phylogenetic Framework for Biophilic Design" (Barbiero and Berto 2021). In fourth place is Sandra Preto (Portugal) with 3 documents, whose text with the highest impact (cited by 2) is "Should an artificial window substitute a natural one?" (Gomes and Preto 2018). In fifth place, Mark Scott Rosenbaum (United States) makes the regional list. Although he should not be on this list because of his affiliation, his production is closely developed and co-authored with Colombian, Spanish, and Peruvian colleagues. His article of greatest impact (cited by 66) is "The restorative potential of shopping malls" (Rosenbaum et al. 2016).

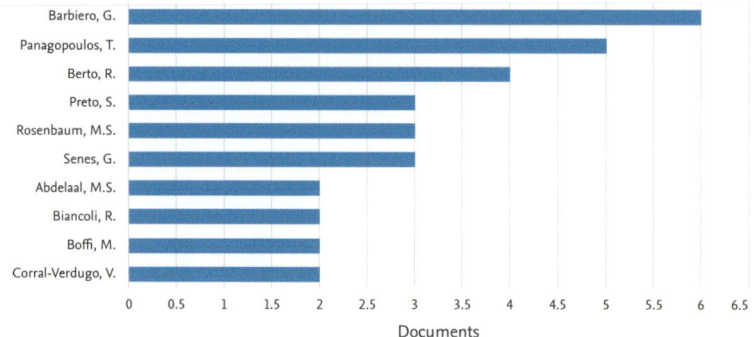

Fig. 3 Most prolific authors according to the number of documents published in the Latin-Mediterranean Region on the topics "Biophilic Design" or "Biophilic" or "Biophilia" registered in the SCOPUS database. Own elaboration based on SCOPUS bibliometric data, 2022

4.1 What Are the Trends in Scientific Production in the Latin-Mediterranean Region Over Time, Based on the Number of Annual Publications?

With an average of 4.8 publications per year published between 1991 and 2021 from the Latin-Mediterranean Region and a total production of 145 documents registered in the SCOPUS database, exponential growth in terms of productivity on the subject can be observed, as shown in Fig. 4. Since 1991, the average number of papers published per year in the Region is one product. In 2011, global warming began to become increasingly noticeable with record high temperatures, extreme weather events on all continents, and a marked increase in CO_2 concentration. It is considered that, as a result of these events, the scientific community grew its interest in Biophilic Design and Biophilia so that the production of scientific articles on the subject begins to increase gradually to reach a production of six papers per year in 2016. In 2019, the world experienced extraordinary weather phenomena, such as unusual heat waves in Europe, unprecedented wildfires in South America and Australia, unseasonal snowstorms in the United States, historic floods in Sri Lanka and Congo, and drought in Victoria, the world's largest waterfall, among other things. In the wake of this, António Guterres, UN Secretary-General, stated at the 25th Conference of the Parties to the United Nations Framework Convention on Climate Change (COP25) held on December 2, 2019, that the climate crisis is threatening human civilization and the world must act before it is too late (VOV world 2019). In addition to this, at the end of this same year, the COVID-19 pandemic began in Asia, confining in the following months up to a third of the world's population. In this context, the production of articles on Biophilic Design and Biophilia in the Latin-Mediterranean Region quadrupled, reaching 25 articles and 35 articles per year, respectively, in 2021 (see Fig. 4).

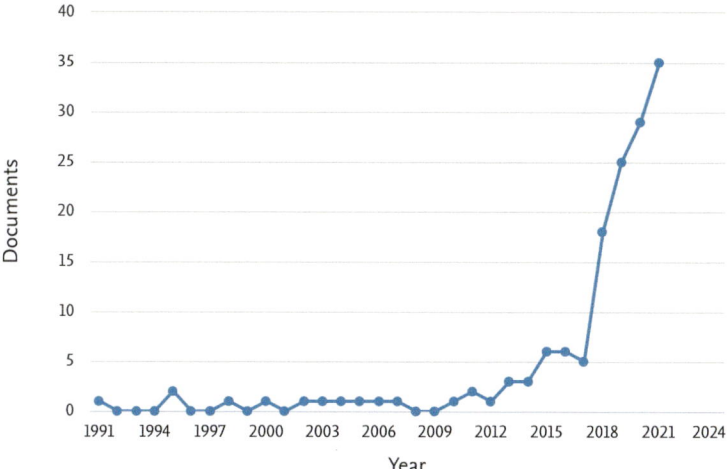

Fig. 4 Documents per year in the Latin-Mediterranean Region (1991–2021) focused on the topics "Biophilic Design" or "Biophilic" or "Biophilia" registered in the SCOPUS database. Own elaboration based on SCOPUS bibliometric data, 2022

In global terms, the United States is the country with the highest production of texts registered in the SCOPUS database on Biophilic Design and Biophilia (274 records), followed by Australia (96 records), the United Kingdom (84 records), and Canada (62 records). In fifth place is the main representative of the Latin-Mediterranean Region, Italy, with 33 publications. Followed within the Region by Spain in the 12th position with 21 records and Portugal (15 records) and Brazil (14 records) in positions 15th and 16th, respectively (Fig. 5).

At the same time, it has been observed that the products published so far in the United States alone on Biophilic Design and Biophilia (274 records) are 1.7 times more than those produced by the sum of all the products published in the Latin-Mediterranean Region (158 records). The aforementioned data show a low impact at a regional level in terms of scientific production, which may be due to the very young age of the studies in comparison with other regions, such as the United States and Canada, and highlights the lack of awareness of its researchers as a Latin-Mediterranean Region, who, as will be seen in the following, tend to develop their work with researchers outside of other locations.

As for the media that disseminate the work of researchers in the Latin-Mediterranean Region, a total of 104 sources were found with texts related to the subject. The journals that publish the most works are mainly three: Frontiers in Psychology (with eight published articles), Sustainability (also with eight published articles), and Advances in Intelligent Systems and Computing (with four published articles) (Fig. 6).

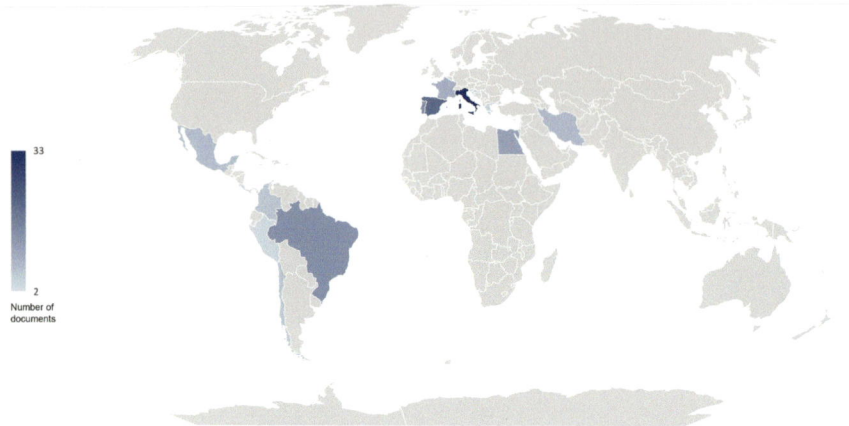

Fig. 5 Distribution of the number of published products according to location within the Latin-Mediterranean Region. Own elaboration based on bibliometric data from SCOPUS, 2022

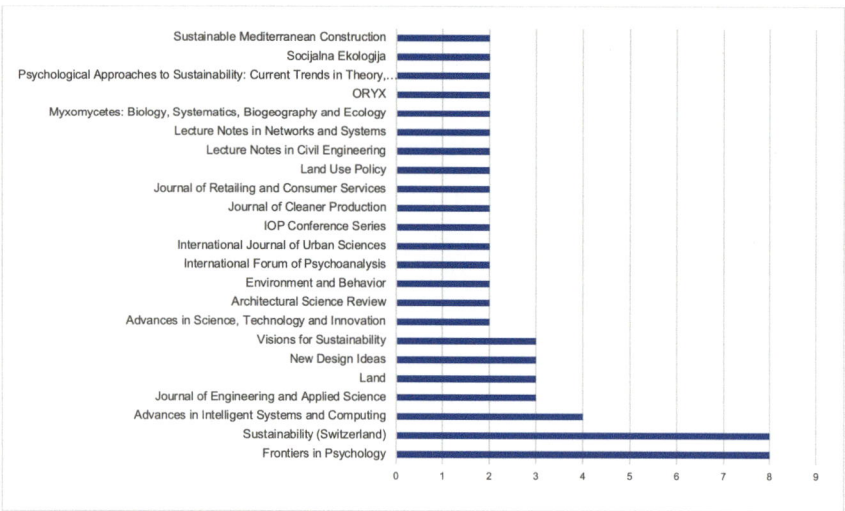

Fig. 6 Resources registered in the SCOPUS database with more than one published article on Biophilic Design and Biophilia produced in the Latin-Mediterranean Region. Own elaboration based on SCOPUS bibliometric data, 2022

4.2 How is the Co-Occurrence Network of Important Terms Extracted from the Specific Literature? How Do They Vary Annually Within the Latin-Mediterranean Region?

To create with VosViewer a term co-occurrence map based on text data about important terms extracted from the Biophilic Design and Biophilia literature, the following criteria were employed: the bibliographic database files of the 145 results obtained with SCOPUS were read. The terms for the analysis were extracted from the fields "Title" and "Abstract", ignoring structured abstract labels and copyright statements. The Full Counting Method was used, establishing as ten the minimum number of occurrences of a term; so, of the 4531 total different terms, only 113 were the ones that met the threshold. For each of the 113 terms, a relevance score was calculated. Based on this score, the most relevant terms were selected. The default choice is to select 60% of the most relevant terms. Therefore, 68 terms were selected. No terms were unselected in the "verify selected terms" process.

Analyzing the term co-occurrence map data shown in Fig. 7, six thematic clusters linked to the terms Building, Emotion, Effect, Model, Park, and Perception were identified.

The first cluster linked to the term Building encompasses topics associated with the terms Architecture, Analysis, Sustainability, Strategy, or Art, aspects that can be considered as specific to the architecture profession. The second cluster centered on the term Emotion includes aspects such as Child, Species, Feeling, Influences, and Intention, concepts much smoother than those that make up the first cluster. In the third cluster, the term Effect stands out along with the words Data, Contact, Workplace, Stress, Recovery, Employee, and Covid. The fourth cluster highlights the term Model and includes aspects such as Change, Use, Interaction, Order, Climate change, and Temperature, which have to do with outdoor or atmospheric aspects. The term Park emerges in the fifth cluster, which includes words such as Contribution, Population, and Connection and is closer to more collaborative ideas. And finally, the sixth cluster focuses on the term Perception, including aspects such as Group, Implementation, and Biophilic urbanism (Fig. 8).

If we analyze the aforementioned terms in more depth and, according to the concepts most repeated annually, the minimum number of occurrences of a term is stipulated as ten to be considered in the analysis, results begin to be obtained starting in 2007, with the term Animal appearing as the one that stands out with 24 occurrences (Fig. 9). The analysis clearly shows a series of trends that vary over the years. In the first years (2007–2013), a number of aspects referred to Animal (24 occurrences), Species (34 occurrences), Emotion (39 occurrences), Intention (19 occurrences), Feeling (14 occurrences), and Case (16 occurrences). These are aspects clearly linked to the understanding of Biophilia or Biophilic Design from the point of view of the organic or organism. From 2014 to 2016, the most repeated term is Child with 46 occurrences, observing a leap from animal to human and from emotional to tangible through works analyzing terms such as Park (30 occurrences), Biodiversity (10 occurrences), and World (19 occurrences). In the period between 2017 and 2018, research focuses on aspects linked to the processual and the construction of theories championed by concepts such as Analysis (37 occurrences), Model (37 occurrences), Knowledge (29 occurrences), Role (29 occurrences), Interaction (28 occurrences), Order (27 occurrences), and Question (27 occurrences).

In 2019 and 2020 the emphasis is on the Implementation (17 occurrences) and Application (28) of Biophilic in tangible problems and places to solve or alleviate

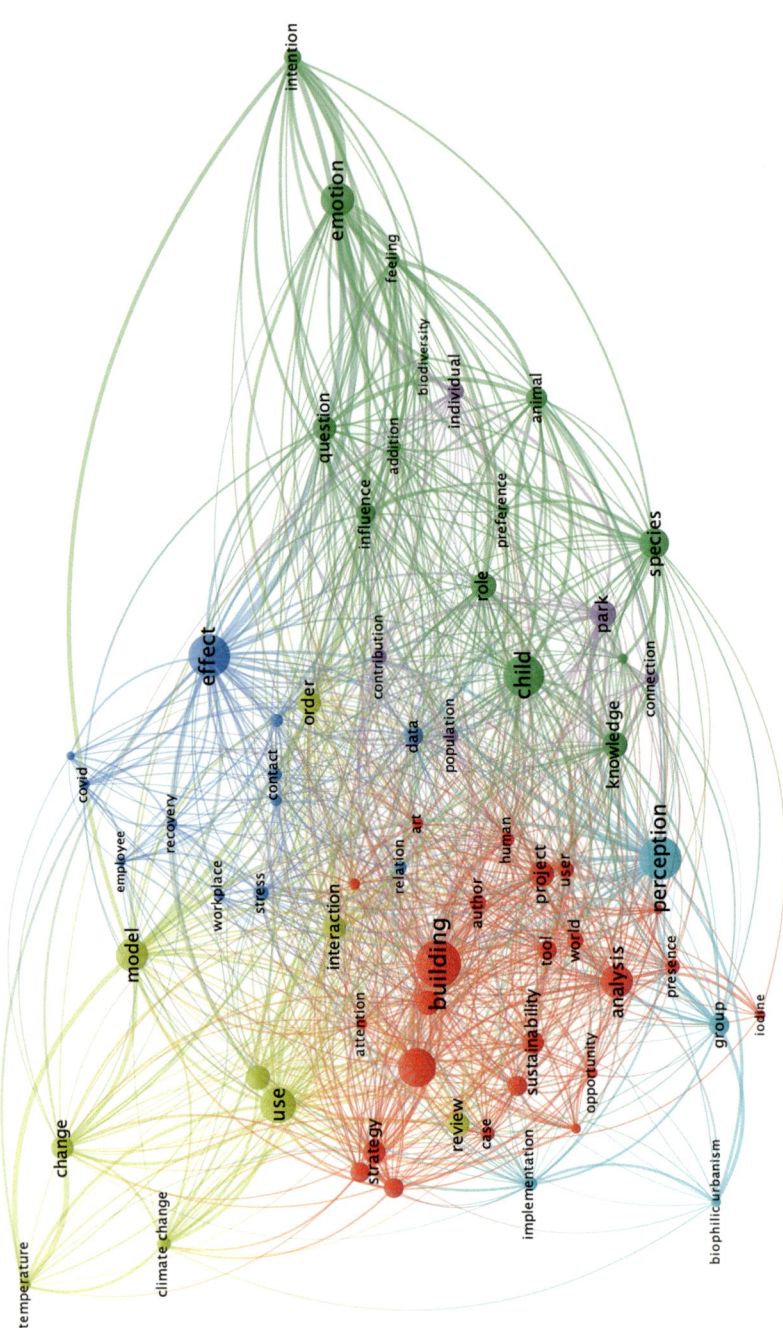

Fig. 7 Term co-occurrence map based on text data about important terms extracted from the Biophilic Design and Biophilia literature from Scopus database ("Title" and "Abstract" fields). Own elaboration with VosViewer, 2022

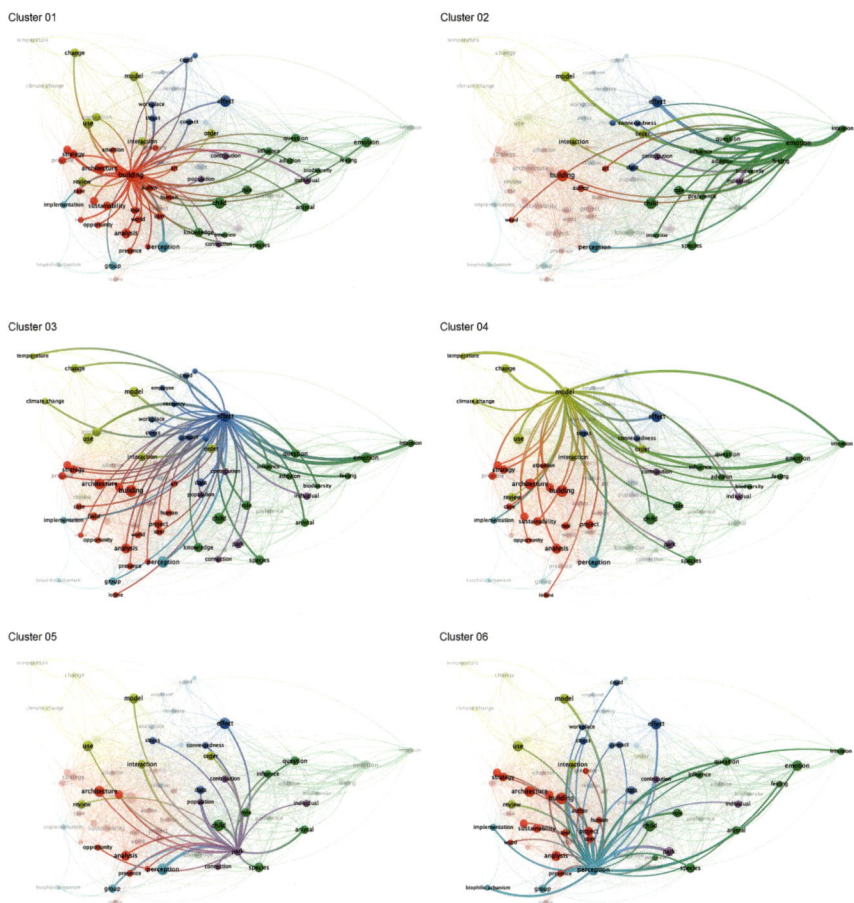

Fig. 8 Term co-occurrence maps based on text data about important terms extracted from the Biophilic Design and Biophilia literature from Scopus database (for more details refer to Fig. 7): Cluster (01) related to the term Building; Cluster (02) related to the term Emotion; Cluster (03) related to the term Effect; Cluster (04) related to the term Model; Cluster (05) related to the term Park; and Cluster (06) related to the term Perception. Own elaboration with VosViewer, 2022

dilemmas, such as stress or the pandemic of Covid-19 in the workplace, among others.

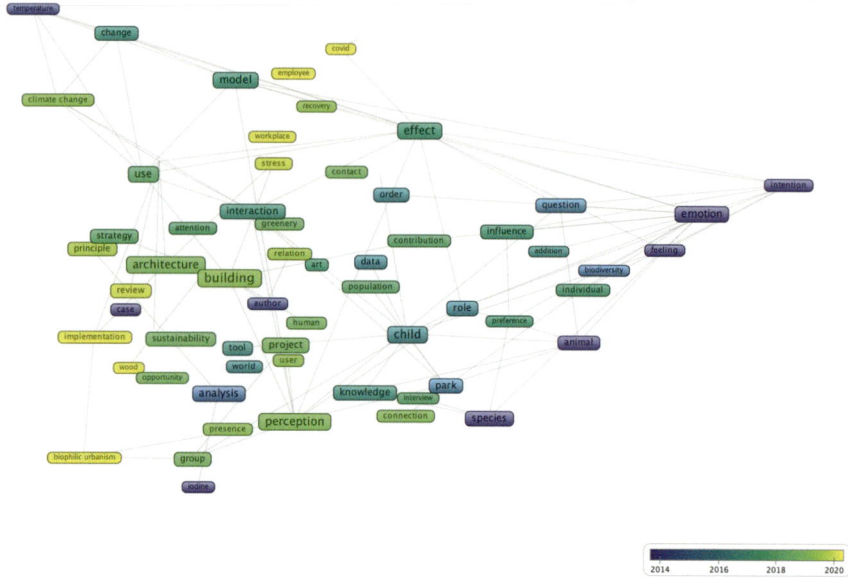

Fig. 9 Map of the most used concepts by year based on text data from Scopus database. Own elaboration with VosViewer, 2022

4.3 From Where, Within the Latin-Mediterranean Region, Are the Insights Being Produced? How is the Network of Linkages Being Built?

To create with VosViewer a term co-occurrence map based on text data on co-authorship of the Biophilia and Biophilic Design literature, the following criteria were followed: The bibliographic database files of the 145 results obtained with SCOPUS were read. The type of analysis was based on co-authorship and its unit of analysis was the field "Country" using the Full Counting Method. Documents co-authored by a large number of countries were not ignored. To be part of the map, the minimum number of documents of a country should be one, and the minimum number of citations of a country should also be one. Of the 45 countries, 43 meet the thresholds. For each of the 43 countries, the total strength of the co-authorship links with other countries was calculated, and finally, the countries with the greatest total link strength were selected (Fig. 10).

With regard to the linkage network in terms of authorship and taking the country as the unit of analysis, it can be observed that of the 43 countries in the Latin-Mediterranean Region, there are six countries that are not connected to any other and therefore are not shown in Fig. 10.

Of the countries with the highest production, it can be observed that Italian researchers are mainly linked in terms of authorship with peers from the United States, followed by colleagues from the United Kingdom, and in third and fourth

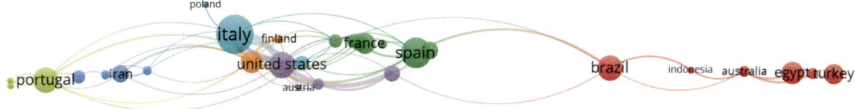

Fig. 10 Co-authorship cluster of the Latin-Mediterranean Region based on text data from Scopus database. Own elaboration with VosViewer, 2022

place researchers from Germany and Finland, all of them countries outside the Latin-Mediterranean Region. Italian collaborations of lesser impact exist with Canada, Poland, Netherlands, and Serbia; and within the Region, they are linked only with Portugal and Spain. As for Spain, the network of linkages is extended without the existence of dominant links, as in the case of Italy. The main co-authors of Spanish researchers within the Latin-Mediterranean Region come from Mexico, Colombia, Chile, Italy, and Brazil. They are also related with similar strength to Japan, Canada, Germany, the United States, and the United Kingdom. With respect to Portugal, the link with the United States is somewhat stronger than with the rest of its network. It collaborates within the Latin-Mediterranean Region with Greece and Iran, and externally with New Zealand, Latvia, Italy, the United Kingdom, and Germany. Finally, Brazil distances itself from the major producers, the United States, Canada, and the United Kingdom, and allies with Australia (the second largest text producer). It establishes its strongest ties with Spain, with links to Indonesia, Colombia, and Germany, at a minor scale. (Fig. 11).

It is interesting to note that both in terms of the volume of production and the connections between co-authors, there is no awareness of the Latin-Mediterranean Region as such among researchers. Therefore, the publication of this book is considered pertinent, as it reflects and makes visible the joint contribution at a regional level and could be the first step to establishing links and relationships between Latin-Mediterranean researches and researchers.

4.4 Which Funding Sponsors are Providing the Economic Stimuli for Research on Biophilic Design in the Latin-Mediterranean Region, Where are they from, and What has been their Funding Tendency Over Time?

The first localized Funding Sponsor that supported research on Biophilic Design and Biophilia in the Region is from 2012: the Interministerial Commission of Science and Technology—CICYT (Spain). It was not until 2017 that another support appeared: the Fundação para a Ciência e a Tecnologia—FCT (Portugal). And it is not until 2018 that the supported articles began to proliferate (9 articles supported in 2018, 9 articles supported in 2019, 16 articles supported in 2020, and 26 articles supported in 2021).

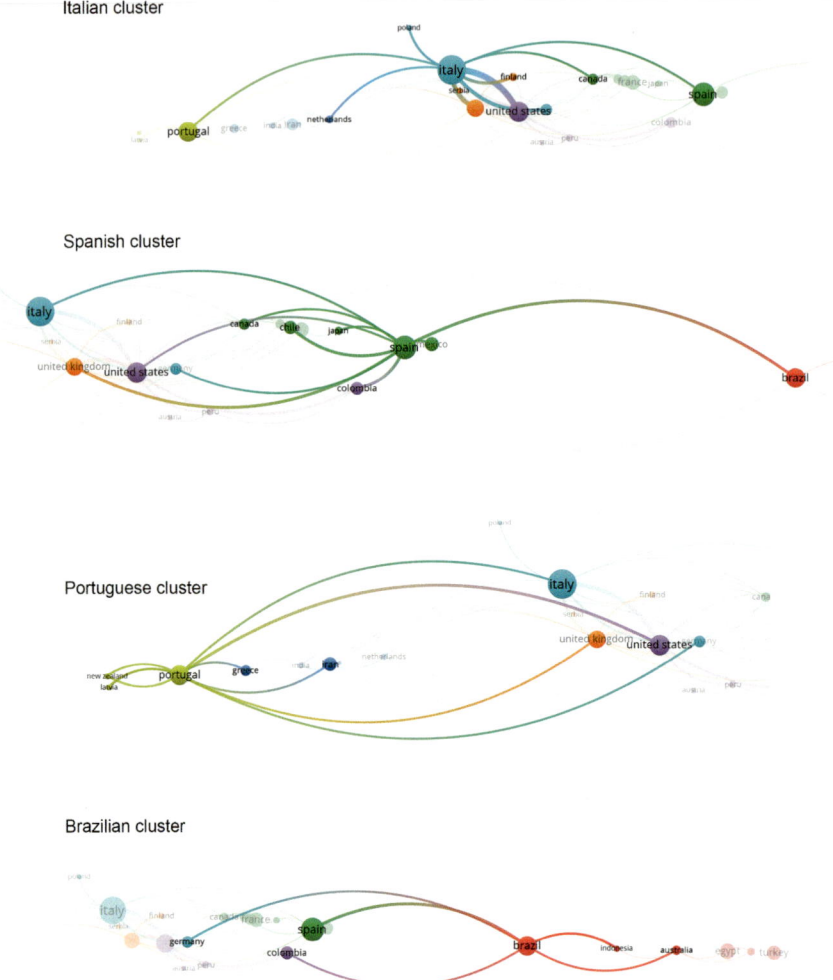

Fig. 11 Co-authorship clusters (for more details refer to Fig. 10): Italian cluster; Spanish cluster; Portuguese cluster; and Brazilian cluster. Own elaboration with VosViewer, 2022

The two main funding sponsors of the identified works are the Fondazione Cariplo of Italy and the Fundação para a Ciência e a Tecnologia of Portugal, (Fig. 12), which are part of the top ten funding sponsors by the number of papers financed worldwide that support research on Biophilic Design and Biophilia. Numerous Latin-Mediterranean texts have been supported by European Funds that are accessible from the Mediterranean sector of the Region, such as the European Cooperation in Science and Technology (COST) or the European Regional Development Fund (ERDF). And although most of the countries in the Latin-Mediterranean Region are linked to their national funds (Coordenação de Aperfeiçoamento de Pessoal de Nível Superior in

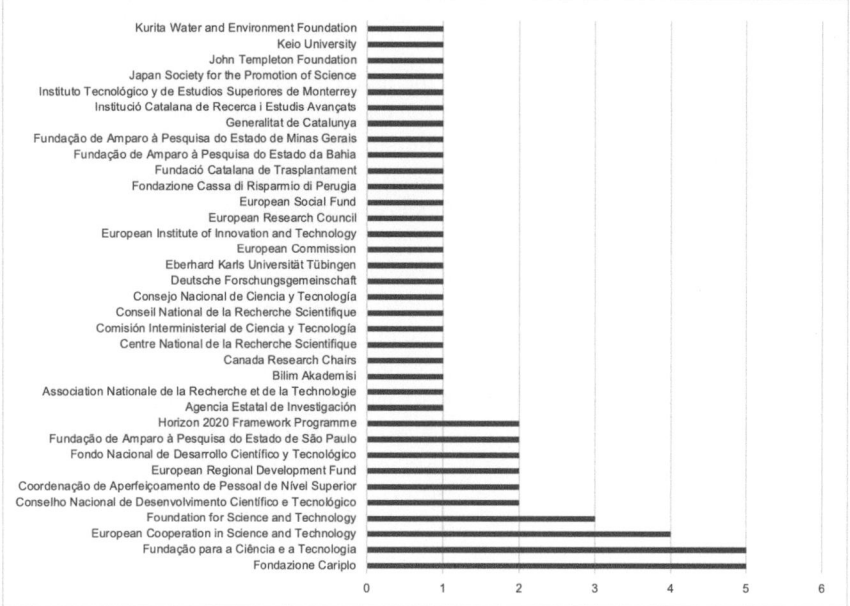

Fig. 12 The 35 main funding sponsors by the number of documents supporting research on Biophilic Design and Biophilia in the Latin-Mediterranean Region. Own elaboration based on bibliometric data from SCOPUS, 2022

the case of Brazil, Fondo Nacional de Desarrollo Científico y Tecnológico in the case of Chile, Bilim Akademisi in the case of Turkey, etc.), alliances exist to access high-impact sponsors outside the Region, such as the Foundation for Science and Technology based in the United Kingdom, and the Canada Research Chairs or the Deutsche Forschungsgemeinschaft, among others.

It is worth mentioning the existence of other minor funds (with only one publication so far) that have supported research on Biophilic Design and Biophilia in the Latin-Mediterranean Region, opening the door to further development and to the recognition of studies in the Region (Fig. 12).

5 Final Reflections and Further Research

As mentioned above, so far only the SCOPUS database has been used for the study. It is considered important for the accuracy of the information to expand the databases used in future works. Web of Science, JSTOR, Mendeley and SciELO, among others, will be added. Special emphasis will be placed on the search for specialized databases or databases with large collections of texts on architecture and art, such as Avery Index to Architectural Periodicals, EBSCO Art and Architecture Source, ProQuest

Art, Design and Architecture Collection, GW Libraries Art and Architecture Source or WGSN, among others.

The authors are aware of the bias in the data presented in this article by using only the SCOPUS database. This is especially remarkable in the study of disciplines in which scientific texts are not traditionally published exclusively and necessarily in indexed journals included in bibliometric databases. Thus, for future work it is expected to incorporate other more inclusive ways of tracking bibliography on Biophilia and Biophilic Design.

This is the first study to consider the territory comprised of the different countries as a region in itself (Latin-Mediterranean Region), and perhaps this lack of consistency as a Region means that there is a lack of collective work done by researchers on the topics of Biophilia and Biophilic Design.

It is important to point out that this work analyzes the studies carried out on the subjects within the Region, so it is considered essential to take a second step in the research by analyzing the studies on the Region regardless of its origin.

The topic is currently of great interest. The trend has been booming since the Covid-19 pandemic and it would be encouraging if this trend can be maintained, given the different benefits that Biophilia and Biophilic Design bring. This has also been reflected in the increased funding available for research on these topics, and although today the representation is mainly Mediterranean, there is an important area of opportunity for Latino researchers to participate and collaborate.

References

Barbiero G, Berto R (2021) Biophilia as evolutionary adaptation: an onto- and phylogenetic framework for biophilic design. Front Psychol 12:700709. https://doi.org/10.3389/fpsyg.2021.700709

Benvenuti S (2014) Wildflower green roofs for urban landscaping, ecological sustainability and biodiversity. Landsc Urban Plan 124:151–161. https://doi.org/10.1016/j.landurbplan.2014.01.004

Berto R, Barbiero G, Barbiero P, Senes G (2018) An individual's connection to nature can affect perceived restorativeness of natural environments. Some observations about biophilia. Behav Sci 8(3):0034. https://doi.org/10.3390/bs8030034

Biancoli R (1995) "Center-to-Center" relatedness between analyst and patient. Int Forum Psychoanal 4(2):105–110. https://doi.org/10.1080/08037069508409529

Blau ML, Luz F, Panagopoulos T (2018) Urban river recovery inspired by nature-based solutions and biophilic design in Albufeira, Portugal. Land 7(4):141. https://doi.org/10.3390/land7040141

Camrass K (2022) Urban regenerative thinking and practice: a. Build Res & Inf 50:339–350. https://doi.org/10.1080/09613218.2021.1922266

Dadvand P, Pujol J, Maciá D, Martínez- Vilavella G, Blanco-Hinojo L, Mortamais M, Alvarez-Pedrerol M, Fenoll R, Esnaola M, Dalmau-Bueno A, López-Vicente M, Jerrett M, Nieuwenhuijsen MJ, Sunyer J, Basagaña X (2018) The association between lifelong greenspace exposure and 3-dimensional brain magnetic resonance imaging in Barcelona schoolchildren. Environ Health Perspect 126(2):027012. https://doi.org/10.1289/EHP1876

Disterheft A, Caeiro SS, Leal Filho W, Azeiteiro UM (2016) The INDICARE-model - measuring and caring about participation in higher education's sustainability assessment. Ecol Ind 63:172–186. https://doi.org/10.1016/j.ecolind.2015.11.057

Erebor EM, Ibem EO, Ezema IC, Sholanke AB (2021) Energy efficiency design strategies in office buildings: a literature review. In: IOP conference series: earth and environmental science. International conference on energy and sustainable environment, 23–25 June 2020, vol 665. IOP Publishing Ltd., Nigeria, p 012025. https://doi.org/10.1088/1755-1315/665/1/012025

FAO (1998) Informe del grupo de trabajo sobre la cuenca del Mediterráneo. FAO, Roma. https://www.fao.org/3/x2095s/x2095s0a.htm

Ferrerira V, Barreira AP, Loures L, Antunes D, Panagopoulos T (2020) Stakeholders' engagement on nature-based solutions: a systematic literature review. Sustainability 12(2):640. https://doi.org/10.3390/su12020640

Fisher KD (2019) An evidence-based biophilic design framework for health and wellbeing. In: Agrawal A, Gupta R (ed) Revisiting the role of architecture for 'surviving' development. 53rd international conference of the architectural science association. Architectural Science Association (ANZAScA), Roorkee, pp 145–154

Gillis K, Gatersleben B (2015) A review of psychological literature on the health and wellbeing benefits of biophilic design. Buildings 5(3):948–963. https://doi.org/10.3390/buildings5030948

Gomes CC, Preto S (2018) Should an artificial window substitute a natural one? Adv Intell Syst Comput 607:247–258; AHFE 2017 international conference on usability and user experience, Los Angeles. https://doi.org/10.1007/978-3-319-60492-3_24

Guntiñas ME, Leirós MC, Trasar-Cepeda C, Gil-Sotres F (2012) Effects of moisture and temperature on net soil nitrogen mineralization: a laboratory study. Eur J Soil Biol 48:73–80. https://doi.org/10.1016/j.ejsobi.2011.07.015

Hanafi M, Naguib M (2013) Bio-regenerative rating technique: a critical review. WIT Trans Ecol Environ 175:233–246. https://doi.org/10.2495/ECO130201

Heymans A, Breadsell J, Morrison GM, Byrne JJ, Eon C (2019) Ecological urban planning and design: a systematic literature review. Sustainability 11(13):3723. https://doi.org/10.3390/su11133723

Hu M, Roberts JD (2020) Connections and divergence between public health and built environment - a scoping review. Urban Sci 4(1):12. https://doi.org/10.3390/urbansci4010012

Hung SH, Chang CY (2021) Health benefits of evidence-based biophilic-designed environments: a review. J People Plants Environ 24(1):1–16. https://doi.org/10.11628/ksppe.2021.24.1.1

Hutton B, Catalá-López F, Moher D (2016) La extensión de la declaración PRISMA para revisiones sistemáticas que incorporan metaanálisis en red: PRISMA-NMA. Med Clin 147(6):262–266. https://doi.org/10.1016/j.medcli.2016.02.025

Kellert SR (2008) Biophilia. In: Fath BD (eds) Encyclopedia of ecology, five-volume set. Elsevier Inc., Netherlands, pp 462–466 https://doi.org/10.1016/B978-008045405-4.00636-4

Khanehshenas F, Habibi P, Zakerian SA (2020) The effect of biophilic design patterns on employee's health and well-being: a systematic review. Iran J Ergon 7(4):1–11. https://doi.org/10.30699/jergon.7.4.1

Landis B (1975) Fromm´s theory of biophilia-necrophilia. Contemp Psychoanal 11(4):418–434. https://doi.org/10.1080/00107530.1975.10745401

Lee EJ, Park SJ (2022) Biophilic experience-based residential hybrid framework. Int J Environ Res Public Health 19(14):8512. https://doi.org/10.3390/ijerph19148512

Martín-López B, Montes C, Benayas J (2007) The non-economic motives behind the willingness to pay for biodiversity conservation. Biol Cons 139(1–2):67–82. https://doi.org/10.1016/j.biocon.2007.06.005

McGee B, Park N-K (2022) Colour, light, and materiality: biophilic interior design presence in research and practice. Interiority 5(1):27–52. https://doi.org/10.7454/in.v5i1.189

Moher D, Liberati A, Tetzlaff J, Altman DG, PRISMA Group (2009) Preferred reporting items for systematic reviews and meta-analyses: the PRISMA statement. PLoS Med 7(6):e1000097. https://doi.org/10.1371/journal.pmed.1000097

Page MJ, McKenzie JE, Bossuyt PM, Boutron I, Hoffmann TC, Mulrow CD et al (2021) The PRISMA 2020 statement: an updated guideline for reporting systematic reviews. BMJ 71:372. https://doi.org/10.1136/bmj.n71

Páramo P, Mejía MA (2004) Urban parks as opportunities for the interaction of children of animals. Rev Latinoamericana de Psicologia 36(1):73–84+i

Rosa CD, Collado S (2019) Experiences in nature and environmental attitudes and behaviors: setting the ground for future research. Front Psychol 10:763. https://doi.org/10.3389/fpsyg.2019.00763

Rosenbaum MS, Otalora ML, Ramírez GC (2016) The restorative potential of shopping malls. J Retail Cosum Serv 31:157–165. https://doi.org/10.1016/j.jretconser.2016.02.011

Russo A, Cirella GT (2018) Modern compact cities: how much greenery do we need? Int J Environ Res Public Health 15(10):2180. https://doi.org/10.3390/ijerph15102180

Tian RC, Nicolas E (1995) Iodine speciation in the northwestern Mediterranean Sea, method and vertical profile. Mar Chem 48(2):151–156. https://doi.org/10.1016/0304-4203(94)00048-I

Tirri C, Swanson H, Meenar M (2021) Finding the "heart" in the green: conducting a bibliometric. Int J Environ Res Public Health 18(18):9435. https://doi.org/10.3390/ijerph18189435

Vojnovic D, Procida G, Gabrielli Favretto L (1991) Chemometric differentiation of raw and commercial milk by trace elements using principal component analysis. Food Addit Contam 8(3):343–349. https://doi.org/10.1080/02652039109373983

VOV world (2019) Los 10 eventos internacionales más destacados en 2019, seleccionados por VOV. VOV Voz de Vietnam, 31 December 2019. https://vovworld.vn/es-ES/enfoque-de-actualidad/los-10-eventos-internacionales-mas-destacados-en-2019-seleccionados-por-vov-814843.vov

Wijesooriya N, Brambilla A (2021) Bridging biophilic design and environmentally sustainable design: a critical review. J Clean Prod 283:124591. https://doi.org/10.1016/j.jclepro.2020.124591

Yassein G, Ebrahiem S (2018) Biophilic design in the built environment to improve well-being: a systematic review of practices. J Urban Res 30(1):128–146. https://doi.org/10.21608/jur.2018.88412

Zare G, Faizi M, Baharvand M, Masnavi MR (2021) A review of biophilic design conception implementation in architecture. J Des Built Environ 21(3):16–36. https://doi.org/10.22452/jdbe.vol21no3.2

Zhong W, Schröder T, Bekkering J (2022) Biophilic design in architecture and its contributions to health, well-being, and sustainability: a critical review. Front Archit Res 11(1):114–141. https://doi.org/10.1016/j.foar.2021.07.006

Rural

Introduction

Overgaard-Porsen et al. analyse the relationship between poverty, its socially derived problems and how rural communities are still governed by patriarchal structures and given migratory dynamics, with the respective damage to family structures, linked to environmental issues, problems such as insufficient water supply, low family incomes, poor housing quality and lack of basic infrastructure, which together generate a low quality of life. The application of a sustainable regeneration model based on business projection, social development, community leadership and social cohesion, Design Thinking methodologies and strategic planning in community architecture design processes is presented. It highlights the importance of how in this model, the public sector, academia, the community and NGOs work side by side and how regenerative designs permeate the entire model, with the idea of improving the environment of communities, regenerating the social fabric. This model allows and encourages the community to take action in the development of their well-being and not just wait for help from the government, making this a virtuous cycle. In order to achieve the best results from the implementation of the model, it is recommended to follow some simple guidelines, such as elaborating a diagnosis according to the Participatory Intervention Model, so that the community achieves a common vision, which guarantees that the decisions taken are those of the community, independently of external interest groups. It is essential to get the municipality involved in the process to access funding and expertise on solving infrastructure problems in the community, with the participation of universities, NOGs and the private sector. It is of utmost importance to provide expertise to the community on how to turn ideas into tangible improvements.

From the work carried out in the community of Tilaco, in the Sierra Gorda Biosphere Reserve, Benitez and Cobreros analyse a reality that affects many rural communities in Mexico, the modification in the way of inhabiting the territory and the way of making and planning housing as a consequence of the effects of migration, which generates problems of habitability, introduces foreign construction systems and even generates a loss of identity. A review and contextualisation of rural housing is carried out from different aspects of planning, construction and use of housing from a regenerative design approach based on ideas from the Living Building Challenge certification, in search of a positive impact on the community and its cultural and bioclimatic aspects. There is a need to develop strategies to provide knowledge on the planning and construction of housing to maintain a balance and harmony between the natural environment and the inhabitants, to improve identity as a sense of belonging, to reduce the environmental impact and to connect with the needs of the inhabitants of the community, listening to their feelings about the perception of housing and the value it has for them. These strategies could help the inhabitants become aware of what is involved in planning a house and of the main elements to consider such as the environment and climate. The inhabitants' use of strategies based on the principles of Regenerative Design and People-Centred Design could be the starting point to find a central point between the construction of the house with local character and a house style brought from another context, using each of the defined categories such as place, water, materials, energy, etc., in order to structure and plan the whole process and with which the inhabitants can feel identified and also help as far as possible to reduce the environmental impact.

Sustainable Development of Rural Communities in Mexico

Rena Overgaard Porsen, Abril Joana Margarita de León Rincón, and Irma Beatriz González Hernández

Abstract According to data from the National Council for the Evaluation of Social Development Policy (CONEVAL 2020), in the south of the state of Nuevo Leon between 60 and 80% of the population lives in poverty and extreme poverty. This economic aspect generates a series of problems: from the social point of view, rural communities are still governed by patriarchal structures and the family structure is determined by migratory dynamics that fragment families leaving them in a continuous state of abandonment. From an environmental point of view, problems such as insufficient water supplies have a direct impact on family income. Finally, poor housing quality and lack of basic infrastructure generates a low quality of life. In this chapter we aim to present a sustainable regeneration model applied to a location of El Refugio de Cerros Blancos, a small village of the municipality of Mier y Noriega, in Nuevo Leon, Mexico. This model is founded on entrepreneurial outreach based on social development, community leadership and social cohesion, methodologies of Design Thinking and strategic planning in the design processes of community-based architecture and the implementation of follow-up strategies.

Keywords Rural sustainability · Public entrepreneurship · Design thinking · Sustainable architecture

R. Overgaard Porsen (✉) · I. B. González Hernández
School of Architecture, Art and Design, Tecnologico de Monterrey, Monterrey, México
e-mail: rporsen@tec.mx

I. B. González Hernández
e-mail: b.hernandez@tec.mx

A. J. Margarita de León Rincón
School of Social Sciences and Government, Tecnologico de Monterrey, Monterrey, México
e-mail: abril.de.leon@tec.mx

© The Author(s), under exclusive license to Springer Nature Switzerland AG 2025
C. Cobreros et al. (eds.), *Regenerative Design*, Cities and Nature,
https://doi.org/10.1007/978-3-031-76890-3_9

1 Introduction

In an international context, sustainability in the framework of rural development in Latin America is combined with the Sustainable Development Goals (SDGs) that derive from the documents signed on September 25th, 2015, where world leaders adopted a set of global goals to eradicate poverty, protect the planet and ensure prosperity for all as part of a new sustainable development agenda (UN 2015). Primarily, sustainable rural development projects all impact separately and together the global goals of "End poverty in all its forms everywhere". Currently, the SDGs have managed to create a whole context for evaluation and study of quality of life in rural communities to evaluate the improvements of the initiatives applied in each community.

The present work is dealing with the sustainable development of rural communities and is based on an applied and empirical methodology to achieve a series of integrated results in the context of the community (Gibbes et al. 2020). To achieve this, it is recommended to work with a range of actors and in a series of parameters that have to do with the physical and sociocultural context of the site.

Within this sustainability framework, the issues of combating poverty as the main strategy to generate a more sustainable development combined with different initiatives of land use and the built context are areas that are seen as particularly important to promote sustainable development in Latin America (Gibbes et al. 2020). his chapter aims to present the application of a model of sustainable regeneration of El Refugio de Cerros Blancos, one of the 17 villages of the municipality of Mier y Noriega, state of Nuevo Leon, Mexico, founded on entrepreneurial outreach based on social development, community leadership and social cohesion, Design Thinking methodologies and strategic planning in the design processes of community-based architecture and the application of monitoring strategies. The objective of this work is to show a possible path to sustainable development through regenerative design by strengthening the social processes of vulnerable rural communities and the application of architectural projects based on principles of inclusion and systemic vision for social transformation.

The relevance of testing and disseminating the results of these projects lies precisely in the impact that they can have on achieving a substantial improvement in the quality of life of rural communities, in line with the sustainable development goals, because the model encourages and strengthens community leadership, building and making visible tangible and intangible products that are achieved by common agreement and in common participation. Architecture can generate social changes of great importance if it is applied consciously and always considering the sociocultural and environmental context of communities. These characteristics are what make the architectural intervention go beyond a sustainable design, as it builds the foundations so that the future of the new generations is better than it is today: regenerative architecture.

However, this project also has some limitations and challenges due to that working in a systematic manner implies mapping, coordinating and even mediating between a network of actors and issues involved in the development of the project, which presents political limitations, in the case of the corresponding municipal or state governments; cultural (belief systems) of rural communities and, even, the barriers of the disciplines involved or the educational models that govern the universities. It should be noted, however, that the choice of the community must provide the greatest possible certainty of managing these interests in favor of a supreme good, the common good.

The variables that will be managed in this intervention are the elements of the participatory intervention model: community cohesion and leadership, strategic planning, participatory methodologies, and social innovation, through Design Thinking. However, it is important to note that rural communities in Mexico have indicators of quality of life in income, education, and access to health services below the cities and the case of the municipality of Mier y Noriega is no exception.

From the application of budgets at the state level, a situation that favors the municipalities located in the Metropolitan Area of Monterrey is clearly seen. The annual budget for the State of Nuevo León for 2022, according to the Secretary of Finance and General Treasurer of the State, is 117 thousand 745 million pesos (Flores 2021) while, in the same period, the budget for the municipality of Mier y Noriega represents only 72 million 25 thousand pesos (León 2021).

At first glance, the differences in these figures seem to be related to the number of inhabitants. However, when performing a simple arithmetic operation, it is observed that, per inhabitant of the state, 20 thousand pesos 357 pesos are received for each of the 5 million 784 thousand inhabitants of the state. On the other hand, in the case of the municipality of Mier y Noriega, the same operation generated the result of the availability of 10 thousand 151 pesos available per inhabitant.

That is why it seems necessary that rural municipalities such as Mier y Noriega, in the case of wanting to be an economically stable community, with a good quality of life and a fixed population, need to introduce methods and techniques to ensure new forms of governance based on active citizen participation; apply strategic planning methods and methodologies to become a "smart city" by strengthening the types of community and digital networks of integrated services (Fernandez Gúell 1997) to optimize the processes and resources of the sustainable administration of rural municipalities.

2 Context of Rural Communities in Southern Nuevo Leon

The Law of Sustainable Rural Development (2021), is regulatory law of Section XX of Article 27 of the Political Constitution of the United Mexican States and is of general observance throughout the Republic, defines sustainable rural development as: "The integral improvement of the social welfare of the population and of economic activities in the territory included outside the cores considered urban in accordance with the applicable provisions, ensuring the permanent conservation of the natural resources, biodiversity and environmental services of said territory".

Likewise, The National Institute of Statistics and Geography (INEGI) defines a rural community as those having 2500 or fewer inhabitants. However, this criterion only obeys the population, and the characteristics of rural communities are remarkably diverse as well. In this regard, the United Nations Economic Commission for Latin America and the Caribbean ECLAC (2021), in the document entitled "Characterization of rural spaces in Mexico based on national statistics", proposes new elements based on the socioeconomic diversity presented by rural communities, in which agricultural activity is often no longer the predominant one, but remittances, for example. This is precisely the case of the municipality of Mier y Noriega, taking into account that its main income is now coming from remittances from family members working in the United States of America.

The geographical and socioeconomic characteristics of the communities of the state of Nuevo Leon marks a variated outlook. The state is situated in the northeastern part of Mexico on the border of Texas, U.S.A. From the economic perspective, Nuevo Leon with its 5.784 million inhabitants in 2020, ranks second in PIB of Mexico with 1.258.572,22 millions of pesos (INEGI 2020).

Nuevo Leon is part of the northeastern border between Mexico and the United States of America. 79% of the population lives in urban locations and 21% in rural areas. In Nuevo León there are 89 urban localities and 4,733 rural localities spread in several municipalities throughout the State, mainly located near the North–South faced federal road system. The main urban metropolitan area is Monterrey with an approximate population of 5,046,743.00 million inhabitants (INEGI 2020).

The natural environment of Nuevo Leon consists of two mayor environmental systems: on the western part we have the Sierra Madre Oriental system occupying 50.9% of the territory and the coastal plain of the North Gulf covering 14.5% of the state territory located on the Eastern part of the state facing the State of Tamaulipas and the Mexican Golf. Internally, Nuevo Leon can be divided in four large areas with Monterrey as a central part, where the northern part has Sabinas Hidalgo as its main town developed along the national road connecting with Texas, also connecting with the border towns Nuevo Laredo and Colombia. In the southeastern part we have the part of the state developed along the national road with small towns like Allende, Montemorelos and Linares. Finally, we have the southwestern part ubicated in the interior on a plateau surrounded by mountains. Mier y Noriega along with municipalities like Aramberri, Zaragoza and Doctor Arroyo are situated in this part, and Mier y Noriega limits directly with the municipality of Doctor Arroyo (See Image 1).

In this natural context, the communities and small towns situated far from the national road system suffer from severe poverty which combined with a poorly developed infrastructure system leads to a series of social and economic problems such as poor education level and leading massive migration rate of young inhabitants seeking opportunities in Monterrey or in the U.S.A.

According to Data Nuevo Leon (a statistical information service of the Secretary of Economy of the State Government of Nuevo Leon 2020), the municipality of Mier y Noriega, counts 7,652.00 (DATA Nuevo León 2022) inhabitants in 2022. The built context of the municipality has the municipal seat called Mier y Noriega, with 1180

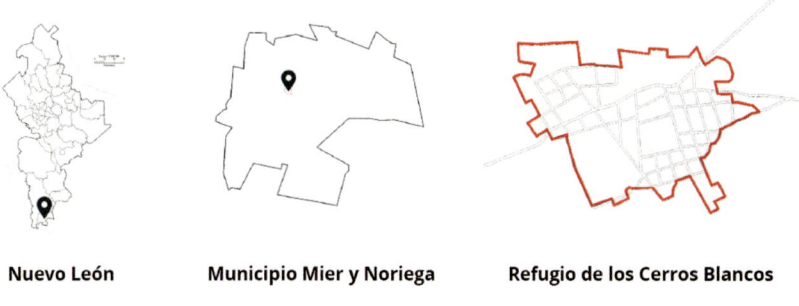

Image 1 The municipality of Mier y Noriega in the State of Nuevo Leon. Location of Mier y Noriega. *Source* Own elaboration

inhabitants, fourteen villages with a population between 100 and 600 inhabitants and three small settlements with a population of less than 100 inhabitants (INEGI 2020).

At the same time, being located as the most southern municipality of Nuevo Leon, Mier y Noriega makes life complicated for its inhabitants, because in addition to the circumstances mentioned above, this municipality is not only not located near a national expressway but on the breach of a plain landsite surrounded by mountains on the eastern southern and western side. This means that every activity that must be developed in Monterrey implicates 5–8 h of transportation time.

3 Participatory Intervention Model

The Sustainable Development of Rural Communities in Mexico model is a holistic model that aims to include a variety actors and subjects; which should be organized within a uniform mesh in four main pillars: (1) Social Cohesion and Community Leadership; (2) Strategic Planning; (3) Participatory Intervention and (4) Social Innovation Methodologies, specifically, Design Thinking (See Image 2). Which is designed to be a way of working centered on the community: its possibilities, its desires, and its skills to generate scenarios that allow neighbors in order to be able to create new development structures (van Es et al. 2015).

The team of investigators who designed and implemented this model are two architects, professors from the "Tecnológico de Monterrey", and the founder of an NGO called Pal Sur del Norte, AC who has 20 years of expertise in processes of social transformation and community organization. Hereinafter, this group will be called the "intervention team" or "the intervention group". Besides this, three more professors have participated in the implementation of the model through field work and the development of the student projects. In this case, interdisciplinarity and the link with the Social Service Directorate of Tecnológico de Monterrey, through the

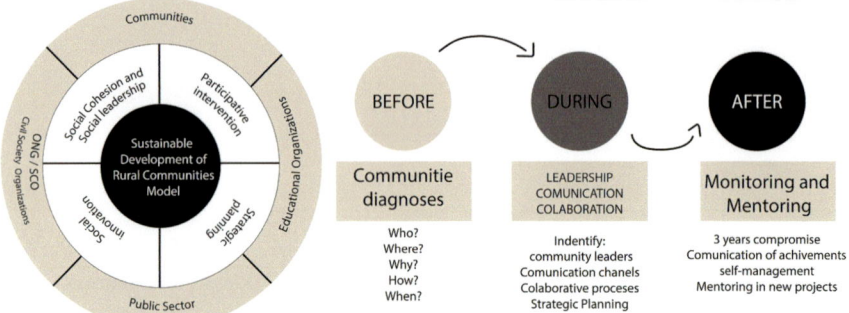

Image 2 Participatory intervention model. *Source* Own elaboration based on 4 years of field work applying participatory methodologies of community intervention, such as theory of change and logical framework

scheme of academic linkage with the community has granted substantial advantages and a great differentiator compared to the rest of the traditional models because we can involve our students in the intervention, sensitizing our future professionals of the social reality in the field.

When the team of investigators decided to enter a community, it is because they already did a previous work of scouting, diagnosis and preparation for the intervention, called Social Cohesion and Community Leadership. While developing the previous diagnosis of feasibility the team used a list of criteria that help if there would be a good response or not (See Annex 1: Criteria for intervening in a community). From this, the intervention team enters a community to work with the sociocultural, geographical, natural context, in order to generate development scenes in which the community can develop in the future, in a sustainable and self-sufficient way. It is important to state that the aim of the intervention group is not a welfare vision, just giving away useless gifts and pearl necklaces and it is made noticeably clear in the first meeting with the community that: "we build together, or we do not do it". The intervention starts from scratch, identifying needs collaboratively with the community, so that the neighbors and the intervention team create the future scenario of human development.

When collaborating with the community there is always the dilemma of what comes out in the detection of needs that identifies the need and those that the intervention team identifies. That is why it is necessary to always maintain the principle of visualizing the latent needs that they (the community) do not see, making field observations in which the team shows them the current state of things, why it is like this, why it would be necessary to change it in terms of the common good. The priority principle for the academic group and the ONG is to think collectively.

Before continuing it is vital to state that what the intervention group produces is not necessarily going to end in a tangible result. It is particularly important that the product has a strong emotional value in the community, that it fosters or recovers a sense of belonging, identity, memory, or elements of economic sustainability. For example, the management of the community center at the village level was initiated by a student from the community and was helped by a commissioner who died months after collecting the signatures. In honor of him and other community members who participated in the workshops on academic strengthening and social cohesion, the community has been mobilized to have its own public space with the best design and the workmanship of the same people who inhabit the village.

The phases of the sustainable community intervention model are described below:

Social cohesion and community leadership

The phase of construction or strengthening of social cohesion and community leadership anticipates the intervention. To carry it out, in turn, the ONG have to organize the following temporary structure:

1. *Before: preparation*

The premise of intervention is that the communities are autonomous, and they are self-determined. Therefore, one of the criteria for intervening is that the community wants to be intervened. As ethical principles of intervention, the following questions must be asked:

1.1. *Why intervene?*

Being a model that is born from the Academy and is based on a model of social service academically linked to the community, the investigators are interested, as academics, in seeking spaces that allow our students to strengthen or develop the disciplinary and transversal competencies necessary to go out to reality and perform as a person, citizen and professional, in an ethical and efficient manner.

Therefore, the intervention has academic interests that must be combined with the real needs of a community. To the question of why to intervene, the answer must always prevail as to why the intervention can improve the quality of life of the community, as the Law on Sustainable Rural Development also says.

1.2. *Where to intervene?*

Choosing the community: do the intervention team choose it or do the community choose the intervention team? Previously, the intervention team must conduct a sociodemographic study to see communication routes, transfer conditions, security, connectivity, base organizations, communication, and support network for the actors involved. In this case, being academic projects linked to the community, it is necessary that there is a support network to prevent and respond as far as possible to the security conditions for students, teachers, allies, drivers, etc. This support network should include an expert in first aid and rescue, academic leaders, the socio-forming organization that facilitated entry into the community, ejido and municipal authorities and community leaders. The trust and support of the community guarantees the stability and continuity of the intervention.

1.3. *Under what perspective to intervene?*

Community intervention must be carried out from the bottom up. That is, without the support of the community it is more difficult to succeed because ownership must be generated. In that sense, it is necessary to work, during the previous phase, the expectations of the community. It is always best to be transparent in defining expectations. We get where we define together in base agreements. These refers to elements in the definition of a joint strategy of work with the community and they have to do with the way in which they perceive themselves and with giving them the necessary tools to strengthen their leadership.

(1) *Community self-perception:* in the interviews and focus groups that the intervention team conducted in the preparation stage, it was possible to identify the way in which the community perceives itself, the origin of that perception, and whether it is possible to work with it or not. If the self-perception of the community is one of victimhood or indifference, originated by the relationship that has historically been established between its authorities of the order that is and the citizenship, it is necessary to determine in citizen training exercises if there is openness to learn, change or modify habits that are not working for them. If there is openness to it and security conditions, then we can proceed. This process is called by the team as "community legitimation". No one should impose a belief system, only establish formative dynamics in which communities are convinced that what the intervention team is proposing can work for them. If that is achieved, everybody is ready to work hand in hand.

(2) *Community self-management*: the next step is for the community to have tools to make decisions and self-manage their needs. The intervention team works on tools for self-knowledge, teamwork, citizen training, leadership and, finally, the establishment of rotating committees to communicate, disseminate, organize and manage needs (See Image 3).

2. *During the intervention*

In this phase, the process that occurs when the first contact with the community is going to be initiated will be explained. It is detailed from the entrance to the community and its legitimization process, to the formulation of community proposals that will give rise to the implementation process.

2.1. *The entrance to the community*

One of the most difficult aspects to achieve is to meet the conditions to be able to enter a community and then, how to attract them, build trust and maintain it to encourage citizen participation. These points are developed below:

2.1.1. *How can the approach to a community be generated?*

Through a person or organization legitimized by the same community. It can be a school, church, family, civil society organization. It is not recommended that it be a municipality or any political organization. The approach to the municipal authorities is given strategically. First, introducing themselves and explaining what they are going to do in the municipality, already legitimized by the community. Then,

Image 3 Focus group with women from the Ejido Refugio de Cerros Blancos, Mier y Noriega, 2018

with strategic alliances that allow both to position themselves, without falling into proselytism or sacrificing their independence.

2.1.2. *How to ensure good communication with the community?*

It is necessary to establish jointly with community leaders the most efficient ways to communicate agreements, progress, achievements, celebrations, etc. In the same way, the academic group must communicate to the parents of the students who accompany them in the intervention. Make an effort to explain to parents everything the team does to control as much as possible so that their children feel safe.

2.1.3. *How to involve community members?*

Through the community institutions and authorities (individual or collective) that they respect. It is mandatory to start from them. Then, relying on their leadership, the intervention team establishes working committees.

2.1.4. *How to maintain the interest?*

Follow-up is key to this. The intervention team must be present all time. The community responds reciprocally. If the intervention team for their support to intervene, they expect the intervention group to be with the community from beginning to end and, at least, 3 years of intermittent follow-up to what was started.

2.1.5. *How to make the intervention visible?*

It is also necessary an approved strategy of public image and communication in social networks between all the actors involved, covering norms and legal requirements, as well as generating ethical content that makes visible the success stories of the community, of the students and the achievements acquired in the face of the challenges faced.

2.2. Proposals and implementation

The proposals came out of the participatory intervention phase, hand in hand with social innovation methodologies such as Design Thinking, which are explained below. In this section, the important thing is that what has been proposed is executed hand in hand with the community involving it in its application, as well as the actors who participated in its design, implementation, and evaluation. It is of utmost importance that the students and teachers who participated in the engagement go to the community to understand their reality and produce something that is truly empathetic and sustainable. The example and congruence between academic leaders in relation to the community will motivate students to give their best. The results in the projects also suggest a lot of difference between a work done with information that someone else brings, versus those who did it based on what they could feel at the intervention site.

2.2.1. The actors

The methodology of actor mapping is one of the core parts of the model (It is not only about identifying those people, organizations and/or institutions that influence, are involved or can influence the development of the work or project, but also to map their interests and the relationships between them. These relationships can determine the success or failure of a work. The stakeholder mapping methodology is one of the core parts of the model. It is not only about identifying those people, organizations and/or institutions that influence, are involved or may influence the development of the work or project, but also about mapping their interests and the relationships that exist between them. These relationships can determine the success or failure of a work. The socio-training organization is in charge of carrying out the necessary steps to establish the links with the actors involved in the project. This stakeholder mapping is also carried out by the organization but agreed upon with the Academy to include the corresponding academic stakeholders and take care of the guidelines under which the participation of students and teachers is inserted in a project. In any intervention model, this mapping of actors must be analyzed to find out with whom it is possible to work or not, depending on the relationships and interests that exist between them. Even so, the premise is that institutional actors should not be left out, that is, the public administration, communal and municipal authorities that have direct interference in the public spaces in which we intervene. If, in addition, there are authorities from other levels of government involved, the corresponding chain of command must be followed with its due protocols. This is the responsibility of the socioforming organization. Generally, for a successful participatory intervention, the actors are part of the following areas:

- *Public*: *in rural areas, the public authorities are: local* assembly, municipality, state and federal government, if applicable, especially for issues of social or financial support for vulnerable areas; or competence of the problem to be solved, such as water.

- *Private*: companies that provide services in the area and can be excellent allies or that affect the intervention area with their economic activity. It can be a favorable actor or quite the opposite, for which it is important to map its position, activities, perception in the community, relationship with authorities, position regarding the intervention.
- *Civil society organizations*: formal and informal associations whose purpose is to address social problems. In the same way as with the company, civil society organizations may or may not be positioned as positive actors in the community. The same process of positioning interests and relationships with the other actors must be carried out, as well as their reputation in compliance and transparency.
- *Local community leaders*: they can be people or groups, such as neighborhood organizations, churches, teachers, etc., that enjoy legitimacy and/or influence among the members of the community.
- *Academy*: is made up of educational institutions linked or interested in being linked to the project, at all educational levels. The experience narrated here was developed in the university environment, but it is not exclusive to this level. Some of the actors that may be present in most upper secondary and higher education organizations are the following

    - *Directions or programs of academic connection with the community*, which guarantee a link according to the Mission of this university, as well as the current educational model, teachers, with their own initiatives or with collective links by School or Department, student groups and volunteer program.
    - *Organizations to carry out the social service of university students*, as well as student groups and volunteer schemes that can be very good allies.
    - *Research centers*, which can also be very good allies to execute and disseminate academic projects linked to the needs of the communities where we intervene.

Once again, managing this network of relationships and interests is the task of the socio-training organization and it facilitates their entry into the Academy so that they can do what is their responsibility: generate, disseminate, and apply their knowledge.

3. *After: monitoring and sustainability*

The monitoring that the NGO gives to the project ensures sustainability. The socio-training organization is in charge of generating the follow-up strategy with the community. In this case, the Academy only maintains communication with the NGO to see how what they created together has been maintained at what stage of development. That is why the choice of the socioformer with whom the academic team will work is very important. In this model, each actor performs an important function, so that there is autonomy on the part of the elements, but also a positive interdependence between them. If the NGO did its job of social cohesion and community leadership well, the Academy will be able to enter without major problems. Annex 2 presents a proposal of Criteria for the selection of socioformers (See Annex 2).

The permanence depends on motivation factors: the municipality, its interest, its follow-up, etc. How does it self-regulate? By giving them tools of self-esteem, agency, leadership and organization, the community leaves self-regulating and knows how

Image 4 Solidarity market activity, in which a student from the exposes her improvement project to the community, 2019

to ask for help when it needs it. Even, what kind of help, to whom, when and what to give in return. How do the NGO let it go? The goal of any intervention is to get out of there and continue to function in an organic and self-sustaining way. Going out is desirable. Do not generate dependency. In Image 4 we can see how the model empowers the community.

4 Territorial Development Strategies

In Mexico planning of natural environments and urban areas developed through a set of legal instruments and plans based on zoning principles strictly related and limited to the political state and municipal boundaries. This legal system works both as ruling and implementing instruments in the planning process for the development of projects and programs in the natural and built context. In practice projects and programs are implemented on municipality grounds meaning that each municipality is supposed to elaborate a plan for the open landscape and/or an urban plan development which will rule the development inside the municipality.

At the same time, the state and national programs for development help and resources are linked to these documents which, nevertheless, in smaller municipalities can be a challenging task to achieve because the legal bases because developing these documents are costly and time consuming. Therefore, most rural municipalities do not possess these documents leading to a poor development scheme, low participation of the general population in the planning processes and a low possibility to achieve public funding to projects. At this point it is important to state that in the process of the academic group, it is sought to involve every actor and potentiality and we therefore work with the methodology of strategic planning because it is easy to use and the methodology includes every type of project and every group of citizens in the community. It is also a planning instrument which is flexible and allows design of long term projects for each community.

As for the specific methodology we follow the theory of Fernandez Güell who point out the following methodology for the process of strategic planning:

A strategic planning must contain a short, medium and long-term vision carried out with a plural legislative character; therefore it is important to carry out several studies to properly understand the important factors of the municipality in order to consider problems and qualities of the natural and built environment and to identify competitive advantages in order to produce an integral vision of the reality of the área in order to be able to generate instruments for an action oriented and flexible implementation strategies for implementation of improvements in the community (Fernandez Gúell 1997).

To achieve these goals, the students and faculty develop projects which offer a intersectoral vision of the municipality's system in the long term as well as identifying trends and anticipating opportunities. Once this analytic part is completed, the priority objectives and focus limited resources on critical issues are being formulated along with the design of the high priority projects needed by the community (Fernandez Gúell 1997).

In the case of planning process for municipalities like Mier y Noriega, students and faculty start developing an analysis of the natural environment, the built environment, and the socioeconomic structure of the municipality, from these results we would make the SWOT analysis allowing to conclude the Strengths, Opportunities, Weaknesses and Threats for future developments for the community, later on these results allowed for the community and the academic group to formulate the strategic goals followed by the process of designing solutions on a short, medium, and long term in order to diminish poverty in the municipality.

This strategic approach allows generating new sceneries for development and understanding of its socio economic context. In the case of Mier y Noriega, if analyzed within the framework of traditional planning, it is a municipality located in the extreme south of the state of Nuevo Leon, where a combination of the poor and insufficient state of the road system and the inclosed form of the territory isolates it from easy access to Monterrey and the rest of the state of Nuevo Leon. With Nuevo Leon being a state strongly focused on Monterrey for education, work and business, for a small municipality like Mier y Noriega, being isolated from Monterrey creates a situation of competitive weakness and lack of education and economic opportunities.

On the other hand, if the future development of Mier y Noriega were to be discussed in relation to a strategic approach, not only taking in account its position inside the state of Nuevo Leon, it could be examined in a context much more flexible scenarios where municipality of Mier y Noriega could be a central location between Matehuala, San Luis Potosí, and Ciudad Victoria, Tamaulipas, and as such being part of a very profitable network connected with the "Potosina Culture".

5 Social Innovation Applied in the Process of Architectural Design

Social innovation refers to the process of developing and implementing new, effective solutions to social and environmental problems (Kouam and Asongu 2022). New proposals or ideas that aim to solve social problems or serve as an alternative to traditional methods of generating ideas, with the intention of creating new relationships between variables and new perspectives to complex situations such as poverty, discrimination, climate change or gender violence, among others.

Brown (2009) points out that "an approach to innovation is needed, which is powerful, effective, accessible, that integrates all aspects of business and society, that people and groups can use it to generate unusual ideas that are implemented and that have an impact." Social innovation is a response to this, it also has as pillar being sustainable and transversal, that is, introducing changes gradually and comprehensively in order to eradicate the problem at its root and not just give a quick and superficial solution or alternative. It intends for these solutions or strategies to be replicable and scalable in different contexts.

In social innovation there are a variety of methodologies such as appreciative inquiry, which allows identifying and building proposals based on achievements and strengths that the community has had through transversal communication and a shared vision between outsiders and inhabitants, building strategies and relationships to achieve shared goals. This methodology was adapted to the process of research and ideation of the territory and users in the process of designing the architectural proposal.

Another of the methodologies used and was complementary to the previous one is: Design Thinking, where Tim Brown of IDEO proposes to reach a solution that is desirable, technically viable and economically profitable or sustainable (See Image 5). Dividing the process into 5 steps:

(1) Empathy, the phase that differentiates a merely academic process and design thinking, is a stage of observation and identification of "Insights" that are obtained through remote and field research to understand the behaviors, dynamics and culture of the group analyzed, going beyond individual perception and identifying the universal elements that can give us a universal perspective.
(2) Definition: In this stage, the information obtained in the previous phase is ordered, systematized and synthesized, the divergent and convergent thinking

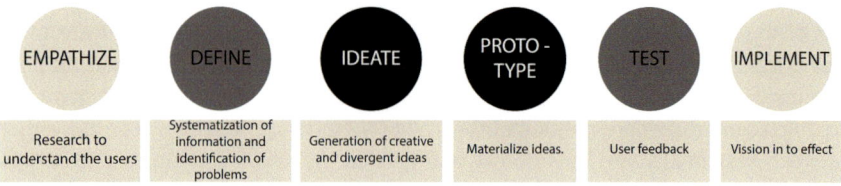

Image 5 Design thinking IDEO phases (Brown 2009)

is used to identify within all the possibilities and probabilities the most feasible and sustainable.

(3) Ideation: is the phase where there is an attitude of experimentation from the holistic understanding of the problem. The ideas can be generated from brainstorming, ideation techniques such SCAMPER, where the main ideas of cases of study are substituted, combined, adapted, modified, changed in purpose and scale and rearranged to enhance inicial ideas and transform it to divergent concept. Also, Morphological analysis can be used in the phase of exploration and 3d dimension (physical 3d models), taking words related to the intention or properties of common things such lighthouses, books, pinwheels in order to give shape and materiality to these ideas.

(4) Prototyping: this phase is necessary, because this not only defines how it looks and works, but what allows it to interact with the design, transform it through trial and error and feedback from potential users and consumers.

(5) Evaluation: uses quantitative and qualitative evaluation tools to identify the "pain" and "gains" of products already on the market. Not all the projects reach the last phase due to lack of funds or interest from those involved. This methodology allows us to see the problem from different disciplinary perspectives and connect for a comprehensive solution. Understanding design as a transversal element and not punctual (Brown 2009, p. 39–108).

Complementing the two previous techniques, the process also uses ethnographic tools and participatory mapping, also social cartographic that involve active participation of the communities, where they pinout their perception of public spaces, knows about their natural environment, landscapes and territory and routes (Alvarez Larrin et al. 2022), with this tools they can identify their more common routes, non-official territory limits and landmarks. The other technique used was storytelling, that allowed an effective communication of the proposals and ideas from the external (the students) to the community. When the communitie use this tools, they help the students to recognize, understand and live first-hand the experiences and culture of the community.

The intention of the integration and adaptation of these methodologies and techniques to the architectural design process is to evolve and migrate to a process where projects are created not only for people, but for and with people.

5.1 Case of the Municipality of Mier y Noriega

The strategic planning process of the municipality of Mier y Noriega was conducted throughout 2022. In the first instance, a variety of strategic proposals were completed for the entire municipality and in a second phase, specific proposals were made for the village named El Refugio de Cerros Blancos.

As part of the initial planning process there were conducted a number of interviews with members of the community, those interviews allowed us to corroborate the hypotheses that were made. When working with communities, it is a basic rule that the community itself identifies its strengths. On the other hand the residents will move on the first opportunity if they feel attached to the place and see possibilities for real changes. In order to understand the later proposals for development and improvement a resume of these interviews are presented below.

1. Strengths

The strengths mentioned by the interviewed population were classified as follows: (1) Sense of belonging: the community openly mentioned pride, attachment and taste for their village. (2) Identity: the population has identity symbols and rituals, such as their dances and religious festivities, of which they are extremely proud. For these festivities, resources are mobilized and managed in a community way. (3) Solidarity: the population shows strong signs of solidarity; each village has been formed by one or several family networks, so that everyone knows each other, has blood ties between them and helps each other. They never leave anyone without food or help in matters of medical emergencies. In addition, as outsiders, the (4) Existence of migratory networks: most go by contract to the United States and return with their relatives to their community; But there is a sector of the population that stays there. The ties between returnees and stayers represent an excellent investment and employment opportunity. (5) There is a demographic bonus of young people who are eager to have a better life and who, in addition, have a lot of capacity for mathematics and English, given the correct training and the opportunities. (6) They have very good teachers, who are enough to motivate and train them to better impact their students; and finally, (7) Female leadership is the transformative force of the community. Women come together to work, clean, educate, help, get groceries and make committees. The potential of the women of these communities and their desire to improve their lives was the determining factor for which we decided to intervene there.

2. Challenges

In order to understand the environment of Mier y Noriega it is important to state that it is semi-desert and has the lowest rainfall in the state of Nuevo Leon The lack of water, therefore, strongly determines the backwardness, vulnerability and poverty of the community. On the other hand, the geographical position of the community combined with a poor road and mobility system leads to isolation and disconnection as the distance, and conditions of the roads in and out of the municipality and between villages represent a challenge to assume in the design of public policies to combat poverty in this municipality, since getting there, from the Metropolitan Area of Monterrey, implies a 5-h trip by car. Without water suitable for planting or optimal ways to market products or receive aid, vulnerability becomes more acute and gives rise to another of the challenges that must be corrected: identifying and accepting the vocation of the land.

The possibility for economic and social development of the community presents important challenges on multiple levels, first of all from its climatic conditions, but also on mobility issues and cultural paradigms that form a strong resistance to social transformation, such as sexism and economic-political assistance. In order to understand the context and the choice of projects the results of the SWOT analysis for Mier y Noriega are described below.

Furthermore, Mier y Noriega is, according to INEGI data (INEGI 2020) despite its vast territory (which covers 997.9 km2, and where the metropolitan area of Monterrey can fit three times). The population reaches 7,652 inhabitants leaving the space heavily underpopulated with 1.21 inhabitants per square kilometer (equivalent to 0.66% of that of Monterrey). According to CONEVAL's 2020 Multidimensional Poverty Measurement, 61.5% of the population lives under severe or moderate poverty conditions. At the same time, in Mier y Noriega people continue to plant cornfields, but the irrigation system is completely temporary, and it seldom rains enough to be achieved. These conditions have been repeated, at least, for the past last 30 years, so by now, it should have been accepted that the development of the community should not depend one hundred percent on farming and agriculture. And instead, identify other options.

At the same time, the conditions of health services are poor: few doctors, few facilities, and infrequent visits by health personnel. Not to mention that, according to CONEVAL, there is currently no welfare infrastructure (Medical Units, Branches of the Welfare Bank or Planned Social Infrastructure Projects).

When discussing the social contexts of the communities and villages in Mier y Noriega, even though women are the engine of social transformation in this area, they have no voice in rural communities. Changing the traditional male dominated gender paradigms that keep women inhibited and impede their development is a priority task, for the good of the entire community.

Another of the great challenges that directly impact the change of the belief system of the communities is the adherence to the paradigm of the old welfare and paternalistic political practices that persist to date in the communities and, finally, in any intervention that is made, internal or external, the care of the environment must be guaranteed.

Unfortunately, by 2030 a reduction of 33.7% of the population is expected to leave the community due to the lack of employment opportunities in the municipality (Image 6).

In Mier y Noriega's analysis, three major problems stand out:

Poor water supply

The average annual rainfall in the municipality of Mier y Noriega is, according to CONAGUA (the public institution that regulates the national water administration in Mexico), 480 mm which compared to the rest of the State which receives 650 mm clearly shows the complication of being a rural and agricultural zone. This turns the municipality into a semi-arid area with a condition that the average annual rainfall is less than the maximum annual potential evaporation, highlighting a water deficit (Tarango 2005). That is why the inhabitants of Mier y Noriega, since its foundation to date, have suffered from the lack of water due to the great droughts that occur. Today communities throughout the municipality receive drinking water thanks to the well called "El Francés" (INEGI 2020).

Image 6 The building context in La Moreña, municipality of Mier y Noriega 2022

Furthermore, 68.5% of the population lacks basic services in housing such as drainage, access to water, fireplace when using wood or charcoal for cooking, and electricity, when in Nuevo Leon the average is 2.3%. At the same time, half of the population in Mier y Noriega lacks drainage and more than one-fourth access to water, because of water scarcity (CONEVAL 2020).

Education lag

At the municipal level, three-fourths of the population has a basic level of education (preschool, primary and secondary) and only 13% study until high school, when in Nuevo León a little more than one-fourth of the population have higher education (INEGI 2020.)This situation generates that poverty patterns become an intergenerational problem, considering that 11.6% of the population over 15 years of age is illiterate, that is, it does not know how to read or write (INEGI 2020).

Poor built context

The built context of the municipality of Mier y Noriega covers the smallest part of the municipality, the main urbanized space is the municipal seat Mier y Noriega which at the same time is the administrative center of the municipality. In addition, there are 14 ejidos or small villages and 3 small settlements (INEGI 2020). The constructive characteristics are buildings built of adobe and cinder blocks. Each detached house is located on a plot of between 2000 and 5000 m^2.

Furthermore, 68.5% of the population lacks basic services in housing such as drainage, access to water, fireplace when using wood or charcoal for cooking, and electricity, when in Nuevo León the average is 2.3%. At the same time, half of the population in Mier y Noriega lacks drainage and more than one-fourth access to water, because of water scarcity (CONEVAL 2020).

All together, as a way to detonate a tangible improval in the quality of life, it is of utmost importance, to generate community participation and local leadership must be strengthened, so that they re-appropriate the issues of public life and participation, with the knowledge of their rights, their legal system, and their execution, as well as the capacity for leadership, management, and follow-up.

All together, as a way to detonate a tangible improval in the quality of life, it is of utmost importance, to generate community participation and local leadership must be strengthened, so that they re-appropriate the issues of public life and participation, with the knowledge of their rights, their legal system, and their execution, as well as the capacity for leadership, management, and follow-up.

As a result of this situation, the goal of the project has been generated followed by a series of strategic objectives:

As a result of this situation, the goal of the project has been generated followed by a series of strategic objectives:

Introduce a sustainable planning system of the municipality that

- In the short term reduce extreme poverty
- In the medium term reduce poverty
- Introduce a circular economy system

To achieve these, a series of specific objectives was developed to contemplate feasible strategies, programs and specific projects that together aim to strengthen the development of the municipality in a model that allows neighbors together with authorities and NGOs to be actively involved in decision-making at all levels for the development of the municipality.

In the strategic planning working group, three major specific strategic objectives were identified:

In the strategic planning working group, three major specific strategic objectives were identified:

In the strategic planning working group, three major specific strategic objectives were identified:

1. Welfare
2. Sustainable territory
3. The constructed context

The set of initiatives should be read as a catalog that can be developed in parallel and/or separately. Think about the time, because it depends on many factors, but thinking about our model, described above, we think that it is necessary to promote quality before quantity, understood in the way in which it is important to carry out projects with a high possibility of success and on an understandable scale—once the first initiatives are successful it is our conviction that other more complex projects will be achieved in less time.

Each of the strategic objectives is described in more detail below.

Welfare

For the residents of the municipality of Mier y Noriega it is of utmost urgency to be able to improve the quality of life and the project foresees a section of improvement of health well-being and education, that is why the introduction of a welfare model is proposed that deals with providing the service, providing spaces and ensuring a competitive level of health personnel in a process carried out among the neighbors, the authorities, NGOs and the private initiative to generate alliances with universities in Monterrey so that their students in the health sector or related careers can carry out their social service or internships in the medical center. Provide medical service places hand in hand with the Ministry of Health, with attractive salaries. Offer spaces worthy of consultation, with facilities and with constant supply of equipment for general practitioner, dentist, and psychologist.

Sustainable territory

According to the authors Murilo Flores (Flores 2007) a sustainable territory is partly achieved

> "Cultural identity as a strategy as a sustainable territory. Precisely one of the great strengths and opportunities is the large amount of territory where the varieties in the landscapes of the municipality range from the arid in the flat parts to the green and wooded in the high parts (2007)." It is also a place with deep roots in the land and in agricultural work. All this reflects in their music, colors, uses and customs. In short, it seems important to take greater advantage of these relationships between the person and the territory.

To initiate this strategic objective, a project proposal has been carried out firstly at the municipal level and secondly a pilot project has been carried out in the municipal seat where the territory is ordered through a series of natural strips developed around the streams seeking to connect ejidos, landscapes and settlements from East to West in the municipality. The use of these spaces will be used for multiple functions such as agricultural research field, water management projects, environmental education field, bicycle transport roads, implementation of continuous reforestation programs and long-term sustainable rural tourism projects (Image 7).

Each settlement and small town have schools and neighborhood commerce, but to solve health issues and more complex procedures and purchases you must go to Doctor Arroyo or this reason a program is proposed to improve the quality of the building context of the towns of the municipality The goal is for each of these plans to be of distinctive design and content to foster the local culture of each community to foster pride and belonging to the place. At the same time, it proposes the application of a technical guide of initiatives, methods, and constructive data to promote the sustainability of housing: how to turn housing into a self-managed structure that produces energy, water, and food management in a sustainable way.

Likewise, a permanent program is proposed where it seeks to generate new buildings for community uses and small industries to promote the strengthening of the economic situation of the neighbors. To generate more efficient construction processes, an adaptive panel construction system and a manual for remodeling projects of existing buildings have been developed.

Sustainable Development of Rural Communities in Mexico 211

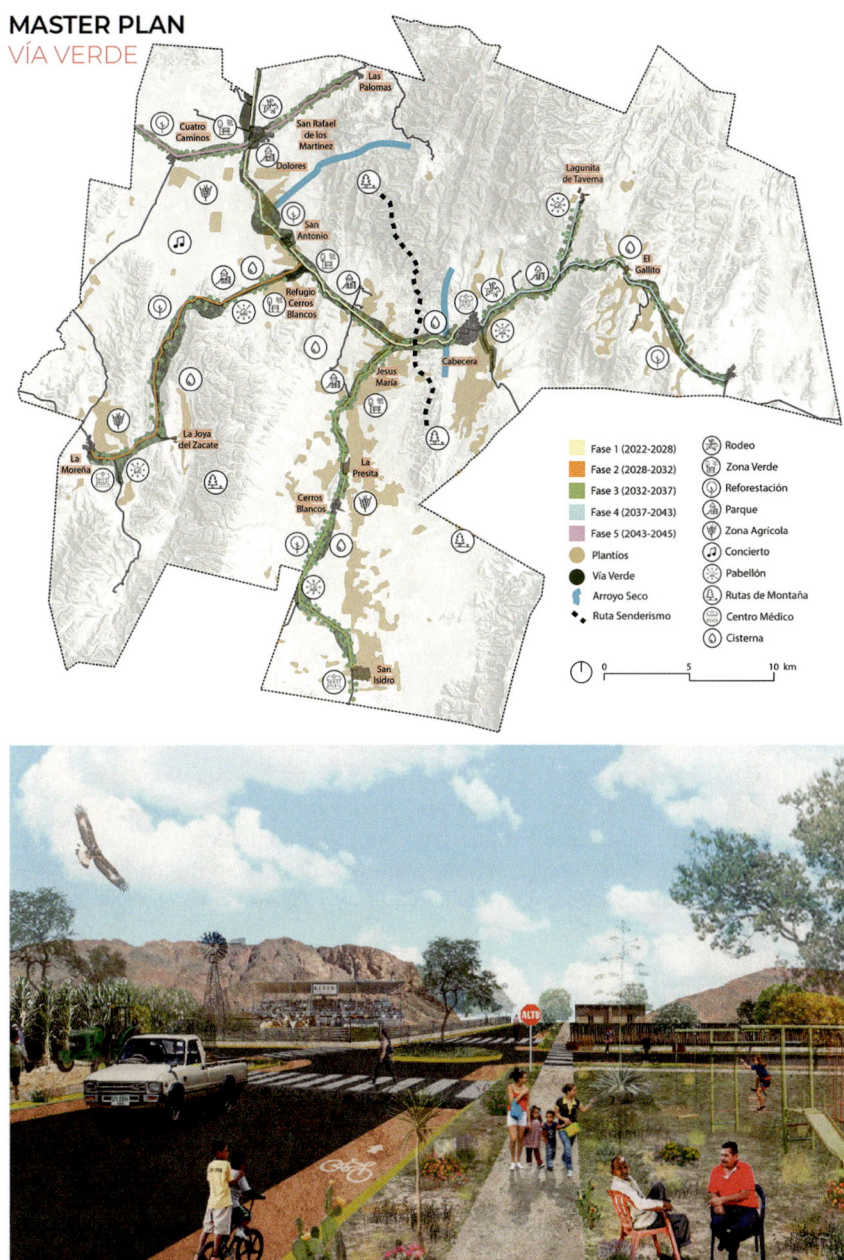

Image 7 *Source* "Paso a Paso, Gota a Gota" project of landscape ordination for the municipality of Mier y Noriega designed by Verónica Anastás, Grecia Arizpe, Carolina Bernal and Rena Porsen

5.2 Case of El Refugio de Cerros Blancos Village of Mier y Noriega

In 2021, the Monterrey Tec engaged in adapting a new study methodology based on disciplinary and transversal skills and seeking that students accomplish professional skills through practical experiences. In the case of the architectural departments of the Tec system, they have focused on the design of community equipment as part of the fourth semester aiming to provide students with hands-on experiences in order to provide a solid understanding of how to include different social and economic backgrounds in the process of designing a building.

During spring 2022, A group of twenty-six students of fourth semester of Architecture of the Tec de Monterrey were divided into groups of 4–5 people, according to Brown (2009) to built smart teams "(…) the members needs to possess a depth skill that allows his or her make a tangible contribution to the outcome", that's why every student has to tell what were their abilities such as visual, digital, conceptualization in order to form smarter and stronger teams.

The students made a remote diagnosis covering natural, constructed and sociodemographic environments. At first, is was difficult for them to found reliable sources because most of rural communities have a lack of coverage of information, CONEVAL and INEGI were the reliable databases that cover El Refugio de Cerros Blancos, to complete their work they used non official information in tourist pages of the municipality, as well as satellite images, related websites, semi structured interviews and non participatory observation when they visited the community. During this visit, the students stayed and lived with members of the community, especially with secondary school students and their families.

In this visit, the students identified who were the community leaders, to their surprise all of them were women. These women acted as coordinators, promoters and finance staff of social courses developed by federal and state social secretariat and organized the community members when necessary. During the stay, the students identified the expectations of the members of the community about the project and the connection that there was between families. This information was obtained through informal chats in breakfast or lunch time in the houses of the community leaders, a series of formal activities also took place, where the project was explained in a nontechnical language and two-way communication. The students had the opportunity of social interaction with the community members at the ceremony dedicated to the Virgin of Guadalupe where the students participated as observants. In all the interactions, the community members were encouraged to express their ideas and concerns.

Staying in the community was a key element of success in a ethnographic point of view and in the development of the project because the students were able to see how was the interaction between family members, see how they live and coexist, what activities they carry out regularly, what spaces they frequent and taboos and prejudices that prevailed in it. The students obtained first-hand information that could not be found or obtained remotely with the interested parties and inhabitants of the community. This experience was complemented with a couple of interviews conducted by the same students to people from the community, photographic survey and measurements of the land in which they were going to work during the semester.

The visit was exhaustive but enriching. The students began to analyze the information collected and obtain "insights" about the community, its needs and expectations. Back in Monterrey, the students identified municipal, community, and terrain strategies that could support an improvement in quality of life, always in relation to the construction of what until now was the idea of a library that little by little was transformed into a "Community Center".

The diagnosis and strategies generated were condensed into a presentation showed to the Pal Sur del Norte association and the finance secretariat of Mier y Noriega representing the authorities and the inhabitants of Refugios de Cerros Blancos,. Areas of opportunity were identified in the strategies showed, pointing out the feasibility and viability, as well sustainability.. Some of the strategies focused on "Reuse of natural resources such as rain, wind and sun, the creation of public spaces of coexistence around the community center, the paving of main avenues, a creation of an educational-cultural node, among other proposals."

After the Strategic Planning phase, the students followed a route based on the stages of Design Thinking for the design process. they systematized the information and translate them to need and wishes to make the architectural program, they explore ideas and emplacement in the terrain, taking in account its main characteristics such as sunlight, climate, prevailing winds that is called "site analysis", also a series of conceptual proposals were generated in 2D (drawings, schemes, sketches, relationship diagrams, etc.) and prototyped in 3D (conceptual models with geometric figures and very general elements where solids and voids).

At this stage, the students made at least three proposals that were related to the strategies they enlisted: the research and diagnosis, the insights obtained by analyzing the information they had and the comments of the leader of the NGO Pal Sur del Norte in the first presentation of the diagnosis. The best proposals of each group were presented to the community in a second visit to El Refugio de Cerros Blancos, complemented by activities and collaborative design strategies such as sticking postit in the spaces they frequently use and their different activities. Other group of students organized an activity to relate daily activities and expectations of the different spaces that made up a community center or a library, complemented with a game where community members could add or subtract elements of a "prototype" model of the project. Other activity was a "chart" of what activities would be carried out in the different spaces of the community center/library. With this activities, the students obtained very interesting results and a greater approach with the community, there was a visible interest in the project.

The students for two weeks adapted the proposals according to the information obtained from the community, taking in account the need for public spaces, coexistence, flexibility to carry out different activities (social and educational) in the same spaces, as well as a space to park their bicycles. Simultaneously to the improvement of the project, executive plans were produced and they made the respective constructive modifications and technical requirements to the space.

Image 8 *Source* "Lófous" designed by Emilio Rodríguez, Janeth Juárez, Natalia Gómez and Natalia Arreola. Directed by Irma Hernandez and Rena Porsen

On May 4, the final presentation was made with the presence of municipal authorities in representation of the mayor of Mier y Noriega, such as the Secretary of Finance of the 2021–2023 administration; representatives of Banobras who would finance the project; by the State Government the Secretary of Equality and Inclusion, Undersecretary of Social Protection and Opportunities, and Nancy Colado representing mobility and Urban Planning Secretariat; Representatives of the community Beatriz Vargas and Ana Cristina Vargas. Each team presented through images, plans, presentation sheets and a model their architectural project showing the benefits of this to the authorities. This team of experts made their decision, having 2 semifinalists. The Lotus project which was contextualized with the architecture and language of the community and the project "the lighthouse" that highlighted the context as an iconic building and more modern language. The winning project was "the lighthouse", which was being developed by the students Ernesto Viveros, Ivanna Aguirre, Kelsey Díaz and José Luis Sánchez, who would continue working together with Arch. Joel Valencia to develop the executive plans and would be in the process of building it (Images 8, 9 and 10).

This project is a success story as a project of academic, governmental, and social linkage due to the interest and contribution of each of the actors, a participatory work and a constant and continuous work by the civil association allied with the community since this was a relevant factor for the community to have an openness and active participation.

Another element that stands out in this process is the direct coexistence with the actors/users who will make use of the community center. In the various visits there was formal and informal coexistence with them, giving space for the opening, to know first-hand, their routes, their beliefs, as well, their aspirations.

A third relevant point for success was the multidisciplinary and specialties of the teachers who advised the students, covering the necessary areas for a comprehensive knowledge of the environment and generation of design proposals.

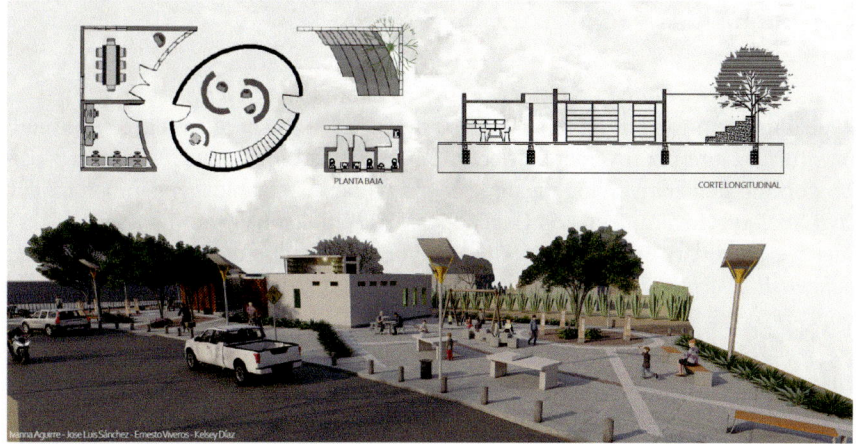

Image 9 *Source* "El Faro" designed by Ivanna Aguirre de León, Kelsey Diaz Villagran, José Sánchez Villaseñor and Ernesto Viveros Alfonzo. Directed by Irma Hernandez and Rena Porsen

Image 10 Construction site May 2023. *Source* Own elaboration

6 Conclusions

One of the most common problems in the communities of Nuevo León is the lack of local financing and the authorities' ignorance of the use of federal and state funds for infrastructure improvement. Therefore, it is of vital importance to have a tripartite working model involving the three actors of the sustainable regeneration model: Civil Society, Government, and Educational Institutions. By working together, they can develop initiatives that channel the development of communities in a sustainable manner. In the past, work has been done separately, yielding results with limited scope, projects that do not fully integrate the needs and expectations of communities, and with few possibilities of implementation due to lack of financing or maintenance. The sustainable regeneration model for the development of rural communities achieves community engagement through the work of NGOs, the translation of students' knowledge and research into sustainable projects or initiatives, and finally, the political power of governmental entities involving both the community itself and public servants to achieve a common good, transforming the current vocation of rural communities into a more active, participatory, and sustainable economy. This reformulates governance, the connection between government and citizens, visualizing the rights and obligations of both manners. This model allows and encourages the community to act in the development of their well-being instead of solely relying on government assistance, thus creating a virtuous circle. At the same time, it's important to mention that the students who participated in these social and architectural processes were positively impacted by developing interdisciplinary, transversal, and social skills that were diminished during the pandemic. They strengthened their awareness by sensitizing themselves to the rural environment and learning how to be agents of social change with an ethical perspective.

Each site is special and has different conditions, but in the case of the municipality of Mier y Noriega it seems important to seek temporary commercial activities such as the development of cultural initiatives, festivals, sporting events and religious festivals etcetera that could generate income without negatively impacting the social context of the communities, the support of an NGO and a trust working in the community would be a viable model to ensure transparency, continuity and relations between the community and the organization of the event.

The implementation of the intervention model for sustainable development of rural communities makes it possible to ensure that new socioeconomic and cultural dynamics can be generated and transform the predatory belief system in the rural area of Mier y Noriega. Those changes bring a different vocation to the countryside, a community role in harmony with the entire ecosystem and a reformulation of governance between government and citizens, visualizing their rights and obligations.

In this model, the public sector, academy, community and ONG's work hand in hand. Regenerative designs permeates in the entire model because all the sectors works with the idea of improving the environment of the communities, regenerating the social net through constant interaction with the community members and involving them in the reconstruction of public spaces and new opportunities for social and economic growth, not only giving them temporary solutions or maintaining their precary conditions. This model allows and encourages the community to take action in the development of their wellbeing not only waiting for government assistance, making this a virtuous circle.

At the same time, it is important to mention that the students who participated in these social and architectural processes were impacted positively on developing interdisciplinary, transversal and social skills that were lessened during the pandemic. They strengthened their awareness skills by the sensibilization of the rural environment and knowing how to be an agent of social change with an ethical perspective.

Throughout the process it has been important to combine the existing methodologies mentioned in this chapter in order to work on parameters which would allow the large variety of participants to be able to develop a common language in order to create more sustainable projects.

Even though only one municipality is presented in this chapter, it is important to notice that the impact on the municipality has been multiple. In the first year the strategic approach of the project achieved extra fundings to the area worth 10% of the annual budget for the municipality. This has allowed to create employment for twenty inhabitants of the El Refugio de Cerros Blancos for six months during the construction of the community center. Looking at the near future the community has achieved funding for better healthcare facilities for the municipality due to be built in 2023, funds to pave the principal roads, encourage the municipality authorities to apply to calls for public funds for infrastructure development and made visible the active participation of the community women in the improvement of public spaces in El Refugio de Cerros Blancos.

In order to achieve success in social-academic-governmental collaboration, the planning of the project has been a multidisciplinary approach in order to achieve more community participation throughout the process and to reach that both neighbors and students could work as one organization throughout the design and construction of the project.

To accomplish the best results of the implementation of the Model for Sustainable Development of Rural Communities it is recommended to follow some simple guidelines to make the model work properly: before starting to work on any intervention, it is important that there has been developed a diagnosis according to the Participatory Intervention Model, in order for the community to achieve a common view on what is needed to improve their everyday life. At the same time, this will guarantee that decisions being made are the ones of the community independently of external interest groups.

To accomplish the best results of the implementation of the Model for Sustainable Development of Rural Communities it is recommended to follow some simple guidelines to make the model work properly: before starting to work on any intervention, it is important that there has been developed a diagnosis according to the Participatory Intervention Model, in order for the community to achieve a common view on what is needed to improve their everyday life. At the same time, this will guarantee that decisions being made are the ones of the community independently of external interest groups.

To accomplish the best results of the implementation of the Model for Sustainable Development of Rural Communities it is recommended to follow some simple guidelines to make the model work properly: before starting to work on any intervention, it is important that there has been developed a diagnosis according to the Participatory Intervention Model, in order for the community to achieve a common view on what is needed to improve their everyday life. At the same time, this will guarantee that decisions being made are the ones of the community independently of external interest groups.

To accomplish the best results of the implementation of the Model for Sustainable Development of Rural Communities it is recommended to follow some simple guidelines to make the model work properly: before starting to work on any intervention, it is important that there has been developed a diagnosis according to the Participatory Intervention Model, in order for the community to achieve a common view on what is needed to improve their everyday life. At the same time, this will guarantee that decisions being made are the ones of the community independently of external interest groups.

At the same time, it is essential to get the municipality involved in the process to gain access to funding and know-how on solving infrastructural problems in the community.

Once these two parts are properly organized. It is the moment to obtain all the possible input on how to design solutions for the problems of the community. In this phase the participation of universities, NOG´s and the private sector is of utmost importance to provide technical know-how to the community on how to lead the ideas into tangible improvements for the community.

During this entire process, from first discussions in the community till the implementation of solutions, it is important to count on an independent organization to coordinate all the phases and actors to assure the most sustainable results for the community.

Annex 1: Criteria for Selecting a Community in Which to Intervene

Prof. Abril de León Rincón

Pal Sur del Norte, AC

1. Evaluate: how did we find out about the community? Who asked for help? If it was your own initiative, touch base first with ejido authorities and people legitimized by the community.
2. If it was not your own initiative, evaluate: how legitimate is the request? Go to the field to tour, validate concerns and needs, map community leaders.

3. Once this is validated, use this check list:
    I. Access to the community
        - Access to expressways and exits YES NO.
        - State of the roads Paved Not paved Very steep.
        - Transportation support and driver to get there, in case the journey is complicated (If you don't have it, it can be very expensive and you won't have the possibility of tracking).
    II. Security
        - Validate news in official media: if there are situations of insecurity that we cannot solve, it is better not to go.
        - Investigate what situations of insecurity arise.
        - Ask those who know the community and its routes.
        - Validate what care measures for health or insecurity emergencies we might have available.
    III. Connectivity and media
        - Do you have internet?
        - Do you have a cell phone signal?
        - Does it have a telephone network?
    IV. Access to services
        - Health: clinics, health centers or similar that can deal with emergencies or provide first aid in a nearby section.
        - Education: schools that can serve as a meeting point and safe space.
        - Water and drainage: access to water (hygiene and consumption).
        - Electricity.
    V. Workspace: Is there a safe and legitimate place for the community or authorities where we can meet with the community and store things in case of giving workshops, for example?
    VI. Lodging: Is there a hotel or safe place in the community or nearby where we can stay?
    VII. Meals: Are there places where we can buy or access necessary meals?
    VIII. Community support network: Is there a support network made up of members of the community and ejidal, educational or municipal authorities that can help us in an emergency or expedite procedures?
    IX. Billing: Ideally, we can find services that are billable for reimbursement, but we know that in the most vulnerable communities, this option is unlikely to exist.

Annex 2: Criteria for Choosing a Socioformador (OSC Partner)

The academic linkage model with the community that is inserted in our participative intervention dynamics requires that the Academy already have the necessary conditions to be able to generate, apply, evaluate and disseminate its knowledge. This task must be carried out by the chosen social training organization, which is part of civil society organizations and is interested in linking with academia because its knowledge and human resources can make an initiative to improve the quality of life of a person a reality community.

Based on the experience of more than 20 years of association, we propose the following criteria for choosing a partnership with a socio-training organization:

1. Have at least 5 years of experience in field work in rural communities.
2. That it is legitimized by the community where it is going to intervene or has the necessary community contacts so that the community allows, facilitates and gets involved in the intervention.
3. To maintain constant communication with all the actors involved.
4. That it is registered with the Ministry of Finance or is incubated in another organization that can obtain and manage financial or in-kind resources.
5. That they have their own economic resources for transportation or support material.
6. That they are willing to adjust to the times and requirements of the educational model.
7. Have a volunteer area.
8. That is willing to carry out efforts of strategic alliances and resources.
9. That he is willing to include the public party in this alliance management.
10. Have a monitoring system per project and communication of results.
11. Who is willing to manage a communication and visibility strategy for the project, approved with all the actors involved (vehicle of collective action).
12. That manages the presence of key authorities for the empowerment and visibility of the project.s

References

Alvarez Larrin A, McCall MK, León Villalobos JM (2022) Mapeo Participativo y Cartografía Social de Conocimientos Culturales, Históricos y Arqueológicos: Recurso Práctico para Profesores y Estudiantes Universitarios. CIGA-UNAM, Ciudad de México. https://doi.org/10.22201/ciga.9786073059121e.2022

Brown T (2009) Change by design: how design thinking transforms organizations and inspires innovation. HarperCollins, New York

CONEVAL (2020) Estadísticas de pobreza en NL (Poverty statistics in NL). https://www.coneval.org.mx/coordinacion/entidades/NuevoLeon/Paginas/principal.aspx

Fernandez Gúell JM (1997) Planificación Estratégica de Ciudades. Gustavo Gili, Barcelona

Flores L (18 de noviembre de 2021) El presupuesto de Nuevo León para el próximo año asciende a 117,745 millones de pesos. El Economista

Flores M (2007) La identidad cultural del territorio como base de una estrategia de desarrollo sostenible. Rev Opera 35–54

Gibbes CH, Inurreta Díaz A, Jimenez-Osorio J (2020) Defining and measuring sustainability: a systematic review of studies in rural Latin America and the Caribbean. Springer Nature, pp 447–468

INEGI (2020). https://www.inegi.org.mx/

Kouam J, Asongu S (2022) Effects of taxation on social innovation and implications for achieving sustainable development goals in developing countries: a literature review. Int J Innov Stud 6(4):1–12

León DN (2022) Data Nuevo León. http://datos.nl.gob.mx/n-l-poblacion-total-y-por-municipio/

León LL (2021) Presupuestos de Ingresos 22, Decreto número 073. Presupuesto de Ingresos para el año de 2022. Monterrey, Nuevo León

Ley de Desarrollo Sustentable (s.f.). https://www.diputados.gob.mx/LeyesBiblio/pdf/235_030621.pdf

Tarango AL (2005) Problemática y alternativas de desarrollo de las zonas áridas y semiáridas de México. Revista Chapingo Serie Zonas Áridas IV(2):17–21. https://www.redalyc.org/pdf/4555/455545052003.pdf

van Es M, Guijt I, Isabel V (2015) Theory of change thinking in practice. Hivos

Regenerative Rural Housing, Case Study: Sierra Gorda Biosphere Reserve in Querétaro, México

Lervin Benitez Álvarez and Carlos Cobreros

Abstract This chapter begins with the analysis of the modification of housing as a consequence of the migratory effects in the community of Tilaco belonging to the Sierra Gorda Biosphere Reserve of Querétaro, presenting problems in its habitability, construction system, adaptation and even loss of identity. We generate the contextualization of housing, defined by several authors as a transcendental space for the development of human beings in the different aspects of their lives. Therefore, the planning, construction and use of housing is an important element in its life cycle. The tools used were: regenerative design, based on the connection between man and nature, maintaining balance; the Living Building Challenge Certification, which sought to have a positive impact on the community and its cultural fabric; and bioclimatic architecture, analyzing the physical and climatic aspects of the community to make the best use of its resources.

Keywords Regenerative Design · Regenerative Development · Rural Housing · Biosphere Reserve

1 Introduction

The community of Tilaco, belonging to the municipality of Landa de Matamoros, within the Biosphere Reserve, is a community that presents migratory movements to neighboring cities and even the United States of America, due to the lack of jobs that allow the heads of families to have an economic income to support the lifestyle of their families; However, when they return to their place of origin after several months or even years, the inhabitants try to replicate the housing of those places in the community of Tilaco and this results in a lack of functionality, generating

---

L. Benitez Álvarez
Craft Vink, Querétaro, Mexico
e-mail: craft.vink@gmail.com

C. Cobreros (✉)
School of Architecture, Art and Design, Tecnologico de Monterrey, Querétaro, Mexico
e-mail: ccobreros@tec.mx

problems in their living, construction system, adaptation and even the loss of identity with respect to the community.

For the development of this chapter, initially we began with the analysis of these migratory effects on the modification of housing, which according to various authors is defined as a transcendental space for the development of human beings in the various aspects of their lives: social, economic, physical, cultural and psychological. The tools or theoretical positions that served as a basis for the development of the research were: regenerative design, Living Building Challenge Certification and bioclimatic architecture.

Propose the development of a regenerative model of rural housing in the Biosphere Reserve of the Sierra Gorda of Querétaro in order to regulate the existing balance between the reserve and the local inhabitants, understanding that the Biosphere Reserve was created for this purpose, as well as the sustainable development of the communities.

Objective

The objective of this chapter is to develop and evaluate strategies for the design of a rural housing model adapted to the context, under the principles of regenerative design and people-centered design, seeking to reduce the environmental, social and economic impact as much as possible.

2 Antecedents

Mexico has a rich cultural heritage that has been shaped in important part by architecture; however, it has begun to be forgotten and is being replaced by constructions that are alien to the environment and cultures in which they are developed (De Tierras y Varas 2009).

As a consequence of migration, the copying of patterns has led to the use of materials, processes and forms that are alien to the surroundings and environmental context, causing the construction of spaces lacking identity, order and harmony, which affect and alter the landscape in which they are located, destroying the relationship between man and territory (Monterrubio Acosta 2001).

Acuña Fernández (2015) considers housing as a necessary element for the existence of the family, the community and society. The Political Constitution of the United Mexican States, in chapter one, article four establishes: "Every family has the right to enjoy decent and dignified housing. The law will establish the necessary instruments and support in order to achieve this objective" (Pardo 2013).

According to the National Housing Commission (CONAVI 2012), the appropriate materials for decent housing are wood, adobe and mostly industrialized materials such as cement, brick, mosaic, asbestos sheets, among others. However, the use of industrialized materials in the construction of buildings and infrastructure generates a high impact on the environment; for example, to produce one ton of cement, 900 kg of carbon dioxide are emitted into the environment (Toca 2017). The International

Energy Agency (IEA 2012) estimates that buildings (commercial, residential and public) emit between 25 and 35% of carbon dioxide to the environment.

With technological advances in the current era, the development of rural housing has been decontextualized and transformed, generating an impact on the environment due to the use of industrialized materials in its construction and a loss of identity as a consequence of the migratory movement that has occurred. Each region is different, characterized by its natural and social factors of origin, which is why rural housing has its own characteristics as a reflection of the context in which it develops, its culture, specifically its identity (Villar Rubio 2001).

The current rural housing in the locality of Tilaco, case study, is mostly made of materials such as cement, wood and sheet metal (De tierra y varas 2009). Based on a sampling carried out in the municipal capital, houses with brick or brick walls account for 82.6% of the total, concrete slab roofs account for 52.7% and cement floors represent 87.1% (INAFED 2006).

The new generations of rural communities that have continued to migrate have begun to promote the loss of identity as a result of copying housing patterns from the context in which they live; trying to replicate them in a rural context from which they come, with the idea that the functioning of these patterns in housing (mainly physical, climatic and cultural aspects) will work properly; However, this can be counterproductive because the lifestyle is not the same and the materials used for the elaboration of the house are not within reach and seeking to obtain them, initially generates an economic expense and with the passage of time ecological damage and the loss of cultural traits as a rural context.

Likewise, the Sierra Gorda has natural and climatic potentials that generate the adequate amount of natural resources for its use in vernacular architecture, considering the construction systems as part of the traditions of living in the sierra; strengthening the principles of regenerative design by considering the use of the elements found in the context to develop a space. Establish the necessary elements to promote the recognition of the population's value of vernacular architecture as a cultural heritage built within the Sierra Gorda and as an example of the traditions that are still maintained; likewise, traditional construction systems are an important point that will allow the transmission of knowledge from generation to generation (De tierra y varas 2009).

According to Rapoport (1972), vernacular architecture is characterized by those dwellings that were built in a certain geographical environment and respond to the physical and cultural space of the context. Correia (2017) considers vernacular architecture to be any private or collective dwelling or building that was built employing traditional technologies (Correia 2017).

Vernacular architecture does not seek to transform nature, but rather, to try to adapt to it by taking advantage of the differences in climate, topography and the various materials that are within reach, used as elements for construction; this type of architecture is developed collectively through the knowledge that has passed from generation to generation rescuing traditions and cultures as a symbol of its identity (De tierra y varas 2009).

This type of architecture is the result of popular culture, it preserves the materials and traditional construction systems of the region appropriate to the context, which is why it constitutes an important heritage that needs to be protected and preserved (Equipo ARQHYS 2012).

2.1 Housing in the Rural Context: México

Housing can be developed in a rural or urban environment, and arises from the need of human beings to be able to develop their daily life, it conditions the context in which it is located, separates the domestic space and the natural environment, generates as main objective well-being and quality of life (MacDonald 2017).

As a physical environment, housing does not only consist of a set of walls organized randomly or in a structured manner, it also corresponds to a cognitive, affective and social element, always seeking an interrelationship between these elements to generate harmony in the dwelling itself.

The rural environment is the territory that has a small number of inhabitants, with agriculture and livestock as the main economic activity. Rural construction began to develop with the use of materials found in the context. Sánchez and Jiménez (2010) defines rural housing as a mainly active and interactive element with the natural, built and community environment. It creates both a cultural heritage and an emotional and cohesive support for family groups, supported or influenced by economic and community activities.

Throughout history, as a result of the constructive activity of man and his relationship with the context, housing has undergone multiple transformations, according to the place and time in which it is located or built, it acquires specific characteristics (Dorantes 2012).

Considering the construction process, rural housing is characterized by the use of artisanal techniques and materials from the natural environment, unlike modern rural housing that uses industrialized materials together with traditional elements and techniques (Quintanar and Rosas 2010).

Rural housing in Mexico has been influenced by two main aspects: the historical, based on the incidence of two cultural traditions: pre-Columbian and Spanish, where the different cultural traits of Mesoamerica and the regional characteristics of Spain are present; and the natural, which includes: the climate that influences the architecture as a protective element for man, who through housing generates microclimates that help his survival; and natural resources, which are used as materials for construction, taken from the natural environment for their use and transformation (Ángeles 2010).

Traditional housing in Mexico has been rapidly transforming as a result of a series of factors such as the urbanization of rural areas, the interruption of popular knowledge, the loss of natural assets and public housing policies that hide the importance of traditional construction systems, promoting the use of industrialized materials, which leads to generate aspirations regarding the concept of dignified and resilient housing (Dorantes 2012).

2.2 Rural Housing PAME: Querétaro

The Pame indigenous people are currently located in the southeast of the state of San Luis Potosí, between 600 and 800 inhabitants are located in the northeast of the state of Querétaro, in the region of Tancoyol, municipality of Jalpan (Zavala 2011).

Most of the Pames settlements are dispersed, that is, the houses are separated by meters from each other. Each plot houses a domestic dwelling, consisting of a house-bedroom and a kitchen; each unit is inhabited by a nuclear family (father, mother and children), which is organized to carry out the daily tasks that allow them to subsist. All of these activities tend to be divided by age and gender; however, with increasing migration to the border and to the United States, it is increasingly common for women and children to organize themselves to perform men's tasks (Ordoñez Cabezas 2004).

As for the typology of the traditional Pame dwelling, it is a square or rectangular jacal, sometimes with one or two rounded sides, with walls of one row of wooden rods, or two rows when they contain stones; the walls are sometimes mud-plastered; plank or adobe walls are also common. The roofs are generally gable roofs covered with palm leaves, although currently they have been changing to galvanized sheet roofs (Rusin 2012).

Rural Eco-Lab (2017) considers that in La Manzanilla and Agua Puerca the spaces are designed according to the needs of the extended family, where the central element is the kitchen, followed by the shingled rod rooms, a resting place for the heads of household. These housing typologies have persisted after having been tested for many having been tried and tested for numerous generations.

A theoretical position that will allow the continuity of the development of the research is the regenerative design whose principles are based on the connection between man and nature.

2.3 Regenerative Design

Regenerative design is based on systems theory oriented towards design processes. The term regeneration alludes to processes that restore, renew or revitalize their own sources of energy and materials, creating sustainable systems that integrate the needs of society with the integrity of nature (Heavengrown 2018).

Regenerative approaches do not seek to reduce destructive impacts, but rather consider that buildings enable social and ecological development, therefore, in order for buildings to serve as an element towards global environmental change, a new link with their surroundings must be established (Colle Raymond 2017).

From the above, there are examples that through the application of regenerative design have managed to develop results at different scales, from the conceptual, starting with biomimicry to artificial intelligence of what is built by man.

2.4 Bioclimatic Architecture

Bioclimatic architecture is responsible for the study of the natural conditions of the environment to take advantage of available natural resources: sun, rain, wind, vegetation, achieve thermal comfort, generating energy savings and reducing the impact on the environment (Hildebrandt Gruppe 2016). The objective of bioclimatic architecture keeps in mind two important principles (Certicalia 2018):

- Construct buildings that are sustainable, energetically self-sufficient and generate 0 emissions.
- Consider the construction system, buildings must have a construction process that is environmentally responsible, as well as the use of non-toxic construction materials.

2.5 Living Building Challenge

International sustainable building certification program, created in 2006 by the non-profit International Living Future Institute. The Institute describes it as a philosophy, advocacy tool, and certification program that promotes state-of-the-art measurement of sustainability in the built environment; it can be applied to development at all scales: new or renovation buildings, infrastructure, landscapes, neighborhoods, and communities (livingfuture.org, 2018).

This certification seeks to encourage and construct buildings with qualities: regenerative, connecting occupants to light, air, food, nature and community; self-sufficient, living buildings produce more energy than they use while collecting and treating water on site; and healthy, creating positive impact on the human and natural fabrics that interact with them.

The Living Building Challenge is composed of seven petals (place, water, energy, health and happiness, materials, equity, and beauty), which in turn are composed of imperatives that in total add up to twenty and must be met to obtain certification (livingfuture.org, 2018). In case only some of the petals are met, certification is obtained at the petal level and if the energy petal is met in combination with some imperatives, NZEB certification is obtained (International Living Future Institute 2016).

2.6 Biosphere Reserve

The United Nations Educational, Scientific and Cultural Organization (UNESCO) in 1971 launched the Man and the Biosphere program, with the objective of establishing the rational use and conservation of biosphere resources to improve the global relationship between humans and the environment by applying natural and social science studies together (Schliep and Stoll-Kleemann 2010).

According to UNESCO (2017), Biosphere Reserves are nationally relevant biogeographic areas supporting science for sustainability with a land area larger than 10 000 ha "[…] special spaces for testing interdisciplinary approaches to understanding and managing changes and interactions between social and ecological systems, including conflict prevention and biodiversity management." (UNESCO 2017).

Within the main characteristics of Biosphere Reserves according to UNESCO (2018) are: achieve conservation, development and logistical support, generate the combination of the various areas with the sole purpose of conservation with others where sustainable development is promoted by the population and business ties of the area, usually with innovative and participatory governance systems, in the process of managing the Reserve, encourage participation among local communities, for the resolution of conflicts related to the use of natural resources, encourage dialogue, integrate cultural and biological diversity of the space, to promote the role of traditional practices or knowledge in ecosystem management.

2.6.1 Sierra Gorda de Querétaro Biosphere Reserve

It is located in the north of the state of Queretaro, representing 32.02% of the total territory of the state, it is made up of the municipalities of: Arroyo Seco, Jalpan de Serra, Landa de Matamoros (covered in its entirety), Peñamiller (69.7%) and Pinal de Amoles (88.03%); it includes 638 localities and the municipal capitals; it has 11 core zones and 1 buffer zone (Unidad de Participación Social, Enlace y Comunicación, INE 2000).

The existence of the reserve in Querétaro arose from a need that the population had: the inhabitants of the area created a social movement called Sierra Gorda Ecological Groups (GESG) to protect and take advantage of the natural resources that existed in the area without depleting them in the future. It was in 1992 when the municipalities mentioned in the previous paragraph were declared the Sierra Gorda Biosphere Reserve (Aguilar 2015).

Currently, the land uses are forestry, agriculture, livestock, housing and conservation (Unidad de Participación Social, Enlace y Comunicación, INE 2000). Within the reserve's territory, land ownership is divided into three types according to data from the 1997 National Agrarian Registry: private with a 69.33% share, communal and ejido, which together account for 30.67% of the land.

The typology of contemporary rural housing in the Sierra Gorda generally responds to rectangular plan volumes, except in some localities where according to the shape of the space is the solution given to the roof, depending on the case: apsidal, double apsidal and circular plan, for which flexible natural materials are used to give the appropriate shape (De tierra y varas 2009).

This change experienced in rural housing in the Sierra Gorda of Querétaro is mainly due to the high rates of migration, as a consequence of the social and economic context in which the population of the Sierra Gorda finds itself, occupational prospects are limited and wages are very low, which is why the inhabitants are forced to leave their communities in search of work, the USA being the main destination (Miller 2001).

3 Methodology

For the development of the research, a methodology consisting of four stages was used: the first corresponds to the planning and information gathering phase; the second consists of the contextualization and analysis phase, where field work was carried out in the case study community: Tilaco, where participatory design tools and the People-Centered Design methodology were used in order to develop the analysis of existing housing, as well as the characteristics that the inhabitants desire or long for when planning and developing a house; The third stage consisted of developing proposed strategies as a result of the analysis and diagnosis of the housing situation in the community based on the principles of regenerative design, the Living Building Challenge Certification, bioclimatic architecture and People Centered Design. Finally, in the fourth stage, the proposed strategies were evaluated with the inhabitants of the community and the Living Building Challenge Certification, seeking the feasibility of using the strategies at the time of planning and developing housing within the community.

It is important to mention that the application of the strategies was positively validated, except in the aspects where joint or team activities are developed, since they consider that this is the weakness they have as inhabitants of the community.

4 Results and Discussions

4.1 Context Analysis

The Tilaco Valley is a rural community that is part of the Sierra Gorda Biosphere Reserve, belonging to the municipality of Landa de Matamoros, within the state of Querétaro, which is currently experiencing the phenomenon of the loss of identity of traditional housing. Located 26 km south of the municipality, located at latitude

21° 09′ 46.0002″ and longitude −99° 11′ 26.9982″, at an altitude of 1100 m above sea level (Giovannelli 2018).

Tilaco is bordered mainly to the north by the Santa Teresita neighborhood 2 km away, to the south by the La Luz neighborhood 2 km away, to the west by the Buenavista community 3 km away and to the east by Otates 6 km away (CIBCEC 2005).

4.2 Physical Environment

The study of the physical environment will provide a broader knowledge about the context in which the research process takes place, allowing a first approach to the identification, prevention and interpretation of the environmental impact of the natural environment.

This section identifies the main characteristics of the natural physical environment such as geomorphology, climate, hydrography, vegetation, fauna and land use, with the purpose of knowing the impact that human activity has had on the environment and understanding how it influences the municipality of Landa de Matamoros.

4.2.1 Geomorphology

The most representative elevations of the municipality are to the north: a group of three mountains known as San Miguel, to the south: a small mountain range called Montañitas, to the east: the hills of La Gacela, El Gato, El Venado, El Tejocote, La Cebadilla and El Grande de San Juan, and to the west: the hills Piedra Encimada, La Barca and Los Orujos. Because it is part of the Sierra Madre Oriental and because of the rugged terrain, the municipality of Landa has different altitudes above sea level. The municipality has some valleys, the most important of which are: Plan de Hongos, Tilaco, Tres Lagunas, Valle de Guadalupe and Acatitlán de Zaragoza (Amador Covarrubias 2015).

The reserve is characterized by the presence of rocks from the Lower Cretaceous of the Mesozoic era, with sedimentary rock and conglomerate lithology. The following predominate in the Tilaco valley and surrounding areas (Rural Eco-Lab 2017): limestone sedimentary rocks (KI[cz]), is found in 55.76% of the population center boundary; limestone, it is white, but the presence of impurities such as clay, iron oxide, among others, give it a cream, reddish or gray tone; and alluvium: found in the valleys of continental depressions, filling the river valleys in the middle or lower parts of its bed, sometimes occupying large areas in the plains that are generated at the foot of the mountains.

4.2.2 Weather

According to Enriqueta García's adaptation, the characteristic climate of the Tilaco locality corresponds to the Semi-warm sub-humid of group C, with average annual temperature greater than 18 °C, with the coldest month below 18 °C and the hottest month above 22 °C. Precipitation in the driest month is less than 40 mm; summer rainfall is less than 43.2 mm and the percentage of winter rainfall is between 5 and 10.2% of the annual total (García 2004).

This subclimate presents particularities, since the relief and geography of the context give it specific characteristics, being in a mountainous environment the air masses and the circulation of the winds change according to the obstructions that it has, in the same way these elements favor the dissipation of the humidity of the environment in a varied way to other areas with the same type of subclimate.

Using a set of graphs obtained from the Meteornom 7 software, the climatic data for the case study were analyzed considering: radiation, temperature and precipitation.

Beginning with the temperature, the minimum temperature is in December and January, which is less than 6 °C, while the maximum temperatures are in April, May and June, ranging from 35° to 36 °C.

The maximum monthly radiation is found in the months of April, May and July above 180 kWh/m^2 and the minimum in the month of December; spring and summer are the months with the greatest discomfort due to heat.

The months with the highest rainfall are September and October, and January is the month with the lowest rainfall, ranging between 55 mm.

According to the stereographic chart obtained through the weather tool, it is possible to clearly appreciate the solar path, which serves as a guide to determine the level of protection against sunlight required in the different months of the year. The months with maximum temperatures (May to August) require protection mainly in the ENE and WNW orientations, and it is also essential to protect the East and West orientations 100%.

On the other hand, in the months with lower temperatures (January, February, November and December) we can allow indirect sunlight to obtain a thermal gain by mass, always taking care that the angle of inclination of radiation does not generate discomfort inside the spaces.

According to the orientation graph, it is possible to determine in a simple way the best location for the openings, as well as for those solid elements that can be used for thermal gain during the night, having the South, South East orientation as the optimal one.

This orientation has the highest percentage of areas in comfort, but also the highest percentage of areas with a lower than average temperature. As for the North orientation, it is important to highlight the need for protection from direct radiation. The prevailing winds come mainly from the south-west, being present in most months of the year.

The behavior of the analyzed climate can be observed on an hourly basis, having as comfort zone the area between 18 and 25 °C and between 80 and 20% relative humidity, initially and without applying any passive technique, we have 33% comfort throughout the year. As for the remaining 67%, it will be necessary to apply passive techniques to provide the necessary comfort and achieve the habitability of the space. We find in general terms, framed by the blue polygon, the points of discomfort due to cold, and in red the points of discomfort due to heat and excessive humidity.

The typical months with the greatest discomfort due to cold (January and December), in which we found 39% of general comfort and 61% of discomfort, without using passive techniques. The temperature ranges between 6 and 30 °C and the relative humidity between 20 and 95%. The temperature of discomfort due to cold ranges between 6 and 17 °C and even in the coldest month we found a minimum of discomfort due to heat between 25 and 30 °C.

Among the passive techniques to be implemented in these months are heat loss through high thermal mass (insulation), internal heat gain and passive gain through high thermal mass (direct radiation). Due to the characteristics of the climate in this area, there is a comfort level of 33.5%, which translates into one third of the target. To reach 90.3%, which represents the target value, a series of strategies are proposed which, based on the climate analysis carried out previously, are the most appropriate. These strategies are composed of:

- Window shading (27.1%): Possible with the generation of eaves, louvers or shades by other means.
- High thermal masses for nights (22%): Can be translated into materials or finishes.
- Evaporative cooling, second state (24.2%): Probably achieved from water bodies.
- Cooling from cross ventilation (27.1%): As its name implies, it could be achieved by taking advantage of the prevailing winds.
- Internal heat gain (14.6%): Theoretically achievable with orientations and other passive techniques.
- High masses of passive direct solar gain (6.3%).

4.2.3 Hydrography

The Sierra Gorda de Querétaro belongs to the Hydrological Region of the Pánuco River (RH-26). This zone is divided into two basins: the Tampaón or Tamuín River, which covers 2,038 km2 and has the Ayutla, Santa María and Jalpan rivers as its main tributaries; and the Montezuma River, which covers 1,532 km^2 of the reserve and has the Extóraz River as its main tributary. 25 of the 66 rivers and streams registered in the entity are found in the protected natural area (Rural Eco-Lab 2017).

The main water resources are minimized to bordos and water pans from which most of the population is supplied. The main bordos are located in the communities of Encino Solo, Landa de Matamoros, Tres Lagunas, Otates, Santa Inés, Malpaís and Acatitlán de Zaragoza. Stored water is obtained from rainwater runoff (INAFED 2010).

The Tilaco valley has the topographic characteristic of being a natural water border during rainy periods, which flows into the Moctezuma River through cavities known as basements, which are located on the slopes of El Sapo hill (Rural Eco-Lab 2017).

Flora and Fauna

In the area belonging to the municipality of Landa de Matamoros, the existing vegetation is distributed as follows: forest 62.8%, scrub 13.3%, pasture 6.7% and jungle 0.9%. For the most part, the municipality is covered by tropical dry forest, in it are pine, red cedar and white cedar; the municipality has a forest importance for its pine, oak, oyamel, white cedar and strawberry tree forests (Rural Eco-Lab 2017).

As for the fauna characteristic of the Reserve, it has a great diversity of wild species; it is composed of 800 species of butterflies, 339 species of birds, 131 species of reptiles and amphibians, 127 species of fungi, 110 species of mammals, and 27 species of fish. Some of the species include: puma, trigrillo, coyote, fox, armadillo, badger, wildcat, rabbit, hare, tlacuache, white-tailed deer, sparrow, cucho, macaw, squirrel, roadrunner, lizard, pigeon, goldfinch, mockingbird, cardinal, crow, parrot, among others (INAFED 2010).

4.2.4 Type and Use of Soil

The following soil types are found in the municipality of Landa de Matamoros: feozemluvic lithosol, luvisol, chromic rendzina and pelic vertisol. There is also a washed red or light-colored clay subsoil, associated with brown or dark reddish brown topsoil, rich in organic matter and nutrients, subsoil with loose or slightly cemented caliche accumulation, and with stony soil, with good air and drainage conditions. Finally, there is a type of subsoil in lumps and accumulation of clay, calcium carbonate or iron, calcareous in all its layers, associated with washed soil of light acidity and clay subsoil rich in lime, with poor drainage and large pieces of rock near the surface (INAFED 2010).

Regarding land use in the municipality of Landa de Matamoros, 24% belongs to residential use representing an area of 256 793.33 m^2, mixed use representing 23% with an area of 242 421.38 m^2, mixed use is divided in turn into residential and commercial (19%), residential and agricultural (70%), residential and livestock (2%) and residential and agribusiness (9%); vacant 16% representing 173 313. 17 m^2, agricultural with 15% corresponding to an extension of 158 513.93 m^2, equipment 12% with an extension of 123 225.46 m^2, agostadero 8% with an extension of 87 960.21 m^2, commerce and livestock with 1% represented by 8 096.50 m^2 and 3 567.82 m^2 respectively (Rural Eco-Lab 2017).

4.3 Socio-economic Study

The town of Tilaco, belonging to the municipality of Landa de Matamoros, has 679 inhabitants, of which there are 334 men and 345 women. Of the total population 229 inhabitants are between 0 and 14 years of age, 166 are between 15 and 29 years old,

195 are between 30 and 59 years old and 89 of them are over 60 years old, likewise, 42 of its inhabitants have some type of disability (SEDESOL 2013).

The population of Tilaco is mostly young, since most of its inhabitants are in the 0 to 30 years of age range. The number of men represents 49.7% and women 20.3%. Within the economically active population, 83.19% of the inhabitants receive between 1 and 2 minimum wages; 6.02% of the population receives between 3 and 4 minimum wages; 3.06% of the population receives between 5 and 6 minimum wages; and 1.20% of the population receives between 7 and 9 minimum wages. The employment structure is mainly based on temporary activities where the most common are day laborers and bricklayers, farmers, cattle ranchers and merchants, and employees in the tertiary sector. The economic base of the town is based on primary and tertiary sector activities (Gobierno de Querétaro 2002).

There is a high flow of emigration by young people, women and men seeking better opportunities for economic development and quality of life. The main destinations are the United States, Monterrey, Mexico City, and the capital of Querétaro. These high rates of emigration in the municipality are mainly due to the lack of employment and the low profitability of agricultural, livestock and forestry activities, as well as the low wages offered to the economically active population (Amador Covarrubias 2015).

Of the population that makes up the town of Tilaco, 30% of the inhabitants emigrate mainly to the USA, returning to the town annually for a stay of approximately 3–4 months, with the city of Querétaro and the CDMX being the second destination of the inhabitants to obtain better economic development and a better education (Gobierno de Querétaro 2002).

The town of Tilaco is organized by a Works Committee (INTEGRA), an Ejidal Commissariat, the Municipal Livestock Association and the Credit to the Word Committee (Gobierno de Querétaro 2002).

It is important to mention that the town of Tilaco is home to the San Francisco de Asís Mission built between 1754 and 1762 by Fray Juan Crespi, which makes it a major tourist center for people who come to the town, thus attracting investment and helping the local economy.

4.4 Architecture

The town of Tilaco has a total of 218 houses, of which 174 are inhabited and 44 are uninhabited, 169 of the houses have concrete pavement and 167 have electricity and the average number of inhabitants per house is 3.9 (INEGI 2010).

According to an analysis of the dwellings found in the community, the following typologies were detected as the main ones (Table 1).

Table 1 Description of housing typology (own elaboration).

Typology number	Description
01	It has stairways at the entrance of the house, most of them are built of red brick or block and flat concrete slabs with a slope for rainwater collection. Square windows, they can vary to the form of semicircular arch, the bedrooms have small dimensions to the north-northwest. The living and dining areas are to the west. Some are on a slope, taking advantage of the development of a basement or cellar
02	One-story houses with front entrance with low enclosures, living and dining areas attached to each other, presenting small spatial dimensions. Flat roofs of galvanized sheet metal. The doors and windows are almost flush with the roof and the facade is oriented to the west
03	Housing with galvanized sheet gable roof formed in one piece, where the living areas are located. Half attached to the ground, taking advantage of its elevation where a cellar is preserved. It presents a compact volume and of similar form to the vernacular constructions. Few windows and of reduced dimensions, minimum overhangs and main facade to the north. They are generally surrounded by vegetation that provides shade to the house
04	Housing of recent construction, with regular volume to 2 waters, main entrance on the north side, bedrooms with the same orientation. Front terrace for recreational activities and generate comfort in summer. Porch-garage with galvanized sheet roofing and wooden trunks structure of trees in the area
05	Two-story house with facade facing northeast. The bedrooms are on the first floor and the living areas are on the first floor. The windows are located on the facade due to the existence of a lateral adjoining building. Galvanized steel sheet roof with rainwater collector. Upper porch on the first level to generate comfort in the rooms. Manufactured with industrialized materials
06	Housing with orthogonal volumetry, consists of two levels with flat roofs for both cases. The main facade is oriented to the south, on the first floor are the main door and rectangular windows. Construction with the use of reinforced concrete columns and creating a terrace towards the back of the building. Rooms in the upper part
07	Built in 1940. Its facade has a modernist influence, with a lateral staircase that gives access to the main entrance, built with local stone. It has a fireplace shared with the kitchen and the living room, it also has a well that collects rainwater. The windows are rectangular and the walls are 50 cm thick. The interior spaces are ample, the distribution is fluid and it has three bedrooms, kitchen and dining room
08	Recent construction presents a combination of architectural styles of European influence. Windows and exterior doors with semicircular arches. Roofs in different slopes and directions without rainwater collector. Wrap-around portico with two entry doors. Convenience area at the front and bedrooms on the first level of the building, there is also a balcony on the first level with access door from the family room
09	Two-story house with lateral entrance porch to the main door. Reinforced concrete gable roof, windows with arched combination and rectangular shape. With influence of North American architecture of Californian type. On the first level the windows are oriented to the south

4.4.1 Building Systems

As a first approximation to the construction systems used in the locality of Tilaco by the inhabitants for the construction of housing in general found in the bibliography, it can be observed that it is characterized by the use of earth, wood and stone. They are simple constructions that are composed of a wide roof, in which habitability and services are mixed. Among the main construction systems are (Rural Eco-Lab 2017):

1. *Adobe and wood.* Adobe settled with mortars made of mud due to the abundance of soils suitable for its elaboration, due to the ease of transformation and construction, the use of these materials presents thermal and acoustic qualities.
2. *Masonry.* Elaborated by stone walls, which form the base for the roof, these walls are load-bearing, the stones are joined together. The transmission of the outside temperature is more direct than that of the other construction systems.
3. *Wood and Tejamanil.* Wood is one of the most abundant products in the Sierra, which is why the inhabitants use it for construction.
4. *Bajareque.* The mud is extracted from the land itself, this material is used to fill the framework formed by reed sticks.
5. *Partition walls and slabs.* These buildings are characterized by the use of a firm floor, partition walls with a combination of cement and earth from the locality and concrete slabs.

4.5 People Centered Design

4.5.1 Surveys

In a visit to the Tilaco Valley, surveys were applied to the inhabitants of the community, the dynamic for the application of the surveys was one for each dwelling and at the same time brief spontaneous interviews were carried out, which were very useful since the people were able to expand more on the subject of housing and their lifestyle.

A total of 52 surveys were applied to the inhabitants of the community, one for each existing dwelling, of which the population of inhabited dwellings corresponds to 174 and with a finite population formula a total of 52 samples were obtained with a confidence level of 95% which were applied to the inhabitants; each survey was composed of 55 questions divided into five blocks to be able to know in depth the characteristics of the community, the dwelling and its inhabitants: I. Territorial domain, II. Physical aspects of the dwelling, III. Services, IV. Daily life and V. Social dimension.

4.5.2 Workshop "the Desire of the House and How to Inhabit It"

The workshop was held with the inhabitants of the community, with which together and through activities we sought to express the characteristics of the desired house and how to live it, being this the reflection of our experiences and knowledge about it; in order to generate strategies that help us to propose a safe, comfortable and durable housing style.

The workshop was carried out through three activities, ending with a group dialogue regarding the main objective of the workshop: housing. The following is a description of each of the phases with their respective activities and the results obtained in each one.

ACTIVITY 01. The poem of desire.

It is a collaborative or group poem. The idea is to generate a series of complementary responses to the phrase "I wish my ..." based on the free flow of feelings and personal associations. The results of the poem will generate a series of characteristics that will assist in the development of strategies for the conception of an ideal dwelling. For this purpose, a moodboard was shown that served as a support to consider characteristics that could be taken into account at the moment of writing the poem, the intention of it being a poem written individually allowed the participants to be moved and open at the moment of writing and expressing their desires towards the need of having a house.

As a result of the diagrams obtained from the poems written by the inhabitants of the community, the following diagram is presented below, which represents the following characteristics: spaces and elements, materials, qualities, characteristics of the spaces and/or house, identity, emotions and sensations (Fig. 1).

After analyzing the poems developed by the workshop participants, it is clear how they continue to have a taste for the use of regional materials such as wood, adobe, palm, reeds, wood, etc. And as with the poem they were able to have memories of their childhood as some of them expressed and remembered that house they lived in with their family and how despite being small it was very cozy and kept them together as that family bond.

On the other hand, it is also noted that they leave aside those memories and continue with the same construction patterns because for them a house made of concrete is better and gives them a status within the community, but despite this they sacrifice because it is also noted that this house is not comfortable.

It is important for them to know that there is no better or worse material, but rather to know which one is suitable for certain conditions and characteristics within the environment and the context in which the house will be located.

When they describe in the poem a house made of adobe and palm, they consider it small, but when they describe it in concrete they think it is big. It is also important to emphasize that within this desired house some of the participants took into account nature, the context, and the animals as an important part of their life development.

Regenerative Rural Housing, Case Study: Sierra Gorda Biosphere …

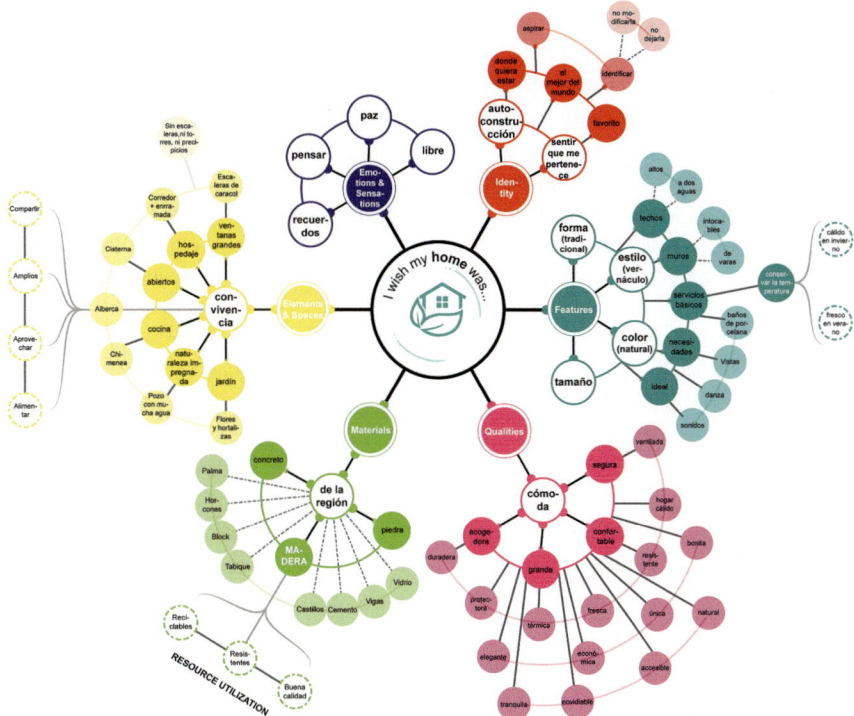

Fig. 1 General diagram representing the desire of the inhabitants of Tilaco: Identity (*red*), characteristics (*turquoise*), qualities (*pink*), materials (*green*), elements and spaces (*yellow*), emotions and sensations (*blue*) (own elaboration, 2019)

In the same way, they identify themselves with aspiring to have a house of which they feel owners, they feel that it belongs to them and with which they generate an emotional bond that allows them to be there always.

ACTIVITY 02. The image of the house.

A collection of photographs representing a wide range of housing types, some of the 10 units represented are typical homes from the case study site and surrounding communities: conventional, vernacular, contemporary, self-built or architect-generated. A player's response to the appearance of a particular type of house will be determined by what they have seen and are familiar with and also by the visual characteristics they associate with the home they expect to live in.

By making decisions about the visual arrangement presented to him, the player is indicating the visual qualities he desires in a house. The player ranks the images in order of their preference, from 1 to 10. The player is asked to describe the particular characteristics of their first two choices and the reasons why they dislike choices 9 and 10.

The participants were presented with a collection of ten images, without giving them any information regarding each of the houses and in which each of the participants ordered them according to the importance they gave them in terms of most liked—least liked. According to the organization of the cards, the following matrix was generated showing the preference of the facades, this was done according to what the inhabitants knew and knew about a house (Fig. 2).

In this activity we also obtained as a result the aspiration of the inhabitants towards large and luxurious houses, arguing that it is because they look well done and give them a social status before others, and for them it is a symbol of progress, although in many cases they see several disadvantages.

It can also be observed in the matrix that the housing with the least acceptance was the one-story stone house and the one with the rooms separated by apparent materials, because most of them argue that it looks small and is not functional, although they accept that because of the materials they are thermal and comfortable houses.

ACTIVITY 03. Build your house.

With the help of a didactic construction tool "CALTIA" we sought to understand the way in which the inhabitants build their homes and the conception they have about

Fig. 2 Façade matrix according to the preference of the inhabitants (own elaboration, 2019)

Fig. 3 Example of desired typology (own elaboration, 2019)

this, considering the primary or basic needs that each one has and based on this how they start with the design and construction process.

The tool consists of a board that contains a grid of perforated squares where the various columns that serve as structural elements are inserted and on which the walls, doors, walls with doors, walls with windows, etc. are inserted. And this is how the typology or form of housing desired by the participants begins to take shape.

Initially, the inhabitants were asked to write down the needs and characteristics they would like to have when building their house, as well as the number of people who would live in it. After generating this list of needs and characteristics, they began to dimension and imagine the house they would like to have, using the Caltia didactic material for architectural design (Fig. 3).

As a result of this activity it was observed and concluded that initially it is important to be able to generate the planning of the house before construction, since several of the inhabitants who participated in this activity argued that this resulted in the construction of a non-functional house, with problems of habitability and thermal comfort, since they built only to take advantage of the resources they had at the time and avoid spending it on other aspects that were not necessary.

Another important point to mention is that the inhabitants who participated in the survey agreed mainly on having spaces for cooking, living and resting, leaving aside other spaces that for them were not important.

4.6 *Diagnostic*

The Tilaco valley belongs to the Biosphere Reserve, so its development is governed by characteristics that maintain a balance between the inhabitants and their environment, promoting conservation and care of the place as part of the reserve to which they belong.

The community's inhabitants are aware that Tilaco belongs to the Biosphere Reserve and they consider that the community is well cared for, it is green, the air is pure, and there is very little contamination; They argue that they are open to dialogue so they can extract resources from the reserve to build their homes and reward them, but they feel that information is lacking. Another disadvantage is that although the community is aware of the reserve, in one way or another, they do the opposite when it comes to housing construction.

The inhabitants identify with the mission as a symbol of transcendence, because it was an important factor that helped the development of the community, and also because they took a fundamental role in the work developed, as well as with the small square because both are places of coexistence and social recreation where they develop their customs and traditions as a community.

If adequate attention is given to the mission, it could promote economic development through tourism, in which the inhabitants develop activities or provide services that help them to support their lifestyle and the development of the community, where as inhabitants they could generate a greater union and integration.

The inhabitants consider that in terms of housing they are only interested in having a house that allows them to develop their basic needs such as sleeping, eating, cooking, living, physiological needs, that provides them with comfort when they live in it and that it is durable, functional and comfortable, regardless of the type of material, but rather they are interested in it being well built and planned.

It is appropriate to use materials such as block, cement, brick to build a house because they consider it to be more resistant, durable, accessible; however, they complain that it does not allow them to live in the house comfortably because it absorbs solar radiation and transmits it to the house, and they also mention that there are problems due to the type of soil in part of the community.

Stone is also considered adequate because it allows the house to be cooler and more comfortable to live in; however, they comment that it is difficult to access and that the same thing happens with wood.

5 Strategy Development

Based on the results obtained in the previous points, the generation of the various strategies based on the regenerative design is started, taking into account the petals of the Living Building Challenge certification, aspects of bioclimatic architecture and People Centered Design, with the purpose of proposing the conception of a regenerative model of rural housing.

For the development of the strategies based on the Living Building Challenge and bioclimatic passive design, cards were generated containing the information shown below (See Fig. 4).

With regard to the strategies based on People-Centered Design, they are presented in a textual manner since they do not require a graphic scheme for their understanding.

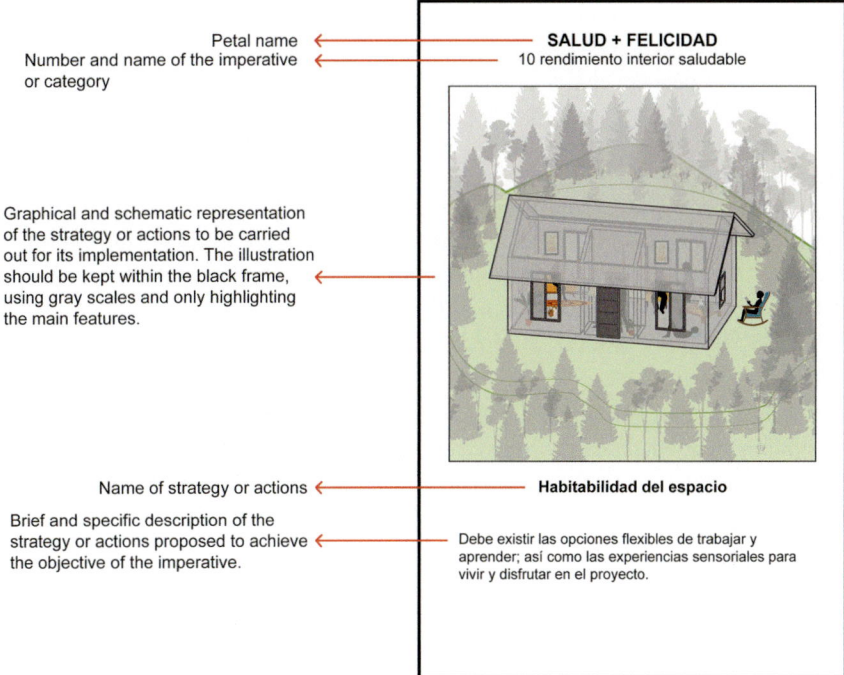

Fig. 4 Worksheet used for the elaboration of strategies (own elaboration, 2019)

5.1 Regenerative Design Strategies Based on Living Building Challenge Certification

The strategies proposed are considered for new housing construction, therefore, for the Living Building Challenge certification, the 7 petals (place, water, energy, health + happiness, materials, equity and beauty) and the imperatives that make up each petal were considered for a total of 20. This certification defines 6 transects according to the Living Future Institute (2019) which are:

- L1. Conservation of natural habitat.
- L2. Rural area.
- L3. Village or campus area.
- L4. General urban area.
- L5. Urban center zone.
- L6. Basic urban zone.

Tilaco is a community that belongs to the Biosphere Reserve and because of the development that has taken place within its nucleus, it is considered in transects L2 and L3 according to the previous classification; it belongs to a rural zone and a village or campus zone because it is mainly composed of land used for agricultural development and food production, small-scale commerce, the number of inhabitants and also because it includes educational and cultural institutions within its nucleus, as well as its dependence on a larger population (in this case it's municipal capital).

5.1.1 Place Petal

This petal is intended to locate sites suitable for building, protecting and restoring a place after it has been developed, encouraging the creation of communities that rely on the pedestrian rather than the automobile, and primarily to look at how man understands and connects with his natural environment. It consists of four imperatives, which are:

1. Site ecology. Virgin green areas, wilderness areas, important farmland, avoid building in flood plains, documenting site conditions, positive contribution, landscape designed to mature and evolve, no use of petrochemicals and decision making and process.
2. Urban agriculture. Area for growing food crops, weekly farmers' market, area for agricultural development, food bank, food supply and resilience.
3. Habitat exchange. Territorial compensation.
4. Life at human scale. Human-scale project, spaces for living and connection, safe cycling, electric vehicles, less parking, vehicle travel, pedestrian routes, public and human transportation, transit subsidy, carpooling and transport impact.

5.1.2 Water Petal

The purpose of this petal is to revalue the use of water, as well as its process: transportation, purification and pumping, and to consider the use of "wastewater" as a nutrient and valuable resource. It consists of two imperatives, which are:

5. Responsible water use. Drinking water, minimizing water use, stormwater, stormwater management and stormwater detention systems.
6. Net positive water. Water supply, water purification, non-potable uses, gray and black water treatment, community installation, supply pumps, water storage and resilience.

5.1.3 Energy Petal

This petal contributes to the creation of renewable energy sources that allow the project to operate year-round in a resilient and carbon pollution-free manner, prioritizing energy efficiency that seeks to reduce wasteful expenditure of energy, resources and money. It consists of two imperatives, which are:

7. Energy + Carbon reduction. Reduced energy consumption, energy calculation, embodied carbon reduction and interior materials with lower carbon footprint.

8. Net positive carbon. Energy supply, total carbon emissions and livable project by resilience.

5.1.4 Health + Happiness Petal

The intention with this petal is to generate healthy spaces allowing the prosperity of the species, looking for the connection between people and their environment and trying to have interior spaces with healthy air and natural light. It consists of three imperatives, which are:

9. Healthy indoor environment. Indoor air quality, no smoking, healthy indoor environment plan, outdoor views and direct ventilation.
10. Healthy indoor performance. Indoor air quality testing, CDPH v1.1–2010 standard method, cleaning protocol: EPA products, access to views and daylight, natural ventilation, temperature control, space habitability and operable windows.
11. Access to nature. People-nature connection, inside-outside interactions and post-occupational evaluation.

5.1.5 Materials Petal

The material petal seeks to generate a material economy that is non-toxic, ecologically restorative and transparent. It consists of five imperatives, which are:

12. Responsible materials. Product Declare, disclose and identify the composition of products, wood extraction for the project, local materials, construction waste, recyclable and compostable organic waste.
13. Red List.
14. Responsible source. Sustainable resource extraction, stone product manufacturers, FSC certification, other wood sourcing and Living Product Challenge.
15. Source of living economy. Promotion of the regional economy and location of material manufacturers.
16. Net positive waste. Waste reduction or disposal, recovered material, material conservation management plan and Design—Construction—Operation—End of life.

5.1.6 Equity Petal

This petal aims to raise equity through the development and creation of fairer and more inclusive communities, seeking the participation, prosperity and maximum potential of its inhabitants. It consists of two imperatives, which are:

17. Universal access. Equal accessibility, public space, people with physical disabilities, quality of fresh air to adjacent buildings, quality of sunlight, quality of natural resources, natural waterways, noise management, noxious emission,

sunlight on facades and roofs, roof shading, waterway edge access, and water appropriation.
18. Inclusion. JUST label and vulnerable or underserved populations.

5.1.7 Beauty Petal

This petal seeks to recognize the need for beauty and the connection to nature as a precursor to caring enough to preserve, conserve and serve the common good. It consists of two imperatives, which are:

19. Beauty + Biophilia. Elements that nurture the human-nature connection, transformation of nature through forms, connection of the project with the place, recreation and celebration, exploration of the site and context, and tracking of biophilia.
20. Education + Inspiration. LBC case report, operations and maintenance manual and accredited professional.

5.2 Bioclimatic Passive Design Strategies. Tilaco, Case Study

Based on the graphs analyzed above, the following strategies were developed, divided into eight categories:

1. Site, context and volumetrics. Cooling and/or shading due to vegetation, attachment, orientation, shape, perforation, settlement and porosity.
2. Glazed openings. Orientation of openings and windows, shading/solar control, movable horizontal louvers, vegetation, pergolas, glass selection and window systems.
3. Covers. Light colors, orientation, inertia and insulation, ventilation and double roofs.
4. Walls. Color, shading, vegetation, cantilevered roofs, inertia and insulation.
5. Ventilation. Cross ventilation, through ducts and temperature differences.
6. Evaporative cooling. Inner courtyard and vegetation.
7. Cold discomfort. Shutters, deciduous vegetation and dehumidification.
8. Complementary. Green roofs.

5.3 Strategies Based on People Centered Design

The strategies proposed in this section arose from the activities carried out in the workshop developed with the community's inhabitants, as well as from the interviews and surveys conducted with them.

1. Identity and appropriation of the inhabitants to housing.
2. Sense of ownership in the design process.

3. Housing creation as a dynamic process.
4. Communal fabric through economic diversity.
5. Participation in project development.
6. Team work.
7. Knowledge of materials and construction processes.
8. To value its constructive processes.
9. Planned self-construction.
10. Housing adaptability.
11. Indoor-outdoor integration.

6 Conclusions

This research project was developed in the community of Tilaco, belonging to the Biosphere Reserve, which is a space that seeks to maintain the balance between man and his environment; as a result of having detected a problem regarding the modification of rural housing caused in part by the constant migration of the inhabitants to nearby cities or to the USA.

When the inhabitants return to the community and have the purchasing power, they acquire their materials and begin the construction of their house, making an exact replica of how they had it in their other place of residence, without considering or planning that the conditions of the environment are different and that probably the functionality of the house will be different; and considering that the house should not be seen as a merely physical space, but as a place that helps and promotes the development of the human being with others of the same species, but also as an element within the environment that should be in harmony.

As a result of these situations, this research project sought to develop strategies that would provide knowledge about the planning and construction of housing and maintain a balance and harmony between the natural environment and the inhabitants, seeking to improve identity, a sense of belonging and reduce environmental impact.

In order to develop these strategies, regenerative design, bioclimatic architecture and the Living Building Challenge certification were taken as main elements. Based on their principles and standards, we sought to connect them with the needs of the inhabitants of the community with whom we worked hand in hand and listened to their feelings regarding the perception of housing and the value it has for them.

These strategies helped the inhabitants of the community to become aware of what is involved in the planning of a house as well as the main elements to take into account such as the environment, climate, materials, etc. and being able to get involved in this process allows them to feel part of it. It should also be noted that during this exploration they conceive the use of different materials for the construction of a house as a social status, thinking that huge houses, concrete, tile floors, etc. are the best visually, but they know that they are not the most comfortable; they also contributed that spaces built with materials from the region are more comfortable and livable and present fewer problems when integrating into the community.

For the development of this research it was important to involve the inhabitants at all times, since they are the main and direct users of the houses and the community, this helped us to know their perspective on the strategies proposed at the time of the evaluation.

That the inhabitants employ these strategies based on the principles of Regenerative Design and People Centered Design, can be the starting point to find a central point between the construction of a local housing and a housing brought from another place where together we can adapt that housing to the context that is required in this case Tilaco, using each of the categories defined as place, water, materials, energy, etc. to be able to structure and plan the whole process and with which the inhabitants can feel identified and with their sense of belonging and also help as far as possible to reduce the environmental impact; developing a self-sufficient housing.

References

Acuña Fernández C (2015) Habitabilidad y política de vivienda en México (Ziccardi A, González Reynoso AE, Mazari M, Lomelí Vanegas L (eds)) (Primera edición). Universidad Nacional Autónoma de México, Coordinación de Humanidades, México, D.F

Aguilar GML (2015) Desarrollo local sustentable en la Reserva de la Biosfera Sierra Gorda (RBSG), Querétaro [Licenciatura]. Universidad Autónoma del Estado de México

Ángeles VJA (2010) Prototipo de vivienda rural sustentable y confortable a base de materiales derivados de la madera y bagazo de coco (Maestro). Instituto politécnico Nacional, México, DF

Amador Covarrubias H (2015) Plan Municipal de Desarrollo 2015–2018 [Municipal]. Municipio de Landa de Matamoros, p 60. http://www.landadematamorosqro.gob.mx/images/transparencia/2016/PLAN%20DE%20DESARROLLO%20MUNICIPAL%20LANDA.pdf

Ayuntamiento de Landa de Matamoros (2006) Querétaro—Landa de Matamoros. Enciclopedia de los Municipios y Delegaciones de México. http://siglo.inafed.gob.mx/enciclopedia/EMM22queretaro/municipios/22010a.html

Certicalia (15 de Marzo de 2018) Obtenido de Certicalia. https://www.certicalia.com/blog/que-es-la-arquitectura-bioclimatica

CIBCEC (2005) Ubicación de la localidad TILACO. http://www.microrregiones.gob.mx/cedulas/localidadesDin/ubicacion/ubicacion.asp?micro=SERRANA&clave=220100048&nomloc=TILACO

Comisión Nacional de Vivienda (CONAVI) (2012) Estadísticas históricas de vivienda. Bases de Datos. México: conavi. Disponible en línea: https://www.conavi.gob.mx/quienes-somos/política-vivienda/estadísticasvivienda

Correia Mariana (2017) s.ma.r.t. Caminos hacia la sostenibilidad. España: Acciona

De tierra y varas: Una muestra de arquitectura vernácula en la Sierra Gorda de Querétaro (2009) Poder Ejecutivo del Estado de Querétaro: Secretaría del Estado Urbano y Obras Públicas

Dorantes CJC (2012) La casa vernácula en tiempos de sustentabilidad: Un estudio cruzado entre la Zona Centro y la Sierra Gorda de Querétaro. (Licenciatura). Universidad Autónoma de Querétaro, Querétaro. Recuperado de. http://filosofia.uaq.mx/nugahu/fils/ant0002.pdf

Equipo ARQHYS (2012) Arquitectura vernácula. Recuperado de. https://www.archys.com/contenidos/vernacula-arquitectura.html

FONHAPO (2009) Diagnóstico de las necesidades y rezago en materia de vivienda de la población en pobreza patrimonial. México, D.F. Recuperado de, p 126. http://cedoc.inmujeres.gob.mx/documentos_download/diagnostico_vivienda.pdf

García E (2004) Modificaciones al sistema de clasificación climática de Köppen (5a ed.). Instituto de Geografía UNAM. http://www.igeograf.unam.mx/sigg/utilidades/docs/pdfs/publicaciones/geo_siglo21/serie_lib/modific_al_sis.pdf

Giovannelli (2018) Tilaco. Nuestro Mexico. http://www.nuestromexico.com/Queretaro-Arteaga/Landa-de-Matamoros/Tilaco/

Gobierno de Querétaro (2002) Plan de Desarrollo Urbano del centro de población de Tilaco, municipio de Landa de Matamoros. Gobierno del Estado, p 14

Heavengrown (2018) Diseño regenerativo. Heavengrown. https://n9.cl/mwvta

Hildebrandt Gruppe (5 de Abril de 2016) Obtenido de Hildebrandt Gruppe. http://www.hildebrandt.cl/en-que-consiste-la-arquitectura-bioclimatica/

INEGI (2010) Indicadores de rezago en la vivienda. http://www.microrregiones.gob.mx/catloc/IndRezViv.aspx?refn=220100048

International Living Future Institute (2016) Living building challenge 4.0

Instituto para el Federalismo y el Desarrollo Municipal (2010) ENCICLOPEDIA DE LOS MUNICIPIOS. Institucional. http://siglo.inafed.gob.mx/enciclopedia/EMM22queretaro/

Colle Raymond J (2017) s.ma.r.t. Caminos hacia la sostenibilidad. Acciona, España

Gobierno del Estado de Querétaro. https://whc.unesco.org/uploads/nominations/1079.pdf

MacDonald J (2017) s.ma.r.t. Caminos hacia la sostenibilidad. Acciona, España

Miller D (2001) Home Possessions, 1era Edición. Oxford, New York

Monterrubio Acosta C (2001) EL OBJETO ARQUITECTÓNICO RURAL Y SU EVOLUCIÓN PRODUCTO DE LA CRÍTICA Y LA AUTOCRÍTICA. En P. I. CYTED-HABYTED, VIVIENDA RURAL. Santiago de Cuba, pp 29, 30, 31

Ordoñez Cabezas G (2004) Pames/Giomar Ordóñez Cabezas. CDI: PNUD, México, 31p (Pueblos indígenas del México contemporáneo) Incluye bibliografía ISBN 970-75-027-8 Recuperado en. https://www.gob.mx/cms/uploads/attachment/file/12576/pames.pdf

Pardo F (2013, julio) Disfrutar de una vivienda digna y decorosa [Consulta]. Recuperado de. https://www.arquine.com/disfrutar-de-unavivienda-digna-y-decorosa/

Quintanar CS, Rosas EOJ (2010, junio) La vivienda rural. Su complejidad y estudio desde diversas disciplinas 30:174–196

Rural Eco-Lab (2017) Apendice. Instisuto Tecnologico de Estudios Superiores de Monterrey

Rusin A (2012, marzo) Los Pames. Tokipametri. http://tokipametri.blogspot.com/2012/03/los-pames.html

Sánchez Quintanar Concepción, Jiménez Rosas Eric O (2010, junio) La vivienda rural. Su complejidad y estudio desde diversas disciplinas 30:174–196

Schliep R, Stoll-Kleemann S (Julio de 2010) ELSEVIER. Obtenido de ELSEVIER. https://www.sciencedirect.com/science/article/pii/S0264837709002087

Secretaria de Desarrollo Social (2013) Catálogo Localidades. http://www.microrregiones.gob.mx/catloc/contenido.aspx?refnac=220100048

Toca Fernández Antonio (2017, octubre) El cambio climático y la industria de la construcción [Noticia]. Obras web. http://obrasweb.mx/construccion/2017/10/31/el-cambio-climatico-y-laindustria-de-la-construccion

UNESCO (2017) Obtenido de UNESCO. http://www.unesco.org/new/en/naturalsciences/environment/ecological-sciences/biosphere-reserves/

UNESCO (28 de Mayo de 2018) UNESCO en Montevideo. Obtenido de UNESCO en Montevideo. http://www.unesco.org/new/es/office-in-montevideo/ciencias-naturales/ecological-sciences/

Unidad de Participación Social, Enlace y Comunicación, INE (2000) Programa de Manejo Reserva de la Biósfera Sierra Gorda (Primera). http://www.paot.org.mx/centro/ine-semarnat/anp/AN15.pdf

Villar Rubio M (2001) LO VALORABLE EN LA VIVIENDA RURAL. En P. I. CYTED-HABYTED, VIVIENDA RURAL. Santiago de Cuba, pp 81, 82, 83

Zavala JF (2011) Los Pames en Querétaro. El oficio de historiar. http://eloficiodehistoriar.com.mx/2011/03/04/los-pames-en-queretaro/

Urban

Introduction

Hernández-Fonseca and Rivero-Peña reflect on how cities in vulnerable, overly complex and uncertain contexts establish urbanisation actions to manage their territory and how any space outside an urbanisation logic is assumed as a space ready to be filled with urban objects. Contrary to the urbanising vision that prevailed in many Latin American cities, the authors propose the landscape approach as a way of defining context-related and open processes of intervention in the place. It considers the social, urban and ecological pre-existences of the place/landscape and the multiple interrelationships of the elements of the territory as a dynamic totality and invites us to rethink the transformation of neighbourhoods as a basis for processes of regeneration and reoccupation of the city in a context of emergency. Urbanising actions, understood as conventional urban development, have not, by themselves, allowed for the consolidation of old city neighbourhoods. By understanding urban decay and unproductivity as negative states, and territory as vacant space, razing has been carried out as a renewal strategy. To face vulnerable, overly complex and uncertain contexts, it is necessary to seek the consolidation of the place/landscape in the medium and long terms. In this way, rather than short-term urban development based on procedures, plans, projects and urban architectural objects, it is important to promote regenerative processes as a basis for intervention. Considering the neighbourhood as a place/landscape, its regeneration must start from the understanding, respect and integration of the pre-existing and the existing as part of the process of occupation and re-occupation of the territory. Landscape guidelines look for the regeneration of the city from a contextual-ecological perspective. Regeneration based on the landscape approach is a proposal open to constant transformation, considering the continuities established in everyday life and aspects valued by the inhabitants. Not everything is defined, and not everything is clear, therefore, it should not be based on a pre-established linear process, but on a long-term process, open and adjustable to the diverse conditions that the place faces over time. To face of vulnerable, overly complex and uncertain contexts, it is necessary to seek the consolidation of the place/landscape in the medium and long term. In this way, rather than short-term urban development based on procedures, plans, projects and urban-architectural objects, it is important to promote regenerative processes as a basis for intervention. Considering the neighbourhood as a place/landscape, its regeneration must start from the understanding, respect and integration of the pre-existing and the existing as part of the process of occupation and re-occupation of the territory. Landscape guidelines look for the regeneration of the city from a contextual-ecological perspective. Regeneration based on the landscape approach is a proposal open to constant transformation, taking into account the continuities established in everyday life and aspects valued by the inhabitants. Not everything is defined, and not everything is clear, therefore, it should not be based on a pre-established linear process, but on a long-term process, open and adjustable to the diverse conditions that the place faces over time.

Haselsteiner introduces ecovillage initiatives from civil society that call for radical and immediate changes, presenting ten case studies from the European and Mediterranean contexts. After the first initiatives of the 1960s and 1970s, which can be classified more as 'back to the countryside' movements, the idea of exemplifying ecological communal ways of life is gaining new momentum due to increasingly evident climate impacts. Ecovillages function as a way of empowering people to find solutions to pressing environmental and social problems away from technocentric positions. From their diversity, the case studies represent a good overview of the objectives, values and practice of regenerative sustainability in which they live, pursuing greater coherence and commitment. The question is left open as to whether and how a transferable model could be developed for society as a whole. Not all people are promotors of regeneration but they cannot be excluded from urban life. Apart from organised ecovillage movements, there are also initiatives by individual architects for ecologically sustainable architecture, which take particular account of local conditions, reflect the needs of the regional population and apply cost- and climate-friendly forms of construction. Although ecovillage initiatives strive for the greatest possible diversity and openness in their social composition, the group of participants is still reduced to a small number of like-minded people from social, artistic or manual professions who bring their knowledge and skills directly to the design of the ecovillage initiative. Even with the restriction that this form of living together will only be practicable and feasible for a small group of people and will only remain niche solutions, they are an essential catalyst and an important stimulus for transformation towards regenerative sustainability. ecovillage movements, there are also initiatives by individual architects for ecologically sustainable architecture, which take particular account of local conditions, reflect the needs of the regional population and apply cost- and climate-friendly forms of construction. Although ecovillage initiatives strive for the greatest possible diversity and openness in their social composition, the group of participants is still reduced to a small number of like-minded people from social, artistic or manual professions who bring their knowledge and skills directly to the design of the ecovillage initiative. Even with the restriction that this form of living together will only be practicable for a small group of people and will only remain niche solutions, they are an essential catalyst and an important stimulus for transformation towards regenerative sustainability.

Gastalver and Weikert introduce the interesting new concept of 'Gastrosophy' to build a new ethical corpus that introduces the perspective of food in urban planning, with two objectives: on the one hand, to raise awareness among city dwellers about how we eat and how and why we should change our eating habits to take better care of our health, our communities and the planet; on the other hand, to involve public actors, administrations and institutions in the development of the city in transition towards a more sustainable, resilient, healthy, biodiverse and supportive system. It is needed an urban-territorial transition to a model in which the metropolis is ruralised, a way of approaching the production of the territory and the city that is framed within what is now understood as regenerative urbanism, where food sovereignty and security is key, through urban and peri-urban agroecology. This transition is based on the planning of metropolitan areas centred on bioregions and medium-sized cities to organise the countryside and the food supply of cities, in the search to transform the urban landscape, deconstructing the twentieth-century city and looking to ruralise peri-urban areas and associate them with cities in order to dissolve the rural–urban dichotomy. The procedure to address this already complex system, the relationship between city and food, has been to venture a concept of Fed Cities in order to formulate the right questions to provide solutions to cities and their relationship with the countryside and food production. It is possible to have the vision of a regenerative present and future of a living food system, specifically in urban, utopian or eutopian environments, in which our food can be supportive, safe, healthy, sustainable, regenerative and tasty.

Mousavi and Godoy propose an approach that focuses on creating a new relationship between natural energy sources and appropriate building materials, repairing the atmosphere of emissions, as well as increasing the well-being and comfort of occupants. The authors contextualise their work in Latin America (LA), where there are many points in common, such as the commitment and advances that have been made in the generation of sustainable building regulations to generate a basis for moving towards regenerative buildings, which could allow a regional vision towards a regenerative future. A path towards regenerative cities is possible using the Positive Energy District model, not only to benefit the environment but also the economy and improve environmental health by reducing the rate of energy consumption of buildings, adopting renewable energy sources, adopting policies that contain guidelines for improving the performance of building systems and applying energy labels. It is possible a transition to carbon-neutral urban areas with the formation of long-term laws in regional planning, from a holistic target approach to building efficiency, active systems and design. In LA are regulatory, economic and technological barriers. Detailed monitoring of building conditions and energy consumption pattern with intensive regulatory and financial support for the implementation of these models are necessary. There is an important role of certifications and their adoption and integration into local regulations. In terms of financing and management, it is important to promote innovative financing mechanisms and to involve strategic actors such as local governments, developers and financial institutions in the planning and implementation of sustainable building policies and practices. It is essential to address sector-specific constraints to the successful implementation of sustainable and regenerative building practices, including addressing the lack of skilled labour and promoting the adoption of innovative technologies. A comprehensive and collaborative approach is needed to address the challenges and opportunities of sustainable and regenerative construction practices in LA.

Ugalde-Monzalvo proposes nature-based solutions, inspired, supported or integrated by nature, being adaptive, efficient, systemic and incremental, that can increase the resilience and sustainability of a city. A model for assessing urban bioclimatic suitability as a solution based on nature's capacity for the building and the built environment itself to collect, accumulate, distribute or dissipate energy that can be applied at different scales and in new projects as well as in the existing built environment is proposed, through the concept of 'urban bioclimatic suitability' (UBA) of a city. It describes its current state, limitations and potential and serves as a measure of the city's ability to meet climatic needs. The UBA integrates different approaches for the systematisation of bioclimatic strategies: (1) those that link the city to the climate and propose generalisations of bioclimatic strategies; (2) those that link human needs to the climate through acclimatisation and comfort indices; and (3) those that link the effects of climate on cities to human needs in terms of constraints, objectives and potentials. The UBA is constituted as the measure of the capacity of a system (city, neighbourhood). It depends on the characteristics of the environment and the needs of the inhabitants and cannot increase indefinitely. Conversely, the inability of the system to meet the needs of the inhabitants can grow indefinitely if the favourable characteristics of the system are degraded or lost and demands increase. If the system is not renewed, the condition of 'unfitness' will grow.

The UBA of a city allows the evaluation of energy consumption; the development of strategies at different scales that promote and facilitate the sustainable development of a locality and the application of nature-based solutions considering the built city as a new nature; the identification of optimal strategies at the right time and place; the implementation of zoned bioclimatic systems at urban level; the generation and application of suitability indices for different cities and the definition of a suitability index by climate types; the development of a sustainability indicator based on urban renewal indices and the monitoring of infrastructures, operation, evaluation and sustainability; the generation and application of suitability indices for different cities and the definition of a suitability index by climate types; the development of a sustainability indicator based on urban renewal indices and the monitoring of infrastructures, operation, evaluation and improvement of applied bioclimatic systems and/or ecotechnologies; as well as the implementation of public policies for urban renewal of the city based on bioclimatic strategies.

Barrio's Regeneration and Reoccupation Process, a Landscape Approach to a Border City

Diana Cecilia Hernández Fonseca and Héctor Rivero Peña

Abstract Nowadays, cities in vulnerable contexts, too complex and uncertain, establish urbanization actions to achieve order in their territory. Rapidly, *urban development* is understood exclusively as *urbanization* and is imposed as a procedure to organize, control, and build a profitable territory. In this way, any space outside an *urbanization* logic is assumed as an *empty* space ready to be *filled* with urban objects. Contrary to the urbanizing vision that prevailed in many cities in Latin America, this chapter proposes the *Landscape approach* as a way to define context-related and open-ended place intervention processes. The Landscape approach considers the social, urban, and ecological pre-existences of the *place/landscape* and the multiple interrelationships of the territory elements as a dynamic *wholeness*. This proposal invites us to rethink the *Barrios* transformation as the basis for city regeneration and reoccupation processes in an emergency context, such as the border city of Ciudad Juárez.

Keywords Regeneration process · Landscape approach · City reoccupation · Barrio · Ciudad Juárez

1 Introduction

In the second half of the 20th century, there is an urgency to urbanize Mexican cities. *The Secretaría de Bienes Nacionales e Inspección Administrativa* (The Secretariat of National Assets and Administrative Control) began in the 1950s with the elaboration of Regulatory Plans for some cities in México under hygienist, functional and aesthetic planning canons. These plans are only "graphic summaries of technical and

---

D. C. Hernández Fonseca (✉) · H. Rivero Peña
Architecture Department, Universidad Autónoma de Ciudad Juárez (UACJ), Ciudad Juárez, México
e-mail: cecilia.hernandez@uacj.mx

H. Rivero Peña
e-mail: hrivero@uacj.mx

© The Author(s), under exclusive license to Springer Nature Switzerland AG 2025
C. Cobreros et al. (eds.), *Regenerative Design*, Cities and Nature,
https://doi.org/10.1007/978-3-031-76890-3_11

social studies" (Sánchez Ruiz 2008, p. 8), which propose modern urban models and incorporate modern urban instruments.

These actions establish an urbanization vision for the growing industrial cities in México. In Ciudad Juarez, an originally agricultural city linked to the Rio Bravo Valley (Rio Grande), the first Regulatory Plan was elaborated by the Architect Rafael Mijares in the early 1950s. It was done in coordination with the *Junta Federal de Mejoras Materiales de Ciudad Juárez* (Secretariat of National Assets and Administrative Control and the Federal Board of Material Improvements of Ciudad Juarez).

By 1984, municipalities in Mexico had the authority to "prepare urban development programs and to apply land use control instruments" (Ugalde 2021, p. 114). This is how it is intended to incorporate a more specific and specialized planning for the development of Mexican cities, to attend to more complex situations, which not only consider their physical aspects; but, theoretically, their economic and social development.

In Ciudad Juarez, when the urban development plans became official, the main goal was to establish land use control for urban growth. An important change to promote urban growth is the *agrarian reform*, carried out in 1992 by President Carlos Salinas de Gortari. The reform modifies Article 27 of the Mexican Constitution, establishing:

> (..) legal modalities to disincorporate ejido land, corresponding to agricultural areas in Mexico, (…) for its subsequent incorporation into the private legal regime, likewise, a series of programs have been produced to promote the ejido land disincorporation (…) and provide it as urban land (territorial reserves), to population centers. (Flores 2008, p. 55).

In the case of Ciudad Juarez enables the so-called *territorial reserves*, later called *growth reserves*.

From that moment, a constant intervention of the territory is carried out with urbanization procedures that do not necessarily consider the territory in which they are imposed. These procedures are generic urban models that cause disarticulation in their implementation, complicating the adaptation to the physical-territorial context and disrupting the social, economic, and cultural dynamics of the place.

Ciudad Juarez, Chihuahua (Mexico) together with El Paso, Texas (USA) conforms a "Transfrontier metropolitan region" (Herzog 1997, p. 2), defined by Herzog as,

> (…) two or more settlement centers located around an international boundary (. …) no longer simply an artifact upon which designers from different corners of the globe practice their profession, (here) the city has become a container increasingly shaped and transformed by such international actors as corporate investors, transnational financial interests, transborder common markets, and cross-national governmental organizations. (1997, p. 2)

Since the *Programa de Industrialización Fronteriza* (Border Industrialization Program) in 1965, the maquiladora industry, assembly companies, were established in the Ciudad Juarez-El Paso Trans-frontier metropolitan region. These industries grew rapidly in the region since the *North American Free Trade Agreement* (NAFTA) in 1994. In this agreement, Ciudad Juarez provides cheap labor, and El Paso provides the services and logistics for this type of industry, generating enormous contrast in the benefits that each city receives.

Maquiladora industry proposes an intensive exploitation of the territory. This has caused an explosive, extensive, and accelerated urban growth of Ciudad Juarez. This intensive exploitation consumes the *growth reserves (territorial reserves)* with vast urbanized areas, dispersed throughout the territory, for the construction of low-income housing neighborhoods for *maquila* workers. These urbanized areas are developed under *Instituto del Fondo Nacional de la Vivienda para los Trabajadores* (National Housing Fund Institute for Workers) credit procedures: financed with public money but developed by private initiative.

Although there is an urban development plan and urbanization provides all the basic urban services, there is no articulation in the occupation of this sector. Housing neighborhoods are surrounded by large extensions of unoccupied land, without urban equipment, and segregated from central areas of the city.

This defines an urban development based on the overexploitation of the territory and its human resources (based on cheap labor). The *Maquila* industry, grouped in industrial clusters, occupies large territorial extensions, consumes excessive water, and produces 85% of environmental pollution, including contamination caused by toxic waste (among which are hydrofluoric acid, paint, varnishes, solvents, epoxides, residual chrome and solvents, which they throw directly down the drain or into garbage dumps close to their location (Centro de Estudios Regionales y Comunicación Alternativa 2009). Despite the lack of regulations or sanctions, the *maquila* industry continues to be the sector that generates the largest number of jobs, corresponding to 62%, offering low and insufficient salaries to generate a sustainable economy.

Ciudad Juarez depends on global investment and economic trends (international treaties, company relocation, and efficient merchandise transportation systems). Its geographical situation has also made it an important location for drugs-arms trafficking and migrant smuggling. The complexity of this situation is reflected in its environmental problems, its high crime rates (considered for several years among the most dangerous cities in the world), its urban backwardness or its gender violence problems. In contrast, the city of El Paso for several years considered the second safest city in the United States.

The "Transfrontier metropolitan region" (Herzog 1997, p. 2), confirms the great interaction between cities, but also the great differences that can occur in the same territory.

> Citizens on both sides of the boundary are increasingly drawn together into a web of north-south relations, in which the dichotomies of "Third World/First World" and "developing/developed" are cast aside as urban neighbors share common transnational living and working spaces. (Herzog 1997, p. 2)

In Ciudad Juarez, it has not been possible to address the social problems caused by the accelerated, complex, and uncertain global processes. In most cases, determine actions at the local level, evidencing the lack of control and the limited local capacity to manage the growth and development of the city (Bauman, July 19, 2001).

2 Empty/Vacant Space

Under an ambiguity/complexity scene, cities have established *urban development* as a goal for its improvement, and have proposed urban plans, urban regulations, or urban projects as a strategy to achieve certainty.

However, in cities with vulnerable contexts, it is common that *urban development* is understood exclusively as *urbanization* and is imposed as a procedure to organize, control, and build a profitable territory.

To make this possible, the non-urbanized or the precariously urbanized areas (areas in abandonment, decay, or in the process of urban degradation) are considered empty or vacant spaces and, in this way, justify their incorporation or reincorporation under *urbanizing actions*. The consideration of empty/vacant also allows to classify a space as dispensable or without urbanization logic (for example, agricultural or natural land), pointing out its availability for urbanization.

The *empty space* (*space to be filled*) and the *vacant space* (*space to be erased*) are based on modern urban paradigms proposed since the late 1920s, with the implementation of zoning and the definition of *land use*. In this way, the territory is annulled as a support of the human being, by considering *empty* what does not have a defined *use*. From this moment on, the territory is considered a *resource* for urbanization, so it will be vital for the city to have sufficient land for urban growth, relegating its organic character and its interaction with the inhabitants. From this perspective, nature within an urban environment is reduced to a *park* or *green space*, which is considered a well-being space (Ruano 1999) with *recreational use*. This is a situation of domination, where the human being is placed above nature and the territorial ecology.

The objective is to adjust the territory to an urbanizing procedure that simplifies its occupation, nullifying, degrading, and making invisible other non-urbanistic processes that are or have been carried out in the territory (environmental, ecological, social, cultural, rural). This has caused the deterioration of the original support territory, which is ignored and destroyed by *urbanization*.

On the other hand, in the second half of the twentieth century, urbanistic studies emphasized urban morphology, based on the analysis of the *city shape*. The urban analysis methodology proposed by Kevin Lynch (1918–1984) in his book *The image of the city* (1960–2008) was a real milestone in morphological studies and opened a very prolific line of research, although it is not always the most appropriate way to understand the complexity of the city.

From Lynch methodological proposal, the city is reduced to its *image* and its *legibility*, emphasizing its ubiquitous character (Menéndez de Luarca and Soria y Puig 1994), separating it from the historical context and the local transformation processes. From this point of view, the urban image is considered a reflection of the conditions of the city and, therefore, to address the image is equivalent to address the city problems.

Under this order of ideas, it is important to achieve a *visual city*, attractive and easy to read, removing what affects the legibility and perception of the so-called *urban landscape*, which "is promoted here as a kind of modern palliative for urbanization" (Marot 1999, p. 47).

In the construction of an *urban image* or in the implementation of an *urbanizing procedure*, the transformations are conceived as autonomous, because of intrinsic circumstances, disjointed from their context.

The *empty* and *vacant* space has prevailed as a validation of urban actions in Ciudad Juarez. Both are based on the premise that everything that is not part of an urbanization procedure complicates or prevents the development of the city.

In a circumstance of empty/vacant space, the *object of urbanization* operates as a colonization way to *complete* or *restore* the city under a controlled procedure, as opposed to the non-urbanized places, the invasive processes, and the urban deterioration.

Ciudad Juarez is an example of sprawling urban growth. Since the end of the 20th century, the urban space has tripled its size, occupying the agricultural valley by establishing a new *territory reserve* to be *filled* with urbanization. In a semi-desert context, like Ciudad Juarez, the original territory is seen as an empty space, a blank canvas ready to be filled, devastating the existing nature and its ecosystems.

At the same time, the reoccupation of abandoned or deteriorated central urban space has been considered. In the so-called Historic Centre of Ciudad Juarez, several complete blocks were demolished to build urban projects and propose a new *urban image*. The old parts of the city have been understood as vacant spaces, ready to be erased and *correctly* reintegrated into an urbanization process, literally destroying what exists in preparation for an urban renewal.

From the urbanizing logic that has predominated, the old, the unproductive, or the decadent are not accepted. It is necessary to suppress in the name of urban development any sign of abandonment, deterioration, decay, or intense use of the city (Fig. 1).

From an urbanizing point of view, these "other spaces", heterotopia spaces for Foucault (2008), or marginalized *barrios*, due to their physical and institutional abandonment, are considered an anomaly, unproductive and out-of-control places, a *non-urban* spaces. However, Solà-Morales draws attention to what he defines as *terrain vague*:

> The photographic images of terrain vague are territorial indications of strangeness itself, and the aesthetic and ethical problems that they pose embrace the problematics of contemporary social life. What is to be done with these enormous voids, with their imprecise limits and vague definition? Art's reaction, as before with "nature" (which is also the presence of the other for the urban citizen), is to preserve these alternative, strange spaces, strangers to the productive efficiency of the city. If in ecology we find the struggle to preserve the unpolluted spaces of a nature mythicized as the unattainable mother, contemporary art seems to fight for the preservation of these other spaces in the interior of the city. Filmmakers, sculptors of instantaneous performances, and photographers seek refuge in the margins of the city precisely when the city offers them an abusive identity, a crushing homogeneity, a freedom under control. The enthusiasm for these vacant spaces-expectant, imprecise, fluctuating - transposed to the urban key, reflects our strangeness in front of the world, in front of our city, before ourselves. (1995, p. 123)

Fig. 1 Space disposed to be *filled,* Colonia *Los Ojitos*, Ciudad Juárez, Chihuahua. *Source* H. Rivero

The goal is to reach an optimal stage over time, without accepting delay, slowness, decadence, or unproductivity as natural states of the city´s transformation process. However, the normal transformation of the city is a constant regeneration in slow and uncoordinated processes of splendor and degradation.

3 Towards Regeneration

Once a territory is occupied, an organizational relationship is established in time, in which the territory is the object of construction (Corboz 1983–2005) and becomes an artifice (Menendez de Luarca and Osorio 2000). Since the presence of human beings on earth, the occupation of the territory has been determined from the relationships established between the inhabitants and their supporting territory and how they transform it, a historical process that shows how cultures are related to nature and their territory (Menendez de Luarca and Soria y Puig 1994; Menéndez de Luarca and Osorio 2000).

Urbanism and architecture do not clearly see the *traumatic* act caused by the implantation of an architectural object (Solà-Morales 2009) and do not understand the city as a permanent process of transformation-permanence, in continuous relationship with the existing context, with the historical time and with its inhabitants. Not considering this is annulling the normal and organic transformation process of the city, eliminating its natural condition and degenerating it.

Fig. 2 Space that can be *correctly* reintegrated, *El Centro*, Ciudad Juárez, Chihuahua (2006). *Source* H. Rivero

It is important to understand that empty/vacant spaces do not exist… what exists is a territory with existences and pre-existences, never neutral (Menéndez de Luarca and Soria y Puig 1994), which makes us see the urgent need to reincorporate the *territory* into the understanding of urban and ecological complexity. In this way, it is essential to understand how the territory is occupied.

The valorization and incorporation of the territory as a key part of the context and as a support for human life, not only as a resource, becomes the guarantee of the present and future city regeneration as a living place. Intolerance to physical deterioration or to ambiguous spaces implies a more serious action: intolerance to social space (Fig. 2).

In a vulnerable context, it is necessary not to think in terms of the development/urbanization of the territory, but in its regeneration. The first is based on immediate actions (infrastructure, projects, objects), and the second is on complex processes in the medium and long term.

To regenerate, therefore, is to reoccupy (rather than renew or clear out) and consolidate its condition as a place. As Mang and Reed defined, "regenerative development and design means the reconnection of human aspirations and activities with the evolution of natural systems, essentially coevolution" (2012, p. 5). "The regenerative paradigm returns to its central position in human life, making it an integral part of the development and design process" (Mang and Reed 2012, p. 8).

From the regeneration paradigm, the city is a context of change, of continuous social construction and persistence, but also a space of deterioration; an urban palimpsest in a constant process of transformation (Corboz 1983–2005). Time (history, process, memory) plays a fundamental role in the configuration and consolidation of the city. In a common urbanistic procedure, where everything must be immediately done and everything must be fully defined and resolved, there is little room for the *indeterminate*, the *adaptable*, and the *in process*.

To incorporate a regeneration and reoccupation process of the city, the first thing that must be considered is to break with the closed paradigms that have prevailed in Latin America's urbanization processes. A change will be required in the way of intervening in the vulnerable city and in the way of rethinking its problems.

4 The Barrio

The recovery of the city's central areas, the so-called *old barrios*, places socially built from everyday life, with their internal logic, offers the opportunity to understand the complexity of the *city's regeneration and transformation process*.

More than a vacant space, the *Barrio* must be recognized as a *place*, defined by Auge (1992–2008) as space for memory, urban experience, and sociability. In these spaces it is possible to find time, the historical connection of urban objects, social articulation, meaning, and regeneration. The *Barrio* as a place, incorporates the changing dimension of the city; and integrates connotations of all kinds: physical, social, environmental, aesthetic, etc. It reflects the complexity of the present-day city.

Norberg Schulz (1979) understands the place as a source of a respective character (*Genius loci*), expressing the need for permanence and *conservation* of the place stability (*Stabilitas Loci*). It is about discovering what already exists. Under this vision, architecture is "committed to the previously existing data of genius loci, history, myths, symbolism and the meaning of a site" (Solà-Morales 2009, p. 41), in opposition to the ubiquity of the global village. However, Solà-Morales comments that the place can no longer be understood as *fons* (*Genius loci*), a static vision that proposes *conservation* as a form of stabilization (*Stabilitas Loci*, according to Norberg-Schulz 1979).

The place not only is transformed, but it is also possible to act on it: It can be intervened in its transformation-permanence. Facilitating the permanence of the place (as a physical and meaningful space for everyday life and memory) implies enabling its constant transformation into contextualized processes. In the *place*, the architecture (the object) is just one more component of the construction of the city.

Therefore, the place brings together habitable experience and physical construction; time and space; events and objects; and "local circumstances and situations" (Corner 1999, p. 4). The place is configured from everyday life and memory. It is a repository of lived, transcendent human moments. A space for significantly shared social activities.

For Sébastien Marot (1999), the change of perspective on the *place* has been initiated by the landscape architects. They began to look at the borders, the indeterminate spaces, and the specificities of the landscape with the aim of reincorporating memory and the sense of place in degraded-devastated sites. The place expresses the intensity of each landscape. James Corner explains about the landscape:

> In this sense, landscape is an ongoing medium of exchange, a medium that is embedded and evolved within the imaginative and material practices of different societies at different times. Over time, landscapes accrue layers with every new representation, and these inevitably thicken and enrich the range of interpretations and possibilities. (1999, p. 5)

From a landscape approach, it can be understood that the *permanence-consolidation* of the place is guaranteed by the possibility of transformation-regeneration processes. It is related to the pre-existences and existences of the context, in an articulated system. The place, as part of the landscape, is an active and open process. It requires for its regeneration an approach in which the narratives of everyday life and memory are integrated. The antecedent and present condition of its territory is strongly articulated to an ecological balance.

The landscape, as a cultural fact, reveals the criteria through which the territory has been occupied. It allows us to know the continuous exchange between those who modify it and the social dynamics of those who inhabit it, as well as the previous practices of territorial construction.

The city can be understood as a *place/landscape*, which needs to be understood to intervene according to its problems. The place-landscape recognizes that the territory is manifested in its landscape, according to its singularities, its dynamics, and the culture and society that occupies it. This means that we are responsible for assuming our intervention in the territory transforming the landscape (Nogué 2008).

The landscape approach is an opportunity to promote regeneration strategies that allow the ecological reoccupation of the *place* by vulnerable communities. "Perhaps, only thing we need to learn is how to look at them from another different point of view." (Marot 1999, p. 56).

From a landscape approach, the *Barrio* offers an opportunity to understand the importance of the relationship between the support territory and its possible intervention from a contextual-ecological vision. This relationship fortifies the synergy between the inhabitant and his place.

5 The Landscape Approach

The *landscape urbanism*, initially proposed by Charles Waldheim (2004) based on his work in Detroit, emerges as a critique to the poor adaptability of urban development to the rapid transformation of cities, demanding an urbanism that integrates the cultural and the environmental as essential aspects. Despite being a recent proposal in urban and architectural studies, its importance lies in the fact that it recognizes that anthropic intervention can no longer be exclusively for the benefit of the human being. Intervention must be also for the benefit of the territory the human being occupies.

Waldheim (2004) proposes *landscape urbanism* as an alternative to intervening abandoned central areas and the expanded periphery (the suburban area) of cities. This responds to his concern about the immense extensions of unoccupied urban territory that result from decreasing densities and the abandonment of the city, a product of the post-industrial process that the city of Detroit and many other North American cities, "areas are not being "returned to nature" but are curious landscapes of indeterminate status" (Waldheim 2004, p. 92).

A relevant aspect to highlight is that, even though Ciudad Juarez is not in a post-industrial process, the *empty* and *abandoned* territory are the most visible symptoms of this supposed *development* based on a productivism paradigm, which generates the same degradation condition that inhibits the regeneration process of the city.

The *landscape approach* based on landscape urbanism, allows, from a large and articulated vision, to define the intervention process, incorporating the social, urban, and ecological pre-existences of the place and its complexity, while inviting us to rethink the concept of *Barrio* as a place. The landscape approach considers *transformation* as the basis for city regeneration and reoccupation processes in an emergency context, such as the border city of Ciudad Juárez.

It is necessary to understand the landscape approach from the occupation of the territory, which lets us know how intervention has historically been done and who has mostly defined its construction. Also, it lets us recognize that the territory, as a living organism, is integrated by a series of biophysical processes related to each other (McHArg 1969–2000), such as water evaporation, wind, soil erosion, waves…which, in principle, occur with or without the human being. They are the result of the natural transformation of physical systems (Brunhes 1910–1920; Corboz 1983–2005), that suffer irreversible changes in their structure because of human intervention.

The landscape approach is a shift of paradigm that takes to the regeneration and reoccupation of the *Barrio* from the understanding of its existing problems and the attention to the city processes, not necessarily with urban projects. Therefore, it is important to recognize, recover, reincorporate, and incorporate complex city processes, which can be conducted simultaneously, adapting them to pre-existences, the existences, and what will exist, in a long-term vision that allows for adjustment.

A similar approximation is the one made by Alexander et al. (1987). Convinced that urban planning does not have a comprehensive vision (*wholeness*), they propose *A New Theory of Urban Design* (Alexander et al. 1987) based on the *dynamic wholeness* of the city:

> Thus, in our view, it is the *process* above all which is responsible for wholeness…not merely the form. If we create a suitable *process* there is some hope that the city might become whole ocean again. If we do not change the process, there is no hope at all. (Alexander et al. 1987, p. 3)

The richness of their proposal consists of considering urban planning as a *process*, which means that it can cover extended periods of time and not a continued path of a resolutive planning project, close to change. Moreover, they assume that the theory itself is open to change, as part of the process, under the understanding that their implementation is in accordance with each context. This reveals the need to integrate multifunctional and versatile visions that promote urban transformation over time.

Therefore, and in accordance with the landscape approach, the problems must be analyzed and understood in an integral way, without isolating the contextual characteristics of the place. The approach must be based on the existent and possible dynamic processes, knowing that when any type of intervention is performed, the condition of the place will be modified in some way.

The implementation of the landscape approach contemplates certain *guidelines*, which are not exclusive, are open-ended and must be adapted to the specific context. In parallel, they have to integrate the *territorial elements* (Hydrography, Edaphology, Geology, Meteorology, Flora, and Fauna) to define intervention processes that promote the possible regeneration and reoccupation of the place.

The *territorial-contextual* elements must be considered as a "Terrestrial whole" (Brunhes 1910–1920. p. 13). Geographical facts are intricately linked and must be studied in their multiple relationships and connections, which are part of the complex characterization of any territory.

Therefore, the relation of a social group organization with its territory allows us to understand the conditions that have supported its historical development or its disappearance.

- *The temporality*, which includes the historical background of the Barrio (as a place-landscape), the natural transformation cycles, and their future possibilities.
- *The valorization and/or revaluation of the landscape.* Based on the understanding that the values can change or that some valorization aspects that had been hidden can be visible again, including the inhabitants' values, the ecosystem values, the scenic values, the cultural values, and the everyday values.
- *The experience of the inhabitant.* This guideline allows a recognition of the Barrio *memory*, considering its relationship with the place-landscape appropriation and, therefore, the interpretation that the inhabitant makes of it.
- *The multiscale analysis of the place-landscape.* Defines intervention processes based on various approaches, from the territorial scale to the architectural scale, to have an integral and complex vision of the regeneration and reoccupation processes.
- *The interaction with the territorial elements.* Necessary for the contextualization of the intervention processes, which are understood from their multiple relationships and ecological impacts, including the anthropic intervention that modifies the dynamics of the territory.

6 Conclusion

The urbanizing actions, understood as urban development, by themselves have not allowed the consolidation of the old Barrios of Ciudad Juárez. By understanding urban deterioration and unproductiveness as negative states, and the territory as a vacant space, razing has been carried out as a strategy for renewal (leaving it empty, as a convenient state for urban action). The nullification of the context, based on establishing a constant *restart* of the city (*tabula rasa*) with the implementation of always different urban plans, compromises the future capacity of the territory as a support for human life.

When facing vulnerable, too complex, and uncertain contexts, it is necessary to look for the consolidation of the place/landscape in the medium and long term. In this way, more than a short-term urban development, based on procedures, plans, projects, and urban-architectural objects, it is important to promote regenerative processes as the basis for the intervention. Considering the Barrio as a place/landscape, its regeneration must start from understanding, respecting, and integrating the pre-existing and the existing as part of the process of occupation and re-occupation of the territory. This process is addressed by considering certain *landscape guidelines* that search for the regeneration of the city from a contextual–ecological perspective.

To regenerate the city, it is necessary to analyze and understand the problems in an integral way, without isolating the contextual characteristics of the place/landscape under study. This results in the need to integrate a multifunctional and versatile vision that promotes the transformation of the Barrio over time.

The regeneration based on the landscape approach is a proposal open to constant transformation, paying attention to the continuities established in daily life and the aspects valued by the *Barrio* inhabitants (beyond an institutional validation).

This approach must be based on the present and future dynamic processes, knowing that by carrying out any type of intervention in the place, its landscape condition will be modified in some way. Therefore, the process in relation to wholeness should be included. Not everything is defined, and not everything is clear, for this reason, it should not be based on a pre-established linear process, but rather on a long-term process, open and adjustable to the various conditions that the place faces over time.

Facilitating the continuity of the place (as a physical space, a space for exchanges and meaning for everyday life, appropriation, and memory), implies enabling its constant transformation, in a commitment to what exists as a prelude to the possible city.

References

Alexander C, Neis H, Anninou A, King I (1987) A new theory of urban design. Oxford University Press. https://archive.org/details/newtheoryofurban00alex

Auge M (2008) Los no lugares. Espacios del Anonimato. Una antropología de la sobremodernidad (M. Mizraji Trad.). Gedisa editores (Original work published in 1992)

Bauman Z (19 de julio de 2001) El desafío ético de la globalización. El País. https://elpais.com/diario/2001/07/20/opinion/995580007_850215.html

Brunhes J (1920) Human Geography. an attempt at a positive classification principles and examples (I. I. LeCompte Trad.). London, Harrap (Original work published in 1910)

Centro de Estudios Regionales y Comunicación Alternativa (CERCA) (2009) Impactos de la maquila en el medio ambiente de Ciudad Juárez. Nóesis. Rev de Ciencias Sociales y Humanidades 18(36):52–77. http://www.redalyc.org/articulo.oa?id=85919840004

Corboz A (2005) El territorio como palimpsesto (L. Manterola y ARTELEKU Trad.) Ediciones UPC (Original work published in 1983)

Corner J (ed) (1999) Recovering landscape. Essay in contemporary landscape architecture. Princeton Architectural Press

Flores RCE (2008) Suelo Ejidal en México. Un acercamiento al origen y destino del suelo ejidal en México, De lo comunal agrario a lo privado urbano [Tesis de doctorado, Escuela Técnica Superior de Arquitectura de Madrid]. Cuadernos de Investigación Urbanística

Foucault M (2008) Topologías. Fractal 48:39–62. https://www.mxfractal.org/RevistaFractal48MichelFoucault.html

Herzog AL (1997) The transfontier metropolis. Harv Des Mag 1:1–5. https://www.harvarddesignmagazine.org/issues/1/the-transfrontier-metropolis

Lynch K (2008) La imagen de la ciudad (E. L. Revol Trad.) Editorial Gustavo Gili, SL (Original work published in 1960)

Mang P, Reed B (2012) Designing from place: a regenerative framework and methodology. Build Res & Inf. Routledge 40(1):23–38. https://doi.org/10.1080/09613218.2012.621341

Marot S (1999) The reclaiming of site. In: Corner J (ed) Recovering landscape. Essays in contemporary landscape architecture. Princeton Architectural Press

McHarg IL (2000) Proyectar con la naturaleza (Trad. P. Fernández, M. San Miguel, A. Centeno y R. Fernández) Gustavo Gili, SA de CV. (Original work published in 1969)

Menéndez de Luarca, Soria y Puig, A (1994) El territorio como artificio cultural: Corografía del Norte de la Península Ibérica. Ciudad y Territorio Estudios Territoriales 99:63–94. https://recyt.fecyt.es/index.php/CyTET/article/view/83925

Menéndez de Luarca JR, Osorio N (2000) La construcción del territorio: Mapa histórico del Noreste de la Península Ibérica. Lunwerg Editores

Nogué J (ed) (2008) El paisaje en la cultura contemporánea. Biblioteca Nueva, S. L

Norberg-Schulz Ch (1979) Genius loci: paesaggio, ambiente, architettura. Gruppo Editoriale Electa

Ruano M (1999) Ecourbanismo. Entornos humanos sostenibles: 60 proyectos. Gustavo Gili

Sánchez Ruiz GG (2008) Preludio en J. V. Arias (ed) Carlos Contreras, Planos Reguladores, 1º ed. Centro de Investigación y Estudios de Posgrado, Facultad de Arquitectura, UNAM, pp 7–22

Solà Morales I (1995) Terrain vague, in anyplace. MIT Press

Solà Morales I (2009) Los artículos de any. Fundación Caja de Arquitectos

Ugalde V (2021) La coordinación institucional del ordenamiento territorial en México. Rev De Geografía Norte Grande 47:105–120. En Línea. https://doi.org/10.4067/S071834022010000300006

Waldheim WC (2004) Detroit: motor city. In: El-Khoury R, Robbins E (eds) Shaping the city: studies in history, theory and urban design. Routledge, Nueva York, pp 77–97

Ecovillages and Transition Towns—What Can We Learn from Them for a Regenerative Future?

Edeltraud Haselsteiner

Abstract The rapidly advancing climate change and its negative consequences require urgent action. Civil society initiatives (grassroots initiatives) are therefore demanding more radical and immediate changes. Following initial initiatives from the 1960s–1970s, which can be more categorized as "back to the country" movements, the idea of exemplifying ecological communal life forms is gaining new impetus due to the increasingly evident climate impacts over the past 10–15 years. Ecovillages operate as a way to empower people to find solutions to pressing environmental and societal problems within social communities and away from technology-centric positions. This chapter pursues the question: What does regenerative sustainability mean in the context of communities, districts, and municipalities? How is this lived and implemented in eco-communities? **Ten case studies from the European and Mediterranean context (France, Greece, Italy, Austria, Slovenia, Spain, and Turkey) are presented and positive as well as critical points are highlighted**.

Keywords Eco-communities · Regenerative cultures and processes · Regenerative water management · Regenerative design and health · Regenerative resource cycles

1 Introduction

Increasingly evident negative impacts due to climate change sharpen the awareness of society as a whole for the urgency of solutions to maintain a functioning ecosystem. Rapid changes in biodiversity and land use, or ever more extreme weather events such as drought and heavy rain with massive consequences for food security and the health of the world's population are just some of the most obvious threats of the climate crisis. The promises that technological development and progress will provide sufficient opportunities to deal with the climate crisis and thus still be able to avert serious problems in a timely manner seem less and less realistic.

E. Haselsteiner (✉)
Urbanity—Architecture, Art, Culture and Literature, Vienna, Austria
e-mail: edeltraud.haselsteiner@aon.at

The discussion moves between two main poles: the belief in technology and "green economic growth" on the one hand and a skeptical position on the other hand, according to which effective solutions can only be found in shrinking and in the economic orientation towards a needs-oriented consumer society. While representatives of the former position dominate the political discussion, there is resistance to the lack of political implementation of necessary measures on the part of grassroots initiatives initiated by civil society (grassroots initiatives). These call for a major transformation of the economic order and orientation towards a social order geared towards renunciation of consumption. Proponents of a lifestyle based on sufficiency and close to nature join together in social groups and communities in order to concretely exemplify and implement the principles of regenerative sustainability in new forms of coexistence. Conceived as cooperatives and communal forms of living, these operate under the generic term "ecovillages".

Following initial initiatives from the 1960s–1970s, which can be more categorized as "back to the land" movements, the idea of exemplifying ecological communal life forms is gaining new impetus due to the increasingly evident climate impacts over the past 10–15 years. Communities like these share social, ecological, and ethical values, such as living in solidarity and based on needs, conserving resources and sharing them fairly, as well as living transparency in grassroots-democratic and consensus-oriented structures and decision-making processes (sociocracy). Ecovillages operate as a way to empower people to find solutions to pressing environmental and societal problems within social communities and away from technology-centric positions.

In this article, questions are discussed along 10 case studies from the European and Mediterranean context (France, Greece, Italy, Austria, Slovenia, Spain, and Turkey): What does regenerative sustainability mean in the context of districts and municipalities and what can we learn from the ecovillage movement for the design of social and societal forms of life for a regenerative future? How is "regenerative sustainability" practiced and implemented in eco-communities? Before doing so, however, the following question will be addressed: What are ecovillages, what connects them, and how did they come about?

2 What Are Ecovillages?

Ecovillages are essentially defined by environmental ideals and high demands on a sustainable way of life. As such, they stand for a paradigmatic change associated with the abandonment of growth and material prosperity goals, especially those linked to the high consumption of fossil fuels. To advance society in the direction of great transformation, they show through concrete examples that a fulfilled life is possible even in industrialized countries with a small ecological footprint (Andreas and Wagner 2012).

Starting with Sir Thomas More's "Utopia" (1518), the "ecovillage movement" sees itself as standing in a long tradition of utopian visions, in that it represents an ideal society that propagates values such as simplicity and equality and thus social stability, peace, and justice reached. The longing for utopias and a fundamental change in civilization is thus also a central unifying characteristic of all globally existing ecovillages. However, the often extreme positions and ways of life raise the question of how far they can be regarded as models that can be transferred and scaled to broader sections of society.

While they try out new forms of equal and grassroots democratic coexistence internally, they often also exercise social and societal commitment externally by offering courses and events for the surrounding neighborhood or those who are interested, getting involved in local politics, culture or art and similar initiatives network worldwide.

The desire to live in an ecovillage often becomes virulent during decisive phases of life change, for example for young parents looking for a village-like structure in which their children can grow up in a protected but very free community and a rural environment, or for older people who want to get involved in a community in a more value-oriented manner after a successful career.

The bond with a way of life close to nature and thus a life in the country results from their values and the pursuit of the greatest possible self-sufficiency and autonomy. Apart from that, however, there are also more urban concepts of ecovillages. The natural way of life and the demand for careful use of the resources of the environment means that existing and now empty buildings are often repaired or, in the case of new buildings, high demands are placed on ecological construction.

What is an ecovillage? Definitions for this are usually broadly based to include space for the most diverse forms of living together and different values. For example, the Global Ecovillage network has established the following definition: "An ecovillage is an intentional, traditional or urban community that is consciously designed through locally owned participatory processes in all four dimensions of sustainability (social, culture, ecology, and economy) to regenerate social and natural environments." (What Is an Ecovillage—Discover Innovative Eco Communities 2013). The demand for "regenerative sustainability" is anchored here as a central requirement.

Equally diverse is the determination of the history of the emergence and the origins of the first ecovillages. Magnusson (2018), for example, speaks of the fourth generation of Swedish ecovillages today and sees their roots as part of the "green movement" against urbanization and often also in connection with the anti-nuclear movement in the 1970s. Other authors (i.e., Mare 2000) go back in history to the 16th century and see the foundation stone laid with Thomas More's story "Utopia" (1518). In it, More describes the state model of the island of Utopia as an alternative to the exploitative and corrupt social conditions in England of his time, characterized by the hierarchical order of the nobility and colonial rule. In Utopia, the fictional image of an ideal society is created, which achieves social stability, peace, and happiness for all members of society with values such as simplicity and equality. Ecovillages as they are forming today can be seen in the tradition of a global "back to the land" movement of the 1960s and 1970s (Mare 2000). This movement was triggered as a reaction to increasing mechanization in all areas of life on the one hand and to growth-oriented modern urban planning, which has often been criticized as misanthropic, especially in Central Europe, on the other (cf. Mitscherlich 1965). Urban planning based on functionalist principles—following the principles of the CIAM (Congrès Internationaux d'Architecture Moderne 1928–1959) with spatial separation of the urban areas for work, living, transport, and leisure—led to monotonous urban landscapes and large suburban housing estates, especially in residential construction, which throughout the 1960s shaped the image of numerous large cities.

3 Regenerative Sustainability in the Context of Neighborhoods and Communities

The concept of regenerative sustainability stands for further development and a paradigm shift in the consideration of sustainability (du Plessis 2012; Mang and Reed 2020; Wahl 2016 and others). This promotes concepts and standards that not only demand stricter targets concerning their impact on the environment but also favor a change from a fragmented to an overall systemic view of the ecological system (Brown 2016; Haselsteiner et al. 2021; Reed 2007).

The first use of the term "regenerative design" dates back to the landscape architect John Tillmann Lyle, who presented a model of a "collaborative interdisciplinary design process" in 1994 (Lyle 1996). Lyle argued against degenerative linear input–output models and proposed a "regenerative cycle model" as an alternative. Based on twelve theoretical and practical principles, essential resources of daily life such as housing, food, water or waste should be kept in a regenerative energy and material cycle and thus enable self-regenerating ecological cycles which regenerate themselves in the built environment. His 12 principles, listed here in sequence, express his fundamental attitude that sustainable development is based on a model of circular processes of the ecosystem.

1. Letting nature do the work
2. Nature as model and context
3. Aggregating, not isolating functions
4. Optimum levels for multiple functions
5. Matching technology and need
6. Using information to replace power
7. Multiple pathways
8. Common solutions to disparate problems
9. Storage as a key to sustainability
10. Form to facilitate flow
11. Form to manifest process
12. Prioritize for sustainability

In a further development of concepts for "regenerative sustainability", especially for the construction sector, the social dimension of sustainability was included more closely, and, based on this, framework concepts for a regenerative design of the built environment were developed (Brown et al. 2018; Cepeliauskaite and Stasiskiene

2020; Haselsteiner et al. 2021). Regenerative sustainability aims to maintain, restore and regenerate a healthy socio-ecological system.

Ecological renewal on the one hand and regenerative social development, on the other hand, were specified based on these principles and interpreted appropriately at the building, neighborhood, or community level. The holistic view of the built environment is also a central aspect. The concepts oppose a one-sided orientation of sustainability towards the energy consumption or CO2 emissions of buildings. These would risk shifting the environmental impact of buildings from one factor to another. Rather, the focus is on adaptation, resilience, and regeneration as well as an "ecological worldview". This worldview is expressed in the application of nature-based solutions, designs, and planning that regard humans as an integral part of nature, as well as a targeted consideration of regional and cultural differences (Cole 2012). Especially the latter, the emphasis on the "place" and the respect for the specific socio-ecological interaction between people and place, is a key distinguishing feature compared to the "Green Design" approach (Mang and Reed 2012). Table 1 lists the regenerative principles selected for this chapter and explains their application using concrete examples.

In the 1990s, Steve Larrick developed a conceptual framework for a system model of living community development, which can also be used for ecovillages as the basis for their overall concept. In it, Larrick puts the ecological development within a community on a level with its social development. Sustainable development moves the living system towards transcendence and ecological and social regenerative development, while negative developments and policies lead to ecological and social collapse and high levels of disorder and entropy. Graphic 1 shows Steve Larrick's model.

4 Regenerative Sustainability in Practice: 10 Case Studies from the European-Mediterranean Context

As already explained in the previous section, ecovillages aim to demonstrate the success of a transformation toward sustainability using a living example. In 1991 the Gaia Trust in Denmark commissioned a study on the current situation and world-leading examples of ecovillages (Mare 2000). It became clear that a wide range of sustainability approaches are integrated into cooperative communities, but that fully functioning ecovillages do not yet exist anywhere (Ecovillages and Sustainable Communities: A Report for Gaia Trust 1991; Mare 2000). As a result, that same year, 1991, social and environmental activists met for the first time to discuss strategies for developing and disseminating the ecovillage vision. In 1994, at a similar meeting, based on this initiative group, the Global Ecovillage Network (GEN) was founded, which since then has been committed to networking and exchange among ecovillages and promoting further dissemination (Global Ecovillage Network—Community for a Regenerative World 2016; Mare 2000). Other national and international networking

Table 1 Regenerative principles: explanation and examples (Haselsteiner et al. 2021)

Regenerative principles: key topics	Description (examples on building/neighborhood/district level)
Place, nature and eco-system	Place-based design approach departing from the recognition that each site is a unique dynamic entity, with the building interacting with green neighborhoods and the environment to regenerate the ecosystem and enable it to evolve into the future (i.e., rebuild soil, adapt to bio/microclimate)
Energy	Restorative and regenerative energy systems, energy as part of a coherent restoration approach aimed at improving ecosystem quality (i.e., effective and sharing of energy, energy storage)
Carbon	Carbon–neutral/climate-positive approaches and carbon-reducing measures to reduce the carbon footprint (i.e., environmentally responsible sources, life cycle assessment)
Water	Reduce the "water footprint", maintenance of natural water cycles, water treatment, reuse, and access to clean drinking water (i.e., water management, net-positive water, rainwater harvesting)
Material and resources	Ecological systems to conserve resources and preserve them for future generations, improve material and resource cycles (i.e., healthy materials, transparent labeling, responsible use, and conservation)
Waste	Zero waste approach, design for disassembly, deconstruction, and flexibility of use (i.e., cradle2cradle approach, upcycling, reuse, and recycling of materials and buildings)
Health and well-being	Reconnect humans with nature, enhance the quality of life and contribute to individual, community, and society health and well-being without exploiting other people, the environment, or future generations (i.e., access to healthy food, ensure air quality, daylight and comfort)
Social equity	Equality, empowerment, and participation of people and countries, fairness in allocating resources, inclusiveness, supporting vulnerable people (i.e., empowerment of women, older and young people, giving disadvantaged groups a voice, no threat associated with food production, globally responsible action in dealing with resources)
Economy	Regenerative, circular and sharing economies, sustainable production and consumption, collaborative, multi-level, place-based economies (i.e., closed-loop business models, energy communities)
Culture and community	Addressing social determinants of health, promoting social cohesion and community identity, enabling accessibility and inclusion (i.e. reintegrating living heritage, empowering rural communities)
Education and inspiration	Enable participation and encourage bottom-up initiatives, eco-literacy awareness and education (i.e. pioneering movements such as permaculture, urban gardening, placemaking, etc., continuous learning and feedback, collaboration, interaction and interdisciplinary planning)
Environment and mobility	Reducing carbon emissions from travel and transport, promoting walking and cycling, walkable cities, rural–urban balance (i.e., carsharing, e-car charging stations, bicycle parking spaces and cycle paths)

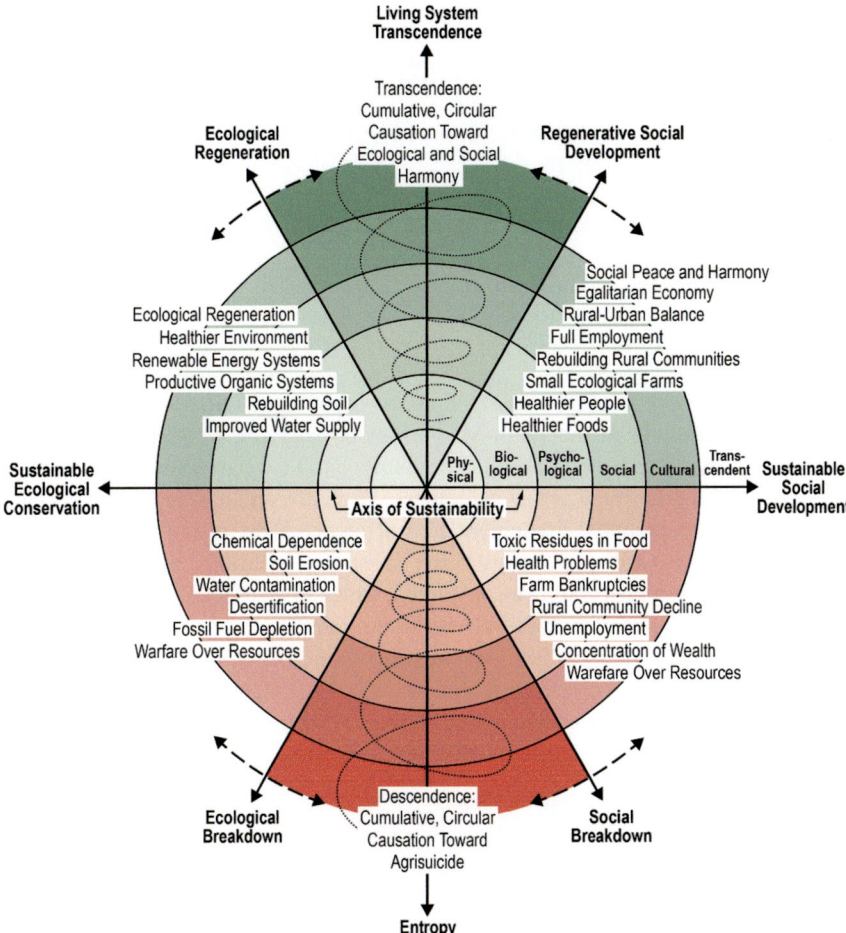

Graphic 1 System model of living community development (Own illustration based on Larrick 1997)

platforms followed, e.g. B. Oasis de Colibris in France, RIVE network in Italy, ECOLISE, European Network of Community-led Initiatives for Sustainable Europe, or Ecovillages Worldwide, an association of ecovillages across Europe.

Ten case studies from the Euro-Mediterranean area were selected for this chapter. The selection was made on the one hand to reflect a certain geographical range within the Euro-Mediterranean area and on the other hand to reflect diversity in terms of size and number of members, ecological goals, and the duration of their existence. Conversely, the restriction to the European-Mediterranean area was firstly to create a certain comparability due to socio-cultural conditions and secondly because of the European-socio-cultural background of the author, which makes it easier for her to classify initiatives in the overall social context. Nevertheless, it should be

Table 2 Subject areas structured according to "regenerative sustainability principles"

Topics	Regenerative sustainability principles		
Architecture and regenerative design	Place, nature and eco-system	Energy	Carbon
Ecological-/sustainability goals	Water	Material and resources	Waste
Common values and practices	Health and well-being	Social equity	Economy
Community management, activities and network	Culture and community	Education and inspiration	Environment and mobility

pointed out at this point that similarly ambitious ecovillage initiatives can be found worldwide, especially in the Latin-Mediterranean area and in Mexico (e.g. Baja Montecito Ecovillage, Tierraluz Eco Community and Proyecto Yum Kaax).

An additional criterion was that sufficient information and data are publicly available to assess the practical implementation of regenerative sustainability goals. It is understandable that due to the restrictive selection conditions, only a very small range of model examples could be considered, which in no case claims to be a representative study. Rather, it is about an exploratory approach to derive possibilities, opportunities, and challenges for a regenerative future of our planet from the ecovillage initiatives and to learn from them. To assess the extent to which regenerative sustainability goals have been and are being lived and implemented in the individual ecovillages, these can be considered along the following four topics:

1. Architecture and Regenerative Design
2. Ecological/Sustainability Goals
3. Common Values and Practices
4. Community Management, Activities, and Network

Each of these subject areas stands for three of the regenerative sustainability principles or goals listed in the table. Table 2 lists these along the 4 categories.

5 Selection of Case Studies

From around 920 initiative groups internationally, which are documented on the Global Ecovillage Network (GEN) platform, 10 projects were selected as case studies for this chapter. To restrict the selection to a realistic number, the search was limited to Europe in the first step. For the European continent, there are still slightly more than 300 (326, as of Nov. 2022) ecovillage initiatives registered as members of the Global Ecovillage Network. From this, in turn, those projects were selected which are located in countries in the Mediterranean region. In the last step, 1–2 projects were selected from each of these countries, which have already existed for several years (at least 5 years), which each form a larger community with several members (>6),

Table 3 Overview of the selected case studies

Name	Location, Country	Founded (Year)	Current members
Meltemi	Greece	1946	184
Il Popolo degli Elfi	Italy	1980	200
Ecovillaggio Torri Superiore	Italy	1989	25
Oasis du Coq à l'Âme	France	2016	45 (30 adults, 15 children)
Hameau des Buis	France	2007	50 (0–90 years)
Arterra Bizimodu	Spain	2013	40
VIDALIA	Spain	2017	25 (17 adult, 8 children)
Sunny Hills	Slovenia	2012/2016	6 founding members (other temporary members)
Güneşköy	Turkey	2000	9 active members (additional supporting members)
Cambium	Austria	2017	85 (65 adults, 20 children)

detailed documentation is available and, on the other hand, specific distinguishing features from the represent projects in the other countries. Care was also taken to select examples from different founding years, i.e. with different durations of their existence. In addition, a project from Austria was selected that is well known to the author and thus provides a good knowledge base for comparison with other initiatives. Table 3 provides an overview of these selected projects.

6 Ecovillages: Visions and Goals of the Selected Case Studies

Within this selection, **Meltemi** is the longest-existing ecovillage movement in Greece. Meltemi was founded in 1946 amid the Greek Civil War by a group of Petroleum Company employees as a retreat to nature. At first, everyone lived in tents. Over the years and with the improvement of the Greek economy, tents have been replaced by small huts for accommodation. From this early era derives the alternative name for the site used by the residents, who affectionately call it Kataskinosi (The Campsite). The people—meanwhile four generations—live here together mainly in the summer. Rules written in the mid-1950s still order coexistence today: Most huts are assigned to specific families who have the right to use them—but nothing more. However, a significant number of homes are not permanently assigned. Renting or letting is not possible, even if it is free of charge. Since Meltemi's history is rooted in the labor movement, these houses are available to workers' union members for their vacations. There is no car traffic within the complex. The residents take care of the care and maintenance of the complex together. The goal of its creation was to save a small part of Greece from devastation, commercialization, and sale to global capital. Its current existence shows that it has succeeded to this day.

Il Popolo degli Elfi is one of the oldest ecological community housing projects in Italy. The community arose from the occupation of a 500-acre property owned by the government. In the countryside, some ruins had been abandoned for decades. The community of 150–200 people consists of an agglomeration of four small villages and fourteen farms, some of which are far apart and can only be reached on foot. The empty stone houses and ruins were renovated on their initiative, the majority of the houses still have no electricity, and heating and cooking are done with wood. The group now has a 20-year usage concession, which is about to expire. They are negotiating with the government to renew their concession. The aim of the community is a way of life that is completely reduced in need and based on self-sufficiency, in harmony with nature. The members see themselves in harmony with the natural cycles and are largely self-sufficient by collecting wild fruits and herbs, growing vegetables, grain, chestnuts, olives, and some livestock. The agricultural products produced are distributed to all villages according to different needs. In addition, the municipality or each village has a collective fund from which those foods and goods that cannot be produced in-house are purchased. The members can decide for themselves how much to contribute to these funds, depending on their possibilities. In addition to self-sufficiency and the renovation of the old buildings, the focus is on art, agriculture, alternative medicine, and communal living.

Also in Italy is the community housing project **Ecovillaggio Torri Superiore.** With its founding year of 1989, it is also one of the older ecovillages. Ecovillage Torri Superiore was created in an abandoned medieval village with traditional stone architecture. The village had to be left for economic, social, and practical reasons in the last 100 years. All buildings are connected on a total of 5 different levels to form a building with 162 rooms. An initiative group of around 25 people restored the buildings using natural building materials and building techniques. Local stones, natural lime, and ecological wall paints were used for plasters and murals, natural insulating materials (cork, coconut fibers) for insulation, and all windows are made of wood. A large part of the energy required for hot water and heating is self-produced using solar collectors and photovoltaic modules. Community life is based on a sufficient use of resources, extensive self-sufficiency with self-produced food based on permaculture, and a lifestyle based on the circular economy. The members provide for themselves collectively and share all meals. Groceries that cannot be produced in-house are purchased from a social cooperative in the nearby town. Fundamental decisions are made by the general assembly using the majority principle, and day-to-day decisions by the resident community, which meets every Monday evening and decides by consensus. With the operation of an eco-guest house, the initiative group offers services in the field of eco-tourism, but also courses and further training programs for organic farming for the public. Half of the buildings belong to the association (used as a publicly accessible cultural center) and a half to members who share 20 restored apartments and take care of the cultural and social activities.

An equally active ecovillage scene exists in France. With around 45 members (30 adults, 15 children), **Oasis du Coq à l'Âme** is one of the largest ecovillage initiatives in France. The cooperative came together in 2016, and in 2021 the group acquired a 25-ha plot of land from a former winery, including agricultural buildings. The members plan to revitalize the existing buildings with the aim of careful use of resources and nature. They see themselves as a "living laboratory" for a way of life based on food autonomy, minimal impact on nature, economical use of energy and resources, extensive independence from fossil fuels, and simple alternative energy solutions. In particular, they work in a cooperative collective with small farmers and employees of agricultural organizations to develop agricultural technologies and practices as a basis for human-scale agroecology. The aim is to jointly develop methods, practices, tools, and machines to regain technical sovereignty and autonomy of action through the reacquisition of knowledge and know-how and to achieve self-sufficiency in organic agriculture. Among other things, a bicycle tractor was developed. The organizational structure follows a cooperative economic model according to the principles of "shared governance" (social and solidarity economy).

Oasis du Coq à l'Âme is a pilot project of the French initiative **Oasis de Colibris** by Pierre Rabhi. Launched some 15 years ago, the campaign challenges the view that solutions must first be sought in economic and technological growth. Rather, they are convinced that the key lies in a fundamental change in society and collective and ecological forms of life and work. The transformation of society is nourished by individual transformation and thus human change and the contribution of each individual, i.e. questioning our way of life, at the heart of change. The initiative sees itself primarily as a networking and support platform for individuals and groups who want to work for an ecological society based on solidarity. However, she also repeatedly sets impulses to remind people that we are all part of the solution and that we as such can contribute to the changes.

The ecovillage **Hameau des Buis** is also a member of the Oasis de Colibris movement. The ecovillage in Berrias-et-Casteljau, France, was founded in 2007. Starting from an old farmhouse with approx. 6 hectares of agricultural land, a residential complex with 20 residential units and huts for around 50 residents was built between 2007 and 2011 in so-called "hands-on construction sites". The plan for the buildings is by Pierre Henry Gomez and was designed according to the principles of bioclimatic architecture. The construction work and the construction of the residential complex were carried out by participants from the ecovillage and volunteers who, under professional supervision, acquired alternative building techniques and the use of natural building materials in a participatory manner. It was built with organic materials from the region, such as straw, clay, wood, and local pozzolan, a rock of volcanic origin. The building concept includes a constructed wetland and green roofs. The former farmhouse has been renovated and is used as a common room. An alternative school for around 12–14 children was also built. The 6ha of green areas are used for agriculture with an orchard, vegetable growing, and goat farming. At the heart of the community is complete food autonomy. To achieve this, there is also a bakery and cheese production. Rainwater recovery, a winter greenhouse, and composting also support their efforts to use locally produced material or energy flows and to manage them in cycles. In addition to permanently assigned housing units, there are additional offers to live or work as volunteers or as woofers as well as training and further education offers on the topic of "living together in regenerative communities". A weekly market and small café serve as a meeting place among residents and their neighbors.

After France with 30 and Germany with 33 registered ecovillages in the Global Ecovillage Network, Spain has by far the most ecovillage initiatives within the EU with around 90 ecovillages. One of these initiatives is **Arterra Bizimodu** with around 40 members. They see themselves as an experiment in sociocracy that explores innovative models of self-government and organizational structures of collective management. For this purpose, a co-living project has been implemented in an old former country hotel since 2013. An important goal for the initiators is to revitalize rural areas, provide housing solutions for alternative ways of life and develop spaces for local products that are considered compatible with their vision. This includes all kinds of workshops for handicraft production, manufactories in connection with regenerative ecological agriculture, but also social or cultural offers. For this multitude of different projects, it is necessary to develop an internal organizational structure that can maintain the flow of communication between them and enable a collectively managed common space for the movement as a whole. In addition, they have around 8000m2 of multi-purpose areas and living space that allows smaller family units space for individual living needs. In addition to ecological goals and tasks that they also share with other ecovillages, they see their greatest challenge in trying out new forms of economy that reflect a balance between the personal and the collective and support the diverse talents of all people who join the project.

The ecovillage **VIDALIA,** also in Spain, was founded in 2017 and currently consists of around 25 people, of which 17 are adults and 8 are children. The initiative is based in Catalonia, in a former textile factory, about an hour's drive from the center of Barcelona. After the textile industry was closed, the majority of residents moved away. All buildings of the former factory are listed as monuments. The entire facility was purchased by a real estate investor who in turn commissioned a group to revitalize, respectfully use, and create a master plan. In 2017, it was handed over to a community initiative for 75 years of use. The group is planning a careful sustainable renovation and a community housing project. For the time being, it is organized as a housing cooperative but plans to set up further organizational and legal framework structures, such as a consumer cooperative or an organizational structure for educational and art projects, to enable broader space for diverse forms of alternative economic activity. In doing so, they are guided by the values of a social and solidarity economy within the framework of degrowth, life in a community as part of nature with little environmental pollution, and a reduced ecological footprint. The sharing of resources is the focus of the first activities, and some exchanges and community initiatives have already been launched, such as a clothing swap shop, a tool exchange for the shared use of equipment and tools, a community laundry to minimize energy and water consumption and at the same time to create a place for social encounters, a car-sharing system and a communal kitchen with dining room where members cook together and share lunch and dinner. The cooperative is

Fig. 1 Ecovillage Cambium, Austria: the former soccer field is now a regenerative vegetable garden (© Sandra Sieczkowski)

still under construction. Work is currently underway on the renovation plan and an ecological energy concept. A start has already been made on creating an orchard and a vegetable garden according to the principles of organic farming and permaculture (Fig. 1).

Sunny Hills in Slovenia is also an ecovillage initiative focusing on the revitalization of abandoned houses in rural areas. A group of six people got together in 2012 and bought up abandoned houses in a small village called Hrvoji, in the Slovenian part of Istria. In 2014 and 2018 the first house—the former vicarage—was renovated, and further renovations are planned. The group is particularly concerned with reviving the natural potential of the site with its traditional building forms and developing sustainable solutions for the renovation in harmony with each other. In addition to preserving the architectural heritage of the dry stone buildings, the buildings are being renovated in an ecological way using natural materials from the region, such as clay plaster. Following a careful use of water, which is very important for this dry region, there are dry composting toilets, the treatment, and reuse of wastewater, and permaculture-based land management. To irrigate the garden, wastewater from the kitchen and bathroom is cleaned in reeds and collected in a reservoir. Heating is using a gasification boiler for heating with wood from the region. The production and use of biochar are also being tested. They define simplicity, joy, community, caring, and diversity as the basic values of their coexistence. Living as simple a lifestyle as possible, they share tools and cars, try to buy only the necessary things, repair, upcycle and recycle whenever possible, produce their food, and cook together.

The group offers different forms of membership. Co-op members, volunteers, and long-term guests are accommodated, live and work together, or help out around the house, with maintenance, gardening, and whatever is on the to-do list. The work is shared between members and volunteers weekly. Sunny Hills is managed both as a cooperative (property owners) and as an association (activities and events). The Sunny Hill association organizes workshops, retreats, conferences, and events.

The **Güneşköy ecovillage** movement started in Turkey in 2000 intending to develop and share sustainable living experiences in rural areas in harmony with nature. For this purpose, a cooperative was founded by nine environmental activists in Ankara and in 2002 a 75-hectare piece of stony land without water was bought from the state. The group began with the basic development of the property, drilled a well, organized a power connection, and erected the first mud huts as housing. Farming began in 2005 and a Community Supported Agriculture model was established. The food produced is distributed among the families involved in the cooperative in Ankara. The cultivation and operation of agriculture follow ecological principles of permaculture, endangered plants are collected and cultivated again, a water-saving system of drip irrigation is used for irrigation, a solar greenhouse and a solar heating and drying system with an underground thermal solar energy storage also serve as a prototype for the difficult climatic conditions of the region and as a model for the local population. They design and implement waste recovery and reuse systems. Even when transporting food, natural means of transport without fossil fuels are used as far as possible. The houses are built from local natural building materials (clay and straw bales). The focus of their initiative is the effort to restore the destroyed natural structures and restore the balance in nature. They try to pass these ideas on to the young generation in the cities in particular. They regularly organize projects and events with young people to get them interested in the topic, strengthen the links between rural and urban areas, and motivate them to set up further initiatives (Weitzhofer-Yurtışık 2012) (Fig. 2).

Cambium is an ecovillage movement and community housing project in rural Austria. The community was formed from the merger of two previously independent initiatives, both in search of new forms of community life in connection with each other and with nature. With the possibility of acquiring the site of a former barracks in Fehring, the two initiatives merged. In 2017, the vacant building of a former military facility was occupied. The total area of 16 ha—building land, land for prospective builders, forest, fields, and meadows has since been gradually developed and managed sustainably. The building is partly renewed individually in self-construction. The group members each have smaller private units for individual design and share large communal zones and workshops, which are also rented out to external users for use. There are also rooms for external seminars and workshops as well as guest rooms. An ecological new building as an extension is planned. Their basic idea is based on the sensible use of forward-looking technologies and at the same time reviving traditional craftsmanship. They see themselves as an experimental field for decentralized renewable energy systems and innovative ecological building constructions (e.g. Strohboid). Around 85 people currently live on the site (65 adults, and 20 children), but expansion is still being considered. The aim is to

Fig. 2 Ecovillage Cambium, Austria: the yurt serves as a place of gathering, mindfulness and stillness (© Sandra Sieczkowski)

create a place to live and work for over 100 people, in which all age groups and different professions, interests, and skills are represented. The group tries to provide itself with a small farm, cooking and eating together twice daily. The organizational structure is based on sociocracy, all residents are organized in working groups that make decisions autonomously by consensus, and a management group made up of representatives from the working groups makes fundamental decisions for the overall project. Communication and feedback with the entire group take place weekly in the plenum. The group attaches great importance to dialogue and networking with the locals. An event center acts as a hub for learning, meeting, and networking (Fig. 3).

7 Architecture and Regenerative Design (Place, Nature, and Eco-System—Energy—Carbon)

All in all, the presented projects show the respectful handling of the place. Almost all projects use an existing building or already built-up areas as the location and starting point of their community initiative. Due to the rapidly increasing proportion of surface sealing and the associated environmental problems such as heat and flooding, it is clear that a fundamental problem is being tackled. Likewise, in all projects, the privately used areas are reduced to a minimum, while rooms for the community are given more space. However, the surrounding natural area has the largest share and

Fig. 3 Ecovillage Cambium, Austria: several tiny houses were built in the workshops (© Sandra Sieczkowski)

therefore priority. The activists demonstrate a very careful, regenerative approach to nature and, like the Güneşköy ecovillage movement, show that even very dry places that are not initially considered fertile can be transformed into living oases and functioning ecosystems.

The economical use of energy and the extensive supply of regenerative energy systems are also among the primary goals of all ecovillages. However, what also distinguishes them from conventional sustainable construction methods is their experimental approach to the further development of technical or non-technical solutions for the most autonomous and efficient possible supply of renewable energy. In the Cambium, the ecovillage in Austria, for example, in addition to a biogas plant and a plant-based sewage treatment plant, a bio-kiln—heating with compost heat—is being tested. The ecovillage Hameau des Buis in France relies on an overall ecological construction according to bioclimatic principles and is testing solar-activated stone walls, among other things. The Güneşköy Agricultural Cooperative in Turkey operates a solar greenhouse or a biofuel tractor (Figs. 4, 5 and 6).

One can also learn from the ecovillages how the overall ecological footprint can be significantly reduced through communal housing and the sharing of resources. As a "living laboratory", Oasis du Coq à l'Âme in France is committed to a way of life with minimal impact on nature and has this also monitored by a scientific committee. The members of the Il Popolo degli Elfi ecovillage initiative in Italy live in houses with minimal comfort, some without electricity (Table 4).

Fig. 4 A-C Ecovillage Cambium, Austria: plant-based sewage treatment plant (© Haselsteiner)

Fig. 5 Ecovillage Cambium, Austria: biogas plant (© Haselsteiner)

8 Ecological-/Sustainability Goals (Water—Material and Resources—Waste)

In their projects, initiators of ecovillages make it clear that water, materials, and resources need to be handled carefully. Rainwater is collected, wastewater is cleaned and reused, and the overall aim is to use water sparingly. Likewise, the building is predominantly made with natural and ecological materials, or existing ones are reused.

The goal of most ecovillages is to achieve food autonomy. For this purpose, smaller to larger organic farms are operated on the principles of permaculture, or, as in the ecovillage Hameau des Buis, a bakery and a cheese dairy are operated. The enthusiasm for experimentation of the initiatives, on the one hand, to deal efficiently with the built-up area and on the other hand to practice ecological agriculture with the greatest possible protection of the soil and in harmony with nature, is also very great in this area. For example, the Oasis du Coq à l'Âme initiative in France focuses on the development of agricultural equipment and tools to become even more autonomous in the ecological management of land.

A central theme of all ecovillages is closed resource cycles. This includes the fact that no waste is produced and that recycling takes place directly in the cycle. Waste is collected and recycled separately, leftover feed is either used to feed the

Fig. 6 Ecovillage Cambium, Austria: bio-kiln—heating with compost heat (© Haselsteiner)

Table 4 Indicators of regenerative sustainability and how they are considered in ecovillage goals (part 1)

Name	Place, nature, and eco-system	Energy	Carbon
Meltemi	++	–	–
Il Popolo degli Elfi	++	+	+
Ecovillaggio Torri Superiore	+	++	++
Oasis du Coq à l'Âme	++	++	++
Hameau des Buis	+	++	+
Arterra Bizimodu	+	-	-
VIDALIA	+	+	+
Sunny Hills	+	–	–
Güneşköy	++	++	–
Cambium	++	++	++

++ *High relevance*
+ *Relevance*
– *No statement*

Table 5 Indicators of regenerative sustainability and how they are considered in ecovillage goals (part 2)

Name	Water	Material and resources	Waste
Meltemi	–	+	–
Il Popolo degli Elfi	–	++	–
Ecovillaggio Torri Superiore	–	++	–
Oasis du Coq à l'Âme	–	++	–
Hameau des Buis	++	++	++
Arterra Bizimodu	–	–	–
VIDALIA	++	–	–
Sunny Hills	++	++	++
Güneşköy	++	++	++
Cambium	++	++	++

++ *High relevance*
+ *Relevance*
– *No statement*

animals or composted, and compost is used in agriculture. Most initiatives, therefore, use composting toilets, thereby enabling an economical and ecological alternative to conventional disposal lines (Table 5).

9 Common Values and Practices (Health and Well-Being—Social Equity—Economy)

As part of their concept, ecovillage initiatives consciously choose a location that enables them to live a healthy life in harmony with nature. This striving for a healthy lifestyle is once again reinforced by self-produced organic food.

What all ecovillages have in common is that they offer to live, work, and living space for people of all ages and different professions, interests, and skills. All initiatives are particularly characterized by this societal and social openness as well as their claim to inclusion and solidarity. Their social commitment often extends beyond the limits of their initiative to socially disadvantaged groups outside the group, e.g. in the integration of asylum seekers.

The search for less growth-oriented economic models leads to the testing of new forms of solidarity economy, the development of a solidarity economy, or offers of the sharing economy. The small companies located in ecovillages are primarily involved in socially or ecologically relevant fields of work. As an economic mainstay, some ecovillages offer tourist services such as room rentals, or they offer the possibility of co-living in connection with volunteer work (Table 6).

Table 6 Indicators of regenerative sustainability and how they are considered in ecovillage goals (part 3)

Name	Health and well-being	Social equity	Economy
Meltemi	++	++	+
Il Popolo degli Elfi	++	++	+
Ecovillaggio Torri Superiore	–	+	+
Oasis du Coq à l'Âme	–	+	++
Hameau des Buis	–	+	++
Arterra Bizimodu	–	+	++
VIDALIA	–	–	++
Sunny Hills	–	–	–
Güneşköy	–	–	++
Cambium	–	–	+

++ *High relevance*
+ *Relevance*
– *No statement*

10 Community Management, Activities, and Network (Culture and Community—Education and Inspiration—Environment and Mobility)

Most ecovillages see themselves in the form of organizations as experimental fields of a living sociocracy, i.e. joint decisions in consensus and the equality of all participants are in the foreground. Living together is based on the idea that communal and individual spaces coexist harmoniously. In addition, community-building activities, such as eating together or celebrating together, are planned as fixed components of everyday life.

Conversely to the striving for autonomy in supply (e.g. food and energy autonomy), ecovillage initiatives see themselves as open systems in the social structure. The integration, communication, and exchange with your immediate neighborhood but also with regional and international networks is of great importance. They try to expand their knowledge and experience together with others and to pass this on again in offers such as workshops and courses.

The initiatives are usually organized as cooperatives and/or associations. Innovative models of self-government are practiced within each group to ensure that the satisfaction of personal and collective needs remains in a flow of giving and receiving. Initiatives that have existed for many years, such as Meltemi in Greece, which has now existed for over 75 years, show that community self-government can also be a success model worth imitating over several generations (Table 7).

Table 7 Indicators of regenerative sustainability and how they are considered in ecovillage goals (part 4)

Name	Culture and community	Education and inspiration	Environment and mobility
Meltemi	++	−	+
Il Popolo degli Elfi	+	−	+
Ecovillaggio Torri Superiore	+	++	−
Oasis du Coq à l'Âme	++	++	+
Hameau des Buis	++	++	−
Arterra Bizimodu	++	++	−
VIDALIA	++	++	++
Sunny Hills	++	++	++
Güneşköy	++	++	++
Cambium	++	++	−

++ High relevance
+ Relevance
− No statement

11 Summary

In order to take up the question: "What can we learn from the ecovillage movement for the design of social and societal forms of life for a regenerative future?" in this last section, goals and visions of the ecovillage initiatives are presented in terms of their alignment with the 17 sustainability goals of the UN. The following table (Table 4) lists the various SDGs in detail and shows how they are addressed in the ecovillages (Table 8).

The results show that there is a high level of accordance with the sustainability goals of the UN on many points. This qualifies these initiatives as a field for learning and experimentation, in order to find solutions to pressing ecological and social problems away from technology-centric positions. As a special feature, not only ecological but also social issues are addressed in ecovillages. In my opinion, this is precisely where the special role of these initiatives lies. Regenerative sustainability means living sustainability in everyday life and aligning our actions accordingly. This is exactly what ecovillages are a suitable model for.

12 Conclusions

The projects presented in the previous sections are just a small sample of all the ecovillages that exist worldwide. Nevertheless, in their diversity, they represent a good overview of goals, values, and the practice of regenerative sustainability in that

Table 8 Sustainable 17 Development Goals (SDGs) and how these are reflected in the goals of ecovillages

17 Sustainable Development Goals (SDGs)	(Main) Regenerative principles and key topics	How this aspect is addressed by ecovillages
1. No Poverty: End poverty in all its forms everywhere	Social equity	**Common values and practices**: basic principles of inclusion, integration and solidarity; commitment to socially disadvantaged groups
2. Zero Hunger: end hunger, achieve food security and improved nutrition, and promote sustainable agriculture	Social equity	**Common values and practices**: strengthening and further development of sustainable agriculture, permaculture and empowerment for self-production of organic food
3. Good Health and Well-Being: ensure healthy lives and promote well-being for all of all ages	Health and well-being	**Common values and practices**: healthy life in harmony with nature
4. Quality Education: Ensure inclusive and equitable quality education and promote lifelong learning opportunities for all	Education and inspiration	**Community management, activities and network**: mutual learning, exchange of knowledge within the community and the neighborhood, further training offers for those interested
5. Gender Equality: achieve gender equality and empower all women and girls	Social equity	**Common values and practices**: equal rights for women, men and people of different genders to live together
6. Clean Water and Sanitation: ensure availability and sustainable management of water and sanitation for all	Water	**Ecological-/Sustainability Goals**: rainwater collection, wastewater treatment and reuse, water conservation
7. Affordable and Clean Energy: ensure access to affordable, reliable, sustainable, and modern energy for all	Energy	**Architecture and Regenerative Design**: site-specific energy and supply concepts integrated into the ecosystem, supply with regenerative energy systems (solar, PV, geothermal energy, experimental energy resources, etc.)
8. Decent Work and Economic Growth: promote sustained, inclusive, and sustainable economic growth, full and productive employment, and decent work for all	Economy	**Common values and practices**: new forms of collaborative work and an economic approach oriented towards the common good and the principles of circular economy
9. Industry, Innovation, and Infrastructure: Build resilient infrastructure, promote inclusive and sustainable industrialization, and foster innovation	Economy; environment and mobility	**Community management, activities and network**: innovation through practical testing and collaborative development, with clear ecological goals and long-term consideration of the impact

(continued)

Table 8 (continued)

17 Sustainable Development Goals (SDGs)	(Main) Regenerative principles and key topics	How this aspect is addressed by ecovillages
10. Reduce Inequalities: reduce inequality within and among countries	Social equity	**Common values and practices**: practicing a sociocratic form of society, equal rights for all persons and opinions
11. Sustainable Cities and Communities: make cities and human settlements inclusive, safe, resilient, and sustainable	Culture and community	**Community management, activities and network**: communal housing and lifestyle as a blueprint for a sustainable form of society
12. Responsible Consumption and Production: ensure sustainable consumption and production patterns	Material and resources; waste	**Ecological-/Sustainability Goals**: high level of self-sufficiency with limited use and sharing of materials and resources, circular economy
13. Climate Action: take urgent action to combat climate change and its impacts	Place, nature and eco-system; carbon; energy	**Architecture and Regenerative Design**: consciously simple way of life with a view to the necessities of climate change and its consequences
14. Life Below Water: conserve and sustainably use the oceans, seas, and marine resources for sustainable development	Water	**Ecological-/Sustainability Goals**: overall ecologically oriented way of life that also takes into account the preservation of water ecosystems
15. Life on Land: protect, restore, and promote the sustainable use of terrestrial ecosystems, sustainably manage forests, combat desertification, and halt and reverse land degradation and halt biodiversity loss	Place, nature and eco-system	**Architecture and Regenerative Design**: design of housing and lifestyle in harmony with the ecosystem (sustainable architecture and food production, supply and disposal integrated into the cycle of nature, biodiversity, etc.)
16. Peace, Justice, and Strong Institutions: promote peaceful and inclusive societies for sustainable development, provide access to justice for all, and build effective, accountable, and inclusive institutions at all levels	Culture and community	**Community management, activities and network**: implementation of communal, participatory and inclusive forms of life with sociocratic decision-making structures, participatory management and accountability (working groups)
17. Partnership for the Goals: strengthen the means of implementation and revitalize the global partnership for sustainable development	Education and inspiration	**Community management, activities and network**: partnerships and networks with all levels of the organization (neighborhood, community, country, global)

they live. There is no doubt that these initiatives pursue all aspects of regenerative sustainability with the utmost consistency and commitment. However, the question remains open as to whether and how a model that can be transferred to society as a whole could be developed from this.

On the other hand, apart from organized ecovillage movements, there are also initiatives by individual architects for ecologically sustainable architecture, which take particular account of the local conditions, take up the needs of the regional population and, for example, implement cost- and climate-friendly construction forms in self-building initiatives together with them. Representatives include Hassan Fathy in Egypt, Shigeru Ban in Japan or Yasmeen Lari in Pakistan.

Although all ecovillage initiatives strive for the greatest possible diversity and openness in their social composition, the group of participants is still reduced—viewed about society as a whole—to a small number of like-minded people. The initiators often come from social, artistic, or manual professions and contribute their knowledge and skills directly to the design of the ecovillage initiative.

Even with the restriction that this form of living together will only be practicable for a small group of people in the future and that these will only remain niche solutions, I still see them as an essential catalyst and important stimulus for the transformation of society towards regenerative sustainability. They show that a different way of life-based on minimal consumption is not only possible but can also lead to a satisfied and "good life".

References

Andreas M, Wagner F (2012) Introduction. RCC Perspect 8:5–8. https://www.jstor.org/stable/26240428

Brown M (2016) Futurestorative: working towards a new sustainability. Riba Publishing

Brown M, Haselsteiner E, Apro D, Kopeva D, Luca E, Pulkkinen K-L, Rizvanolli BV et al (2018) Sustainability, restorative to regenerative. An exploration in progressing a paradigm shift in built environment thinking, from sustainability to restorative sustainability and on to regenerative sustainability. COST Action CA16114 RESTORE. https://www.eurestore.eu/wp-content/uploads/2018/05/RESTORE_booklet_print_END.pdf

Cepeliauskaite G, Stasiskiene Z (2020) The framework of the principles of sustainable urban ecosystems development and functioning. Sustainability 12(2), Article 2. https://doi.org/10.3390/su12020720

Cole RJ (2012) Regenerative design and development: current theory and practice. Build. Res. & Inf. 40(1):1–6. https://doi.org/10.1080/09613218.2012.617516

du Plessis C (2012) Towards a regenerative paradigm for the built environment. Build. Res. & Inf. 40(1):7–22. https://doi.org/10.1080/09613218.2012.628548

Ecovillages and sustainable communities: a report for Gaia trust (1991) Context Institute

Global ecovillage network—community for a regenerative world (2016) Global ecovillage network. https://ecovillage.org/. Accessed 10 June 2016

Haselsteiner E, Rizvanolli BV, Villoria Sáez P, Kontovourkis O (2021) Drivers and barriers leading to a successful paradigm shift toward regenerative neighborhoods. Sustainability 13(9), Article 9. https://doi.org/10.3390/su13095179

Larrick S (1997) A living systems model for assessing and promoting the sustainability of communities. In: Annual conference of the community development society. https://eric.ed.gov/?id=ED417338. Accessed 27 June 1997

Lyle JT (1996) Regenerative design for sustainable development. Wiley

Magnusson D (2018) Going back to the roots: the fourth generation of Swedish eco-villages. Scott. Geogr. J. 134(3–4):122–140. https://doi.org/10.1080/14702541.2018.1465199

Mang P, Reed B (2012) Designing from place: a regenerative framework and methodology. Build. Res. & Inf. 40(1):23–38. https://doi.org/10.1080/09613218.2012.621341

Mang P, Reed B (2020) Regenerative development and design. In: Loftness V (ed) Sustainable built environments. Springer US, pp 115–141. https://doi.org/10.1007/978-1-0716-0684-1_303

Mare, E. C. (2000). *A Concise History of the Global Ecovillage Movement*. Village Design Institute.

Meltemi (n.d.) Global ecovillage network. https://ecovillage.org/project/meltemi/. Accessed 9 Dec 2022

Mitscherlich A (1965) Die Unwirtlichkeit unserer Städte. Suhrkamp Verlag, Anstiftung zum Unfrieden

Reed B (2007) Shifting from 'sustainability' to regeneration. Build. Res. & Inf. 35(6):674–680. https://doi.org/10.1080/09613210701475753

Wahl DC (2016) Designing regenerative cultures. Triarchy Press

Weitzhofer-Yurtışık T (2012) An ecovillage-initiative and its interact: "developing a village" in Central Anatolia [Diplomarbeit]. Universität Wien

What is an ecovillage—discover innovative eco communities (2013) Global ecovillage network. https://ecovillage.org/projects/what-is-an-ecovillage/. Accessed 28 Sept 2013

Links and further information on the ecovillage initiatives

Meltemi, Greece. https://ecovillage.org/project/meltemi/
Il Popolo degli Elfi, Italy. https://ecovillage.org/project/il-popolo-elfico/
Ecovillaggio Torri Superiore, Italy. https://php7.torri-superiore.org
Oasis du Coq à l'Âme, France. https://www.oasis-ducoqalame.com
Hameau des Buis, France. https://hameaudesbuis.org
Arterra Bizimodu, Spain. https://arterrabizimodu.org
VIDALIA, Spain. https://www.vidalia.coop
Sunny Hills, Slovenia. https://soncnigrici-istra.eu/
Güneşköy, Turkey. https://www.guneskoy.org.tr/
Cambium, Austria. https://www.cambium.at

Gastrosophy—Fed Cities: The Perspective of Food in Urban Planning

Jaime Gastalver López-Pazo and Lilian Weikert García

Abstract With *Gastrosophy: Fed Cities*, LaPlasita—a collective focused on architecture and the design and execution of cultural projects- carries out projects, investigates, analyses, and proposes tools to improve the relationship between food and cities. We apply the concept of *Gastrosophy* to build a new ethical corpus to introduce the perspective of food in urban planning. Working on *Gastrosophy* and urban planning, it pursues two objectives: on the one hand, to make city dwellers aware of how we eat and how and why we should change our eating habits to better care for our health, our communities, and the planet; on the other hand, involve public actors, administrations and institutions in the development of the transition city towards a more sustainable, resilient, healthier, more biodiverse and supportive system. We design an urban-territorial transition that moves from the 20th'century city development model to another one in which to ruralise the metropolis, as Cerdà promoted "we must ruralise the urban and urbanise the rural" (Cerdà 1867). This way of approaching the production of territory and the city is within the parameters of what is now understood as regenerative urbanism. We, practitioners, apply utopian thinking to strengthen critical thinking and imagine possible better scenarios. From the field to the table, as a framework to develop different awareness-raising activities, with art as a mediator and urban planning projects.

Keywords Gastrosophy · Sustainability · Utopia · Urban_Planning · Food_Sovereignty

---

J. Gastalver López-Pazo (✉) · L. Weikert García
laPlasita, Sevilla, Spain
e-mail: jgstlvr@hotmail.com

L. Weikert García
e-mail: liweikert@gmail.com

© The Author(s), under exclusive license to Springer Nature Switzerland AG 2025
C. Cobreros et al. (eds.), *Regenerative Design, Cities and Nature*,
https://doi.org/10.1007/978-3-031-76890-3_13

1 What Do We Investigate?

We investigate the food chain from a holistic perspective. Developing the concept of *Gastrosophy* (see Fig. 1), we work in the theoric framework of geographers as Eliseo Reclus. Food impacts different socio-economic and environmental aspects.

How do we feed cities? How do we feed the city we want? These poetic and political questions are the subject of this research.

According to the UN World Cities report (UN 2022): "we are witnessing a world that will continue to urbanise over the next three decades—from 56% in 2021 to 68% in 2050". One of the challenges that cities face is the fair and accessible supply of healthy food—included in the Sustainable Development Goals (SDG)—while regenerating and caring for planet Earth. We must seek a territorial balance and avoid the loss of ancestral and traditional knowledge and know-how that disappeared due to the migration from rural to urban areas. Our proposal following Cerdà is to ruralise the city. We believe that architects and urban planners can engage in this task to improve the life of city dwellers. As the UN World Cities report points out, the fact that the land covered by cities, towns and semi-dense areas is still under 10% can be seen as a space for improvement. Our research proposes redesigning the urban–rural continuum from the perspective of the relationship of the food system to the urban areas (Fig. 2).

In this regenerative urban and territorial transition, food is key. Food belongs to our everyday life and is directly related to climate change. Food production and

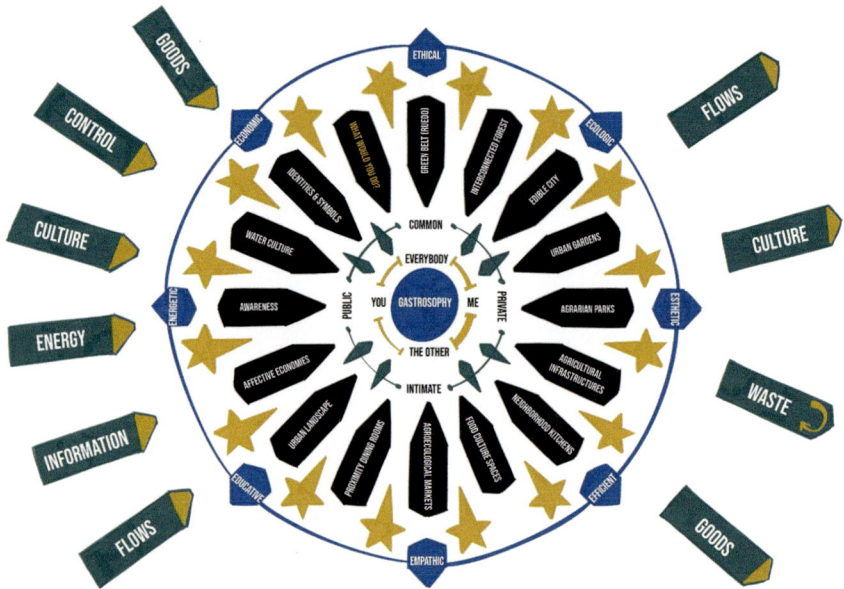

Fig. 1 Diagram representing the concept of gastrosophy

Fig. 2 Ruralise the city. Urban orchard in el Carmel, Barcelona

delivery significantly affect environmental issues. However, at the same time, climate change is already impacting agriculture via heavy rainfall, droughts and other extreme weather events that make farming challenging and unpredictable. Many cities rely on food imports from regions where the climate is becoming less stable (Fig. 3).

The Iberian Peninsula is a global change hot spot. It is desertifying. The rural world is emptying, and the country-city dichotomy recedes dayly into an overwhelming abyss. The city mistreats the countryside, from which it feeds and which it despises. It is necessary to work towards food sovereignty to break this metabolic gap.

Fed cities, an action research, studies and projects practices and tools that can offer cities and the countryside a healthy, safe, sustainable, supportive, and tasty urban food system. We work locally, in the metropolitan area of Seville, as a laboratory where we can experiment and imagine other possible scenarios. However, at the same time, this research can be extrapolated to different European and Mediterranean cities due

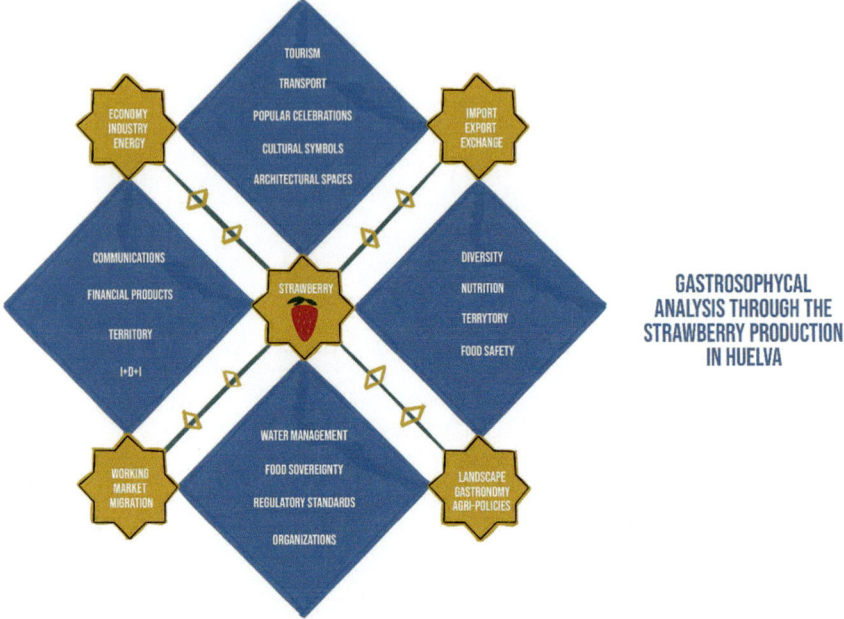

Fig. 3 Diagram representing the gastrosophical case study of the Huelva´s strawberry

to their socio-territorial organisation. We start from local symptomatology (Seville) to make an Iberian and European diagnosis.

Our praxis is based on this question: How to feed more healthily the million and a half people who inhabit the metropolitan area of the city of Seville while taking care of the planet? *Fed Cities* explores a city in transition where agroecology is closer to urban processes, where it is understood that the countryside is not a mere pantry but is much more than a territory at the service of the city. *Fed Cities* works on utopias that ruralise the city, working on the new urban–rural continuum (UN 2022). *Fed Cities*, through regenerative urbanism, bring citizens closer to food production, from which dwellers are detached nowadays. Food reaches our tables almost by magic, without us often being aware of the complex system that sustains this path (Steel 2020).

What attitude do we urbanites want to have with the countryside? Thinking and acting with urban and peri-urban agroecology is a direct way to imagine new possible worlds, more inclusive and more socially and ecologically sustainable. Moreover, it allows us to dismantle the urban models of the 20th century and build with other critical and creative parameters and instruments.

Here we will list some of the references, ideas, tools, experiences, and practices we play with to shape and test our theory.

2 What Problems Have We Detected?

In the summer of 2008 in Spain, lorry drivers and fishermen mainly supported a strike due to increasing fuel prices. Lorry drivers stopped, and supermarkets ran out of supplies in a few days, threatening the city´s food supply. As urban planners and citizens, we were surprised by the speed with which "food" disappeared from our reach. We detected an obvious fact: food is not produced in the city. The city, where most people live, is far from the primary sector. Urbanites have detached from their food production. In the history of cities, they had close resources to face situations of shortage: Intramural and extramural orchards. It is not the same today. Already in 1921, a "planned railroad strike made real the possibility that a city of the size of New York, dependent on distant food supply and losing nearby farmland, would be a risk if food supplies were cut off" (Verhoeven and Wiskerke 2018).

Food is a complex system. So we frame the relationship between food and cities by focusing on three problems:

- The city is a food production desert. In cities, food production is anecdotal. A hitch in the supply chain generates complex situations of shortages.
- The current global-industrial food system harms the planet and people. Locally based agroecological consumption enhances our spirit, social justice and mitigates climate change.
- The countryside-city relationship is unbalanced, with the city dominating the countryside, which affects the house (=Eco), knowledge-heritage, lands, and the sustainable balance between living beings. The city devours, dominates, and does not recognise the many ecosystemic values of the countryside, as the city understands the rural as its mere pantry. Ruralising the urban will improve this imbalance.

Since 2008 we have varied several times the research methodology. The first challenge was to think how much bread was eaten in Seville, with 700,000 inhabitants, in a year. How much flour and water did it take to produce that bread? How many tons of wheat had to be harvested for that flour? Finally, how much land was needed to grow that wheat? We then looked at how to relocate that land spatially. Once the game was over, we analysed how to reshape urban planning to establish new relationships between food and the city and how to transform dwellers´ habits.

During this exercise, we played with various numbers and maps without any scientific purpose but to raise awareness about the relationship between the city and food. We realised that the perspective of food could become part of the constructive heritage of urbanism.

Since then, food has been gaining notoriety. Grassroots movements arise, like the Via Campesina demanding food sovereignty. In 2015, the Milan Charter[1] emerged.

---

[1] https://www.foodpolicymilano.org/wp-content/uploads/2015/10/Milan-Urban-Food-Policy-Pact-_SPA.pdf.

3 What is the State of the Art of This Topic?

In 2015 the Milan Charter emerged in the frame of Milan's World Expo, "Feeding the Planet, Energy for Life". This charter is the first international protocol that encourages cities to develop sustainable food systems that guarantee healthy and accessible food for the entire population, protect biodiversity and reduce food waste. It is a protocol that each municipality can adhere to and propose municipal policies concerning food. *Fed Cities* works with the Milan Charter, which more than 300 cities have already committed.

In the Cilento region, in southern Italy, the cradle of the Mediterranean diet, years before (2003–09), the Bioregion concept was defined. A bioregion has been defined as "a geographical area where farmers, the public, tourist operators, associations and public authorities agree for the sustainable management of local resources, based on the principles and model of organic farming and the agroecological best practices, in order to boost the economic and socio-cultural development of their community."[2] Each bioregion is identified by a lifestyle, food, human relations, and a characteristic nature. Thus, agricultural production has a higher value and is traditional, which is better appreciated by the market. The concept of bioregion is essential to think about the correct scale in the planning of the associated municipalities. The bioregions join the International Network of Eco Regions (IN.N.E.R). This organisation developed in 2021 a charter to support the development of new bioregions and a tool to support territories in this transition (Fig. 4).

The European Commission also promotes food in cities and the Territory-metropolis relationship through various policy strategies. Beyond the Common Agricultural Policy (CAP), there are other strategies like the "From Farm to Fork" strategy for a fair, healthy, and environmentally friendly food system or the Food 2030 research and innovation policy "to achieve a resilient food system and ensure everyone has enough affordable, nutritious food to lead a healthy life".[3]

The Organisation of the United Nations (UN) is a pioneer in these matters with a global perspective. UN has oriented its food discourse through the Food and Agriculture Organisation (FAO) not only to combat hunger but also with the concern of food supply in cities, seeing that the world is urbanising and that the majority of the world population tends to leave the countryside and live in cities and megacities.

Interdisciplinary projects already underway connect the city, the countryside, food, and people. A benchmark in Spain is the collective Inland—Campo Adentro, which, in addition to research, carries out training and awareness praxis while developing direct action projects. Examples are the sheep's flock in Madrid centre or the shepherds' school in the Sierra de Guadarrama. There are also other references from various disciplines, such as geography with Jaime Izquierdo and his "agropolitan city":

---

[2] https://www.ecoregion.info/.

[3] Food 2030. https://research-and-innovation.ec.europa.eu/research-area/environment/bioeconomy/food-systems/food-2030_en.

Gastrosophy—Fed Cities: The Perspective of Food in Urban Planning

Fig. 4 Mediterranean diet walkshop by LaPlasita

From its [the city] foundation until the Industrial Revolution, it maintained a symbiotic relationship with the countryside through a round trip path that linked the peasants' lands with the food market. Today that relationship is broken, and it is necessary to rebuild it. The agropolitan city is the one that recovers, rehabilitates, and updates the relationship between the city and the countryside through new objectives and innovative management instruments. (Izquierdo Vallina 2019).[4]

Other authors, like the anthropologist Manuel Delgado or the philosophers Daniel Innerarity and Marina Garcés, are also closely related to public space.

Carolyn Steel, an architect, is one of the first researchers to address this issue holistically as she comments in the foreword to her book Hungry City that this is "an unusual and revolutionary study that examines how the modern production of food has damaged the balance of human existence" (Steel 2020).

Some projects unite art, science, and the peasantry, such as the "Terra, Teca, Traca" festival. Also, various co-financed projects by the Carasso Foundation[5] in Spain and France work with education, art, and food.

In 2023 both Spain's contribution to the Venice Architecture Biennale with the "Foodscapes"[6] project and Bilbao's Urbanbat[7] festival "How does food shape our cities?" focus on this issue.

Finally, there are research forums where local experiences of good practices are shared, and work is done to overcome the poor and problematic country-city/rural–urban dichotomy. One example is Spain's Network of Municipalities for Agroecology[8].

4 Feeding Cities Without Eating up the Planet

Food shapes cities and shapes the countryside that supplies them. Food is a crucial element in our lives regarding survival and how it shapes a society's culture and identity.

> The feeding of cities arguably has a greater social and physical impact on us and our planet than anything else we do. Yet few of us living in modern cities are conscious of the process. Food arrives on our plates as if by magic, and we rarely stop to wonder how it might have got there.
>
> When you think that every day for a city the size of London, enough food for thirty million meals must be produced, imported, sold, cooked, eaten, and disposed of and that something similar must happen every day for every city on earth, it's remarkable that cities get fed at all. Food shapes cities, and through them, it moulds us—along with the countryside and oceans that feed us. (Steel 2020).

---

[4] Quote translated by the author.
[5] https://www.fondationcarasso.org/es/.
[6] https://www.accioncultural.es/es/18-edicion-de-la-bienal-de-arquitectura-de-venecia-2023.
[7] https://12festival.urbanbat.org/.
[8] https://www.municipiosagroeco.red/.

Even so, the most important issue today is energy. Even though it is at the centre of the discussion, food is not fundamental. The end of the carbon era takes away the prevailing discourse and resources. All public aid focuses on energy saving and digital transformation, not food.

Fernández Galiano writes in the magazine Arquitectura Viva "At a time when energy debates are more pertinent than ever, food remains in the background. Yet how we produce, distribute, and consume it mobilises our societies, shapes our metropolises, and transforms our geographies more radically than any other energy source."[9]

Nevertheless, many architecture and urban facilities directly or indirectly relate to food: agricultural production facilities, slaughters, industries, logistics, warehouses, markets, digesters, and treatment plants. Also, food is more than that. Let us think about the Mediterranean diet, which is a way of being in the street and relating to others, a lifestyle.

Therefore, it is necessary to put food in the foreground. Food is the energy of life.

More than 30 years ago, technicians, politicians, and citizens demanded and introduced mobility into the thinking of the city and its planning, and today cycle paths, public transport planning, and other sustainable actions are essential in urban planning. For more than 20 years, the environmental and ecological perspective has come into play when designing our cities. Feminist struggles have highlighted the essentials of a gender perspective, which makes the city safer and more inclusive and works with the ethics of care, putting productive and reproductive priorities on the same level. In addition, for some years now, the city, the landscape, and the territory have been worked on from a food perspective, which seeks to stop cities from being so vulnerable and dependent regarding food. For this reason, *Fed Cities* has a double objective: to introduce the perspective of food in urban planning and to raise awareness among citizens and administrations on these issues in order to transform the city and the countryside.

How are urban populations fed today? Is this food system sustainable? What problems exist today in the relationship between the city and food? Is it sustainable how we inhabit cities? Is it possible to continue feeding the cities urbi et orbi over time? When shortages arise, how are urban populations fed? How has the agri-food industry been organised to feed the 6 billion people who live in cities? At what price do we feed ourselves? Is it ethical the way we eat? Has urbanism been thought of with these parameters? Do architects and urban planners have the tools to face these issues? How can we make cities healthier while taking care of the planet? From an urban and food perspective, how can we make city dwellers freer, more autonomous, and capable of deciding on their cities and environment? How can we organise cities more fairly without destroying the countryside?

At the same time, why do we see the countryside only as a food producer, as the city pantry? What does the city owe to the countryside? How should the relationship be established in the transition city? How to break the dichotomous countryside-city

---

[9] https://arquitecturaviva.com/articulos/foodscapes-pabellon-de-espana-para-la-bienal-de-arquitectura-de-venecia-2023.

relationship? How to transform this dialectic of productive-reproductive work, where we mistakenly believe that the visible city is progress and the invisible countryside is backwardness?

The twentieth century has experienced a human and cosmic disconnection with the culture of the countryside. We are using it, exploiting it, degrading it, and despising it. Cities gobble up everything that can produce energy, whatever its form. The extended world conurbation we live in takes advantage of this infinite metabolic cycle.

The problem is that food in cities needs to be more sustainable, healthy, sovereign, or supportive. We propose to reverse this lifestyle with the project *Gastrosophy: Fed Cities*.

- *Fed Cities* is a concept, a think tank, an ideas lab, research, and projects to see how to transform society and cities from the food perspective. We rethink how we eat in cities to promote a city in transition that improves life and the environment.
- *Fed Cities* is a hands-on work that provides solutions to the three problems on which we have decided to focus and that define European and Mediterranean cities: Connecting the city to food production in the urban and peri-urban environment, improving public health and the planet by changing food systems and harmonise the territorial balance and the countryside-city relationship.

This research tries to create a new narrative to explain how we can organise cities, the countryside, and our habits in a healthier, more environmentally and socially sustainable way. We will focus on improving cities through food to make them fairer, more inclusive, beautiful, bio-sustainable, eco-sustainable, intense, and playful.

5 Methodologies

We work with two methodologies, a regulatory framework and a concept.

Arts-based research refers to a research approach that uses the arts as the primary means of investigating, exploring, and communicating ideas, concepts, or research questions. Unlike traditional research methodologies focusing on scientific methods, arts-based research relies on creativity, artistic expression, and subjective interpretation. For this reason, our artistic practices are an investigation of the problem to be dealt with. We propose utopia as a methodology as an artistic practice. We apply utopian thinking to reinforce critical thinking to imagine possible scenarios, search for planning proposals in the city and the region, and to figure out how we turn the urban issue into a rural agape.

Scientific research from regenerative urban planning focuses on the study and development of approaches and strategies that promote the regeneration and improvement of urban areas in physical, socio-economic, and environmental terms. The concepts of bioregion and agricultural parks are of vital importance here. *Fed Cities* is the design and evaluation of strategies from the food perspective in a broad sense (*Gastrosophy*) of adaptation to climate change.

We expanded the regenerative urbanism research beyond Urbs and added the concepts of Civitas and Polis. The Urbs is related to the physical and structural aspects of the city. *Fed Cities* further incorporates Civitas, the form of the political and social organisation of the community working with the notion of identity and belonging, implying the notion of citizenship, and the term Polis, as a city-state, where citizens have an active role in Polis decisions.

The Milan Charter is the framework that we work with. Since this can set the starting point to review laws, regulations, and government programs, as well as identify the actors and institutions involved in implementing food policies. It involves collecting qualitative and quantitative data, conducting interviews with key stakeholders, conducting field observations, and analysing the effectiveness of measures taken to promote sustainable food systems (Fig. 5).

Applying the Milan Charter, we can propose solutions that affect the following:

- Governance requires coordination and collaboration between the administration and the actors involved. It seeks to promote participation and transparency and design flexible and adaptable governance.
- Health is achieved by promoting a sustainable and healthy diet in cities and encouraging the production and consumption of nutritious, diversified, locally sourced foods. It seeks to reduce the consumption of processed and ultra-processed foods.
- Socio-economic equity implies ensuring that all people have equal opportunities to access healthy and nutritious food, regardless of income level and place of

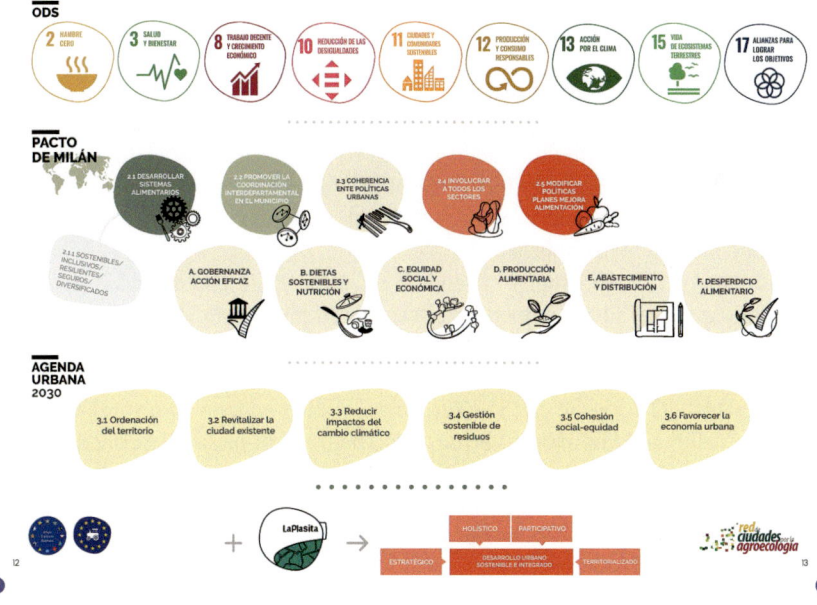

Fig. 5 Frameworks as resources

residence. As well as seeing a market opportunity in these new ways of shaping cities and territory.
- Sustainable food production implies promoting local food production, efficiently using natural resources, properly managing food waste, and adopting sustainable agricultural and fishing practices.
- Distribution and food supply can improve the resilience of urban food systems in the face of climatic, economic, and social challenges. This implies promoting the diversification of food sources, adaptation to climate change, strengthening the capacity to respond to crises and disasters, and promoting local and regional food systems.
- Circular economy and food waste make it possible to reduce the pressure on natural resources, minimising associated environmental impacts and taking advantage of the value of the food produced, contributing to greater economic and environmental sustainability in food systems. Food waste is a global challenge, as at least a third of the food produced is wasted yearly.

Finally, we introduce the concept of *Gastrosophy*. *Fed Cities* takes this obsolete word, coined by Eugen von Vaerst in the 19th century in his book "Gastrosophie oder die Lehre von den Freuden der Tafel" (Gastrosophy or the teaching of the pleasures of the table). We redefine and resignify the term to analyse and design the new forms of relationship that we must establish between countryside-city-food-people-planet. *Fed Cities* plays with this concept in a semiotic way because we believe that new solutions need new thoughts and language as a tool that can help.

6 What is *Gastrosophy*? Some Fragments

Gastrosophy is a way of thinking-acting, a way of looking and associating, a thought-action tool that acts against the imbalance of the current industrial food system since it provides a holistic, hedonistic vision, ethical, aesthetic, ecological, economical, empathetic and energetic approach.

Gastrosophy is: a philosophy of food: We can therefore understand *Gastrosophy* as the philosophy of food claimed by Nietzsche in his Gay Science "Are the moral effects of food known? Is there a philosophy of food? Nothing proves the non-existence of such a philosophy better than the constant agitation for and against vegetarianism!" (Nietzsche 2018). *Gastrosophy* must address issues related to the problem of truth and knowledge. It must also be thought of and described from the every day and with everyone; that is, it is a knowledge of the common. *Gastrosophy* is an ethic, a practical philosophy. Furthermore, it is an aesthetic, a kind of art. For this reason, politics and poetry go hand in hand in *Gastrosophy*, the search for truth, and the unveiling of beauty.

A theory of caring around food: We understand *Gastrosophy* as a theory of caring regarding food. Theories seek to describe, explain, predict, and control phenomena to allow knowledge acquisition to improve practices. Kristen Swanson raised the

"Theory of Caring" in 1991. The core of this theory is that "humans cannot be treated as objects and cannot be separated from self, others, nature, and the larger workforce." The author herself conceives caring as an educational way of relating to a valued being towards whom a personal commitment and responsibility is felt. That is why feminists adopted this theory of caring to speak, among other things, of the productive and reproductive division of labour. With *Gastrosophy,* we adapt the caring concept to the food system to improve its praxis and daily life in the long term. We act from the landscapes to the after-dinner meal, from the value of the work of farmers, ranchers, and fishermen to the knowledge they have about production spaces and ancestral knowledge, from the immense universe of agricultural work to the reuse of waste, from biodiversity to our way of consumption and the time that we dedicate to shopping and cooking. Susan Sontag affirms that care guarantees the survival of the species and its continuity.

A Sophia: From ancient Greek, the use of Sophia has usually been translated as wisdom. It meant intelligence or ability. From the Greek arose the Latin root sap-from which the words *sapere*—to know and *sapore*—to taste and *sapientia*—wisdom arise. In Spanish, to know (saber), to have taste (saber), to be tasty (sabroso), and to be sagacious, to be cultured, to have the ability, to have news, to be cunning and everything with a particular flavour all these things mean to know (saber). All this means wisdom. Can we then start this philosophy of food with these wickers? Can knowledge (saber) and flavour (sabor) be bases that give rise to gastrosophical knowledge? For Aristotle, Sophia appears when the degrees of knowledge separate from sensations. That is, a universal appears in the face of singularity "knowledge separates from the taste of things. Knowledge and flavour are two opposites like the individual and the universal", says he in his Metaphysics. Through *Gastrosophy,* we want to join these two ways of knowledge.

A genealogy of flavour: We must review genealogy and history to construct that *Gastrosophy.* History reflects the power relations that have made the emergence of this or that true discourse possible. "Genealogy makes visible the web of power through which a true discourse emerges." "The emergency", says Foucault, "designates a place of confrontation" (Foucault 1969). The emergency, then, points to a space of struggle and its battles. *Gastrosophy* will face making the city and confront the industrial food system.

A science of appetites, a hedonistic historical line: as a starting point, Charles Fourier seems appropriate to build a society of Harmony from the two original pleasures –love and food—(Fourier 1806). We consider Epicurus, Kropotkin, Reclus, and many others who thought to take care of self, others, and the planet. *Gastrosophy* collects the historical line of more than two thousand years of monadism and epicurism.

A commensalism: *Gastrosophy* treats the fact of eating together as an artistic action, as a revolutionary act, as day-to-day management, eating, cooking, inviting, enjoying a long table, after dinner, being in the kitchen, picking edible flowers, knowing which mushrooms are cookable and which are not.

All of this is *Gastrosophy.*

7 The Hypothesis We Propose

The hypotheses we launch aim to radically transform the city model and work towards a transition city. The hypotheses are a triple helix to transform the urban space:

- Planning hypotheses centred on the bioregions of metropolitan areas and medium-sized cities to organise the countryside and the cities' food supply. Faced with the reality that cities are food production deserts, we propose to supply municipalities, from their closest surroundings, from their bioregion, producing 35% of fresh seasonal products locally and 15% of animal proteins for local consumption.
- Hypotheses designed to transform the urban landscape, deconstructing the city of the 20th century. Ruralise peri-urban areas and associate them with cities to dissolve the country-city dichotomy that has been going on for several centuries. We want to recompose the productive-reproductive dialectic and understand the countryside as much more than an infinite pantry. The countryside and the city are inseparable and must be considered in a balance of forces.
- *Fed Cities* as regenerative urbanism can be a way to carry out these urban transformations, prioritising the regeneration of urban spaces and promoting locally based agroecological food systems (Fig. 6).

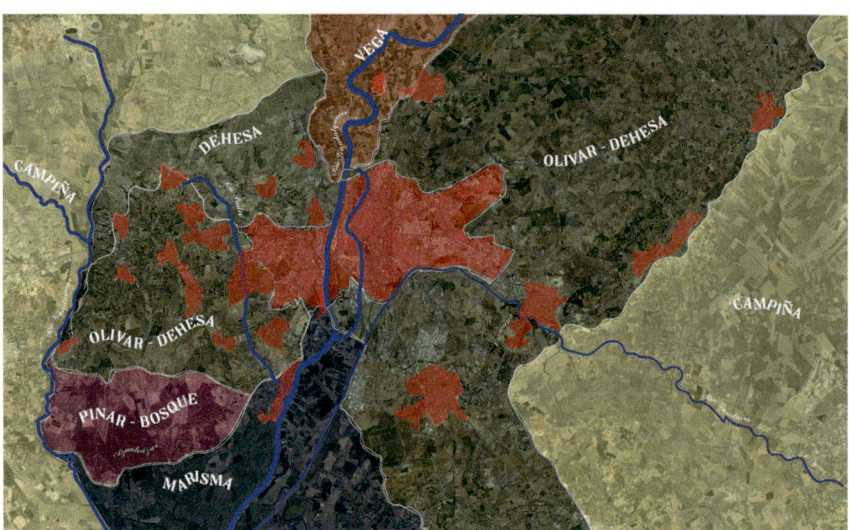

Fig. 6 Productive landscapes of the Seville's bioregion

8 Who Are the Stakeholders, and How Do They Work?

Many stakeholders are involved in this *Fed Cities* process with diverse interests and desires, many conflicting. In that situation, how to act? How to collaborate? How to agree? In the last twenty years, researh has been developed on food, production, sales, consumption processes, modes of new economies, labour situations of migrants, and work on local economies. Radical-and necessary-movements and concepts have emerged, such as food sovereignty, promoted along the Vía Campesina in South America.

Detecting the stakeholders in each municipality is the first of the methodological actions to initiate changes in municipal policies. Carry out the sociogram, establish a leading group of interested parties, and build the project by aligning the agents and the different interests that must lead it.

This will be the first part of the transition. The processes must be participatory, and the citizens must see that their work and engagement have decision-making power. Decisions should be above political changes and not depend on volunteering, social militancy, or funds with a due date. The aim is to create a solid and sustainable structure with its resources and the power to execute.

In addition to urban-territorial problems, there are problems with a political dimension of governance:

- The lack of skills and knowledge of the municipal corporations and the inadequacy of tools to make decisions and act to introduce the perspective of food in their policies. The Milan Charter and experiences in other territories already provide tools and best practices that can be adopted.
- The absence of cooperation between municipalities and the different stakeholders involved. Creating spaces for these other types of governance is possible and necessary. The Bioregion concept can help in create these shared spaces.
- The power of the food lobbies, against which it is challenging to fight. Alliances should be sought with companies in the sector that are committed to the transition and with the alternative business ecosystem.
- The low social awareness of these issues, which can be worked through sociocultural innovation projects that promote collective and critical reflection to transform consumption habits.

9 Some Materials for Urban Planning

We could sing to the 20th-century city with the book of lamentations in our hands:
"Even jackals present their udders/cities/to nurse their young; but the daughter of my people has become cruel like the ostriches of the desert. The tongue of the creatures sticks to the palate because of thirst; small children ask for bread, and no

one distributes it to them. Those who ate exquisite delicacies faint in the streets; those who had been raised in purple embrace the trash."[10]

Our praxis works for a deconstruction of the city, for a transition city that allows the occupation by the countryside and by a particular nature, which allows it to return to inhabit with other species. The Iberian Peninsula is organised with more than 10,000 municipalities of very different scales, and there is an extensive series of medium-sized cities with fewer than 100,000 inhabitants that, we believe, are still capable of generating "an air that makes them free" as Marx said of Los Burgos in the fifteenth century. There are also two large metropolitan areas with more than 4 million people, another two with more than a million inhabitants and some close to a million people.

For this, some of the tools with which we work and serve for decision-making, some concepts that regenerative urbanism should use from the perspective of food are:

9.1 Political Materials

- Common and Void: Understand the concepts of private, intimate, public, and communal. Promote the common within the cities and the territory. Legislation and community work is needed. Generate autonomous free spaces: TAZ, Autonomous Temporary Zone. This attitude that Hakim Bey (1991) devised in 1991 helps prepare us as citizens and build a transition city. We need to generate capital-free present places. These are temporary places that can become revolutionary. Cities and the countryside have vacant lots, voids, and places where the TAZ can be developed. The occupation of these voids and potential places is essential today to move towards another city model. Once citizens occupy them, they must take care of them. Avoid the dispossession to which the current life system submits us. We can deal with deserting that society and building parallel worlds, and we can think with Agamben about putting friendship at the centre of politics.
- Food sovereignty is the right of people to define their own policies and sustainable strategies for the production, supply, and consumption of food based on small and medium production and not on agro-extractivism. Via Campesina, an international movement, promotes this definition (1996). This movement comprises peasant organisations, rural women, indigenous communities, and landless workers who defend family and sustainable agriculture, prioritising economies and local markets and managing the resources in the hands of those who work. Land, water, seeds, livestock, biodiversity and knowledge are among these resources.

---

[10] Lamentations 4:3 https://biblehub.com/lamentations/4-3.htm.

9.2 Economic Materials

- Public purchases: Governments must become aware and make public purchases that have the potential to be a tool to promote positive changes in urban food systems by promoting more sustainable practices, supporting local producers, and improving the quality of dwellers' life through a healthy diet. Specific requirements must be established in the bidding documents. Public procurement can address different aspects of the city and food, such as acquiring nutritious and quality food for public services such as schools, hospitals, or community kitchens. It may also include the contracting of catering services for public events or the acquisition of equipment and technologies related to the production and distribution of food. The administration may even dare to buy land to preserve it from urbanisation and build an agricultural park.
- The price of food: This is one of the keys. Although agroecological food may have a higher price than conventional food, this is due to the potentialities mentioned before, such as their higher nutritional value, quality and flavour, and low environmental impact. However, weaknesses such as the scale of production, lower yield, and limited access may influence its price and limit its availability on the market. There is also the issue that ecological and health costs should be passed on to conventional food prices. If so, these conventional foods would be more expensive.
- This additional cost of agroecological foods may reflect the environmental benefits and preservation of natural resources that they provide. Therefore, the choice of agroecological foods can be an investment in the sustainability and conservation of the environment in the long term.

9.3 Territorial Materials

- Geography and bioregion: are two essential concepts to move towards *Fed Cities*. Understanding what the morphology, geology, climate, geographical features, vegetation, and fauna are like and how all of this is related defines a bioregion that is, for us, a peri-urban territory of cities that will serve as a provider and allow planning territorial, economic and ecological aspects of agricultural production and new landscapes. We usually work within a radius of 30 km so that we can walk calmly in a day.
- Agricultural hinterland and forest: Every municipality must be surrounded by a productive ring of communal lands (pro-commons of Elionor Ostrom) that surrounds the city and protects this land from construction. This zone must be a strip of urban area to produce and generate a biodiverse landscape. These forms are in the historical line of Patrick Geddes' approaches to the garden city. A forest surrounds this ring. Cities must have forests again. These elements can be developed under the umbrella of the agricultural park.

9.4 Productive Urban Materials

- Agricultural parks: Are a figure of urban and territorial planning that permit the protection of fertile urban and peri-urban fields. The parks can be articulated around the water systems of each territory, connecting other green and productive areas to unite the public and common spaces of the cities. Agricultural parks have an environmental, productive, recreation and governance dimension. These spaces must be places of biodiversity that can be private or communal. We can combine the concept of agricultural parks with the next topic of productive urban landscapes.
- Continous productive urban landscapes: We can design productive spaces that function as public places in those city's voids called the Terrain Vague. They could be places designed for grazing, as in la Casa de Campo in Madrid.[11]
- Family farms: It is necessary to promote orchards and traditional and historic farms in the cities, and vocational training with courses for future urban farmers and shepherds that can carry out sustainable businesses. These orchards and family farms must be associated and have cooperatives where they can share. Promoting a new generation of farmers, in this case, in the peri-urban space, is necessary. All this will require new architecture and urban planning. To include the figure of the orchard within the urban plans as essential equipment, such as a school or a park, is crucial.
- Farms and urban grazing: having spaces in cities and their surroundings designed to coexist with livestock, grazing and beekeeping. Using animals to manage and maintain urban green spaces and carry out livestock provides a series of benefits of an ecological nature, biodiversity maintenance, soil quality improvement, environmental education promotion and organic and locally based animal protein.
- Urban gardens: They are the element that is currently most active in cities. There are school gardens where pupils learn these ways of being. It is necessary to ensure pupils can eat the food they grow as part of the daily school menu. Currently, in Spain, the law does not allow it. There are also gardens for elderly people, which have a more socialisation function, preventing, for instance, health issues of this population sector.
- Agroecological markets: If we promote the production of agroecological products, we also have to ensure that the producers can sell their products without intermediaries. Agroecological food to be accessible means that citizens should have access to buy this food. Districts must have at least one agroecological market and prepare squares and public spaces to hold these markets. The Mediterranean city has a significant tradition of food markets that must be reinforced. The weekly market is installed in so-called "herb squares" in Italian cities.
- Food hubs and consumption groups: Many families and individuals organise themselves with what is known as consumption groups, where they buy directly from small peri-urban farms, have places to distribute the baskets and serve as a space for discussion on food-related matters.

---

[11] Project of Inland https://inland.org/product/pastoreo-urbano/

- Communal kitchens, ovens and composting: Districts can own communal kitchens and ovens so neighbours can share resources to cook, bake and eat together. These spaces can be associated with the food hubs and be the place to hold training activities, workshops and talks. Shared composting areas can facilitate the management of waste.

10   Gastrosophical Direct Action

SOME ART PRACTICES OF ART-BASED RESEARCH

Fed Cities is exploratory action research. We analyse, propose, experiment, learn and transform with projects that we carry out to confront our knowledge, thoughts and practices. We generate a desiring economy in each project and try to make proposals for social innovation related to food and the city, which transit within the field of art.

CASAURANTE PROJECT/2014–2020/For six years, we transformed the home living room on Fridays into a food house (home + restaurant = casaurante). Open to the neighbourhood, we cooked a menu of starters, three courses and seasonal desserts. We recovered obsolete ingredients and traditional dishes that are in danger of extinction while creating a space for relationships between friends and strangers.[12] This experience allowed us to meet producers from the metropolitan area with whom we have established relationships since we cooked their products. Casaurante also allowed us to meet many people who came home to eat. With this artistic practice of commensalism, we have acted as intermediaries between producers and consumers.

MEDITERRANEAN DIET/February 2019/We have participated with working groups from various European countries: Portugal, Spain, France, Italy, Greece, and Bosnia, to define and work on the Mediterranean diet. The project seeks to promote the innovative capacities of the Mediterranean region to ensure its sustainable growth while promoting transnational activity between clusters and networks of key sectors in the Mediterranean area. One of the meetings was in Seville, and laPlasita took care of the organisation. We designed a walk through the city, visiting various places in the historic centre, which we understood represented the Mediterranean spirit. We walked through diverse places like a convent for cloistered nuns to taste their sweets and cakes and to a squatted urban garden where children from nearby schools play. We also visited a young bakery and a popular dining room where they cook traditional paellas. The day ended in our studio, where we offered a tasting menu, tapas type, where we made a gastrosophical journey through the history of the different cultures that have been building the city of Seville for more than two thousand years. The cultural exchange enriches the narrative. Cooking the food allowed us to study and reflect on the introduction to the city of different products, species and ways of eating and cooking.

---

[12] https://ciudadalimentada.wordpress.com/portfolio/casaurante-proyecto/

On the other hand, we turn a tourist walk into a gastrosophical experience. Here is how to get down to work with artistic practices. Walking and cooking provide a series of lessons of great value for constructing (collective) knowledge.

THE GREAT TEXTILE, THINKING WITH OUR HANDS/April 2019—Present/We not only think with our heads. Building with our hands also gives us the ability to think. *The Great Textile* is a 120m^2 collective fabric artwork representing the metropolitan area of Seville, its five agricultural landscapes and a utopian vision of what the territory could be like (Fig. 7).

This piece is an artistic representation of the *Fed Cities* research project. It has been sewn in collaboration with students from a vocational school. We proposed an innovation program in project-based learning. The piece works as a communication machine to publicise the topic it deals with. Based on this growing utopian piece, we are doing reverse research for the *Fed Cities* project, slowly giving the research a scientific aspect. *Fed Cities* is then also an arts-based research. With the support

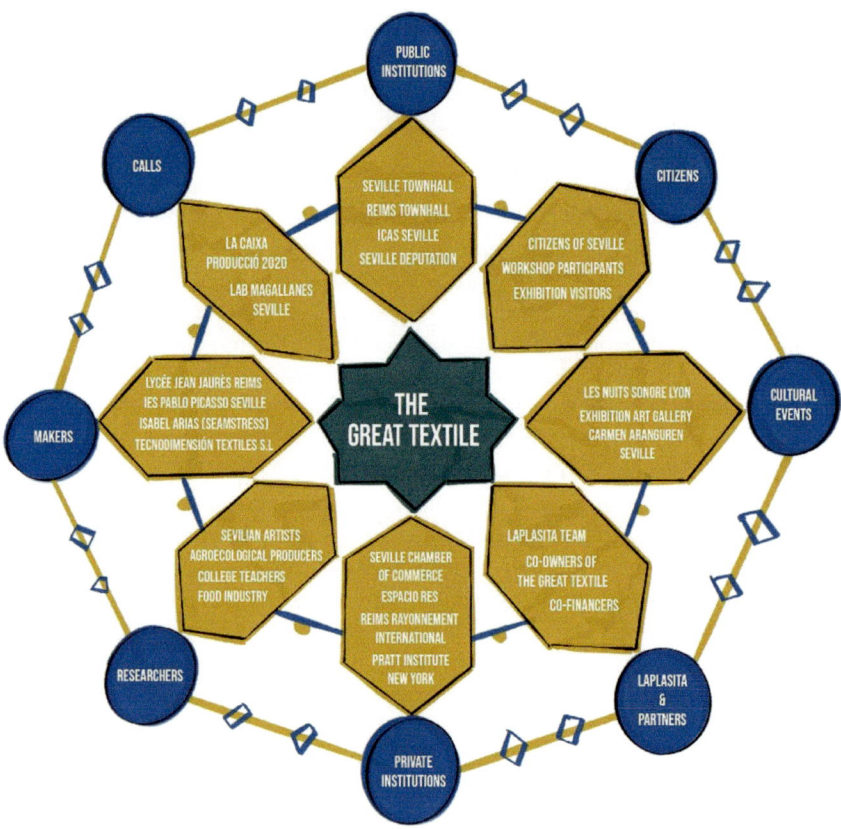

Fig. 7 The Great Textile, flow chart

of different groups of people, we continue sewing the Great Textile. It is a work in progress (Fig. 8).

WALKING AS AESTHETIC PRACTICE—WALK ALONG THE PUDIO RIVER, THINKING WITH YOUR FEET/March 2021/We think with the heads, hands and feet. We understand the action of walking as an aesthetic practice. While sewing and building the participatory fabric artwork *The Great Textile*, we explored

Fig. 8 The great textile, participatory fabric artwork

Fig. 9 The Great Textile, presentation as participatory artwork

the Pudio River, which articulates the metropolitan area territory and can be an essential piece of the agricultural park we propose as part of *Fed Cities*. It is crucial to experience the place, meet its inhabitants and learn about its heritage. From this walk, a small documentary emerged.[13]

CULTURE LABORATORY, FED CITIES/October 2020–May 2021/Right after the confinement due to the Covid-19 pandemic, we started an arts lab residency at the Royal Artillery factory in Seville to continue working on our *Fed Cities* project. There we managed to create a network of stakeholders working on the issue of food and cities from different disciplines. This network allowed us to work more broadly, sitting at the same table with researchers, producers, urban planners, early childhood educators, cooks, poets, artists, and many others who enriched our knowledge. This arts laboratory ended with the exhibition of *The Great Textile* and the organisation of the 1st Gastrosophy Congress.[14] Over three days, 21 experts (researchers, producers, activists, food providers, and artists) shared their experiences and discussed *Fed Cities* (Figs. 9, 10 and 11).

FED FARO/November 2021/It is important to leave the Seville laboratory and be able to apply knowledge to other places, in this case, to Faro in Portugal. In collaboration with the municipality, the tourism and cooking universities and the Tertulia Algarvia[15] foundation, we designed, organised and carried out a workshop to analyse and experience the food production spaces of Faro. Fed Faro consisted of a trip from the Formosa estuary, passing through the food market and agricultural

---

[13] https://youtu.be/4on9V7fPiyk.

[14] https://www.diariodesevilla.es/opinion/articulos/Berlin-Artilleria_0_1579342164.html.

[15] https://www.tertulia-algarvia.pt/es.

Gastrosophy—Fed Cities: The Perspective of Food in Urban Planning

Fig. 10 The Great Textile, work in progress in Girona

farms to the mountains, its fruits, and the olive groves, and ending with a cooking session and dinner at Tertulia Algarvia. After the session, all the participants became aware of the rich resources of the city and the countryside. They had never put on the glasses of looking at the city as a place that produces food, landscapes, and culture. Eating is culture! (Fig. 12).

UTOPIA, FED AGAPES/November 2021–October 2022/We organise dinners with official and unofficial researchers, once a month. Celebrating around a table of food is always a great way to connect with others. Our goal was to democratise the ability to imagine better futures by reflecting on how we eat in cities, what our food systems are like, and our relationship with the countryside. Since we all have some relationship with food, we all eat, almost all of us shop, and some cook. In this way, by reflecting on food sovereignty, we can all participate and propose solutions for the future of food in the city. Applying a creative methodology, we collectively reflect on *Fed Cities*. Sitting around the table, too, we can become aware of the rich diversity of our community. At the table, each diner sews and prints their utopia on the lunch napkin. These utopias will make up a new participatory work of fabric art (Fig. 13).

OPEN INNOVATION: ARTISTIC RESIDENCE/October 2023/Call for three artists to participate in the *Fed Cities* project. This *artist*-in-residence aims to enrich the research with the experience, expertise and perception of the artists and their work.

Fig. 11 The Great Textile, exhibition in Seville

11 Use Case: Bormujos Alimentado—Fed Bormujos

The last project we like to explore is an experience that merges some of the tools and practices we explained: Bormujos Alimentado—Fed Bormujos.

Bormujos is a municipality in the Metropolitan Area of Seville, which belongs to the Aljarafe cornice. A territory that was agricultural for centuries, with a predominance of olive groves and beautiful farmhouses since Arab times. With the turn of the millennium, this territory has become a commuter town of the capital—12 km away—creating a conurbation. It has 20,000 inhabitants, and semi-detached houses characterise its morphology. It still maintains a tradition of wineries where they step on the grape, producing *Mosto,* a typical wine of the region.

We presented our *Fed Cities* project to the municipal corporation, convinced them, and we were hired by the city council, through a minor work contract, to develop a

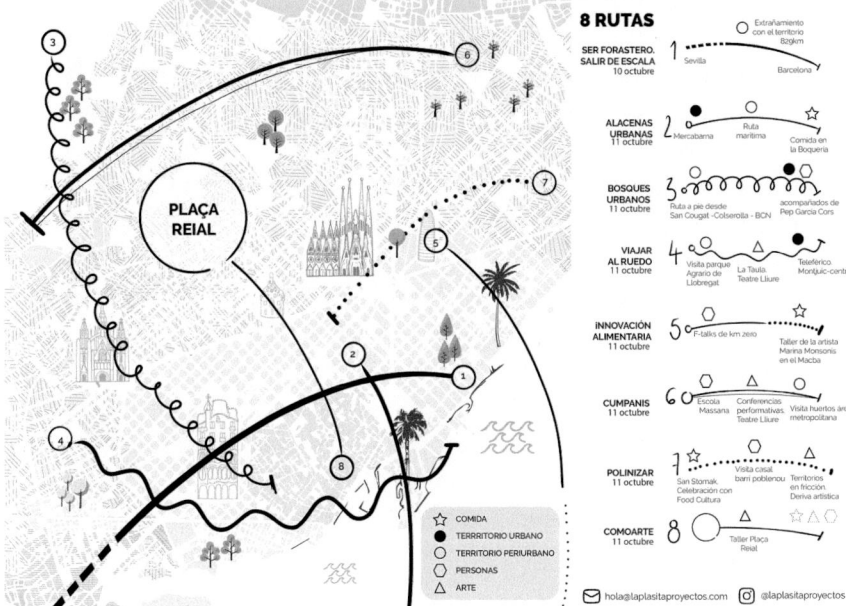

Fig. 12 Fed Barcelona, proposal for 8 walkscapes

strategic *gastrosophical* study, which proposes introducing the perspective of food within municipal policies and being able to move towards a transition city, a healthier city.

We have worked on different aspects, some of them based on the lines of action of the Milan Charter. The final result is a report and several actions carried out during the year (Fig. 14).

With the motor group, we made progress in other lines, such as:

- GOVERNANCE/Governance to improve city government, improve interdepartmental relations, promote participation, and have a transversal strategic plan for municipal food policies. ACTIONS: Analysis of the municipal government—What is already being done in terms of food policies? + Realisation of a sociogram of interested parties (Stakeholders) + Design of a plan of awareness, training, and participation strategies + Signature and adherence to the Milan Charter as a framework + Affiliation to the network of cities for agroecology + Definition and implementation of transversal food policies.
- URBANISTICS/The Milan Charter, in the fourth of its recommended actions, says "promote food production". Urban planning can contribute in this aspect: from its planning figures or territorial organisational capacity, protecting soils and transforming landscapes to ruralise the city and have productive peri-urban areas. ACTIONS: "Eleven parks: One forest" Study of parks, green areas and public spaces and improvement plan through small actions of urban acupuncture

Fig. 13 Ágapes, collective reflection on food and city

and commissioning as productive spaces + Expansion of existing urban gardens from 80 to 120, giving the possibility to the most vulnerable families to exploit the orchard for their consumption + Proposal for an Agrarian Park in the General urban development plan, around the existing streams, 275 ha for local production + Management of water resources in collaboration with other administrations + The Pudio river, as articulator of the Aljarafe + Envision Utopia, as we would like the town to be (Fig. 15).

Fig. 14 Bormujos Alimentado, poster

- DIRECT ACTIONS/We always work with direct actions to show the theoretical approaches or first experience, and then theorise. Activities provide awareness and necessary participation in learning and the transformation of our daily habits to move towards another city model far away from the harmful parameters of the 20th century, generators of large amounts of carbon footprint. ACTIONS: Promote the current Monthly Agroecological Market + Canning Workshop in one of the parks of the municipality: the residents of the town are invited to bring glass jars to package the last vegetables from the garden to have homemade

GRUPO SOCIOMOTOR
GOBERNANZA

PREPARAR UN SOCIOGRAMA Y CONSTRUIR UN GRUPO MOTOR
QUE ACTÚE COMO RESPONSABLE DE LAS ACCIONES A DESARROLLAR

OBJETIVOS DE DESARROLLO SOSTENIBLE

| 01 | 02 | 03 | 04 | 05 | 06 | 07 | 08 | 09 |
| 10 | 11 | 12 | 13 | 14 | 15 | 16 | **17** |

Localización:
Municipio de Bormujos, Sevilla

Fecha / Estado:
Proceso

ODS Principal:
17. Alianzas para lograr los objetivos

Agentes intervinientes:
Técnicos ayto. - Sociedad civil - LaPlasita

DESCRIPCIÓN / OBJETIVOS:

Todo proyecto necesita de energía deseante y desbordante para llevarse a cabo. Para ello es indispensable un equipo pequeño y hábil que empuje y dirija este proyecto complejo. Se necesitará por un lado recursos humanos que deseen y entiendan esta transformación urbana y vital a la vez que el municipio debe proveer de recursos para apoyar al grupo motor.

DURACIÓN:
2 meses

POSIBLES BENEFICIARIOS:
Proyecto de Bormujos Alimentado - La Gobernanza

LIMITACIONES:
El grupo motor debe surgir de forma natural y debe ser plural... Debe estar compuesto por personal del ayto. y miembros de la sociedad civil críticos y propositivos. Otro requisito es que debemos todos de pensar en grande y para todos, no solo desde los intereses particulares de cada uno.

Resultados esperados:
Disponer de un grupo de unas 5, 7 personas (40 /50% personal del ayto.) y con un mediador cuando sea necesario.

Recomendaciones para replicabilidad:
La experiencia de otros grupos sería muy interesante para poder aprender. Organizar reuniones con el grupo de Valencia, de Sevilla, de Madrid...

TEMPORAL:

1º Promover entre la sociedad la participación de personas que estén dispuestas a invertir su tiempo en esta labor.

Fig. 15 Bormujos Alimentado, report fiche

preserves and enjoy tomatoes, peppers, aubergines and other garden products in winter + Healthy breakfast in the agroecological market, where producers provide the ingredients and activities related to health, nutrition and sustainability are organised + Compost bins workshop on composting where, in addition to explaining what it is and its usefulness, we built three compost bins with the participants, bins that we placed in three points of the city so that the neighbours can use them. The idea is that the compost is then used in the urban gardens of the municipality, promoting circularity.

- MUNICIPAL POLICIES/In the current social organisation, everything public arises from the state in all its forms. Therefore without the municipal capacity to promote policies that impact the municipality, it is difficult to advance on these issues (beyond autonomy actions and self-managed that can be developed in parallel). ACTIONS: Health Council: associating health centres and food + School canteens, promoting that pupils eat agroecological and participate in what they eat, and even with workshops where they plant, collect and cook + Entrepreneurship. Promote the local food industry, foster the creation of cooperatives and promote the productive labour market, as well as have a small co-working of startups related to food + Community kitchen and cooking workshops inside a municipal building with an annual social program + where the Food Hub can be, where knowledge and exchange are promoted + Community wood oven, in one of the parks for the construction and management of a citizen assembly of this oven + Involve local restaurants, which promote local agroecological food on their menus (promote it, especially during the celebration of the Tapas week).

The weaknesses are apparent, and they have already been commented on. There needs to be more collaborative and participatory culture. There still needs to be more knowledge about the usefulness of this regenerative food urbanism. In addition, the administration times are prolonged, the regulations do not help, and the political parties have specific interests.

This work, developed during 2022, will be expanded in the second semester of 2025 to 10 municipalities. For three years, we will prepare and work with the administrations of these municipalities to promote the creation of the Pudio River Agrarian Park. A park of more than 2,500 ha is to be articulated around the stream to produce food for the 120,000 people who live in those ten municipalities and which is the first of the eight agricultural parks that the city of Seville should have in the next fifteen years, to move to that other way of making the city (Fig. 16).

12 We Are Saying Goodbye

The *Fed Cities* research falls within the category of Participatory Action Research. Our procedure to approach this already complex system, that is, the relation between city and food, has been:

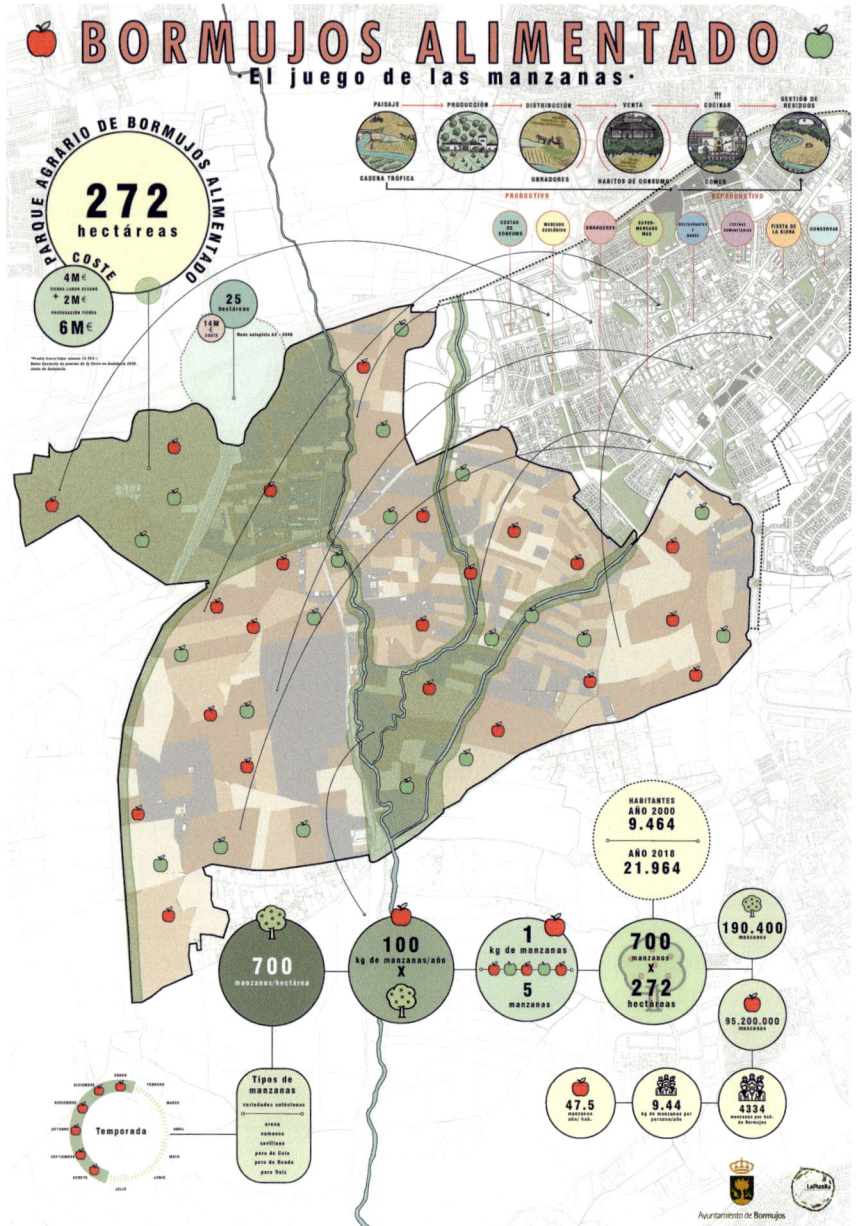

Fig. 16 Bormujos Alimentado, game of apples

- Invent a word, GASTROSOPHY. Build a semiotics around it.
- Invent a *Fed Cities* concept to ask the right questions to provide solutions that solve problems in cities and their relationship with the countryside and food production.
- Think with the head, with the hands and with the feet, collectively and collaboratively.
- Design actions that impact in different ways:
  - By gathering information, the investigation feeds back when exposing itself: thought and action;
  - By raising awareness of society with action as a way of communicating;
  - By creating a relational corpus of interested parties (stakeholders) for "hands-on work".
  - By setting the change in motion, the transformation.

Involving different actors in research, both to collect data and knowledge, as well as to transmit learning and develop awareness-raising activities, is complex. So the times and processes are lengthened. At the same time, actions need resources that are only sometimes here and now.

On the one hand, since food is a topic that everyone can reflect on, talk about, and look for alternatives since we all eat, almost all of us buy, and some cook, it allows us to develop different projects linked to research in an organic way. On the other hand, since it is a sensitive issue and requires a political stance, this interrelationship between the different actors and organisations is complex.

We want to launch at least one prototype of research action activating the health councils contemplated in the 1988 Spanish health law and which have only been implemented in a few Spanish cities to holistically treat the lines of action that mark the Milan Charter.

Other lines of research could lead to the study, analysis and project of the organisation of the Iberian Peninsula as bioregions and regions to make a complete study and seek solutions to depopulation in the national territory, as well as in the creation of production networks and distribution of farm products in the cities, providing a solution to the atomisation of small local agroecological producers. A project that, with our resources, would take two lives but that we will not stop doing for it.

With *Fed Cities,* we want to imagine a present and future of food, specifically in urban, utopian, or eutopian environments, in which our food could be supportive, safe, healthy, sustainable and tasty. Since so many actors, organisations and decision-making centres come into play in this transition towards a better city from the point of view of food, putting this research into practice takes work. Nevertheless, small steps can be taken, small-scale changes that, in due course, may significantly impact the solution to our three primary research questions.

However, many interests move and are very broad and multiple areas in which this problem unfolds, so the process sometimes becomes uncertain, ambiguous and cumbersome.

Wanting to transform the current world food system is fighting against giants, but like that verse by Ossip Mandelstam, we must do without the need for anything.

"There is *grass* in the *streets* of *Petersburg*, the first runner-sprouts of the virgin *forest* that *will cover* the space of *contemporary cities*." (Mandelstamm 2005).

References

Bey H (1991) T.A.Z Temporarily Autonomous Zone, Enclave de Libros Ediciones
Cerdà I (1867) Teoría general de la urbanización
Foucault M (1969) The archaeology of knowledge, Editions Gallimard
Fourier C (1806) Theory of the four movements, Barral, Barcelona (1974)
Izquierdo Vallina J (2019) The agropolitan city, KRK Ediciones
López García D (2015) Producir alimentos, reproducir comunidad, Libros en Acción
Mandelstamm O (2005) Sobre la naturaleza de la palabra y otros ensayos. Árdora Ediciones
Nietzsche F (2018) The gay science, Agapea (Original work published 1882)
Steel C (2020) Hungry city, Capitan Swing
UN (2022) World cities report 2022
Verhoeven S, Wiskerke JSC (2018) Flourishing foodscapes - design for city-region food systems

Magazines:

Soberania alimentaria
a+t magazine is this rural?
Arquitectura Viva. https://arquitecturaviva.com/articulos/foodscapes-pabellon-de-espana-para-la-bienal-de-arquitectura-de-venecia-2023

Building Energy Legislation and Incentives, a Latin American Vision Towards Regenerative Design

Seyedehniloufar Mousavi and Caribay Godoy-Rangel

Abstract Global climate concerns have largely contributed to developing sustainable solutions throughout different industries. In the building sector, one of the most polluting industries due to its high energy consumption, designers generally seek to reduce negative impacts on the environment by trying to cover the pillars of sustainability. However, this may no longer be enough, and it is necessary to generate positive impacts that can reverse the impact of buildings on climate change. This new approach changes the expression from less bad to better. Positive Carbon/Energy Building (PCB/PEB) is one of the approaches that focuses on creating new relationship between natural energy sources and appropriate construction materials, repairing atmosphere from emissions, as well as increasing well-being and comfort of occupants. This approach implies a positive energy conversion where all the aspects involved in the design allow to generate more energy than is needed, through renewable sources, by the building itself. In order to apply these strategies in the societies, it is necessary to have a legislative framework sensitive to new needs and aimed at promoting the integration of different measures, not only where energy efficiency measures are required but also the generation of positive energy from renewable sources. In Latin America, despite the variety of socio-economic situations and climatic differences, there are many points in common. One of them is the commitment and progress that has been made around the generation of sustainable building regulations that allow generating a base to move towards regenerative buildings. This chapter seeks to map out a regional vision of current sustainable building legislation and incentive plans to motivate citizens in shaping a regenerative future, by focusing on the energy aspect.

Keywords Regulation · Regenerative buildings · Positive energy building · Renewable energy · Latin America

S. Mousavi
Faculty of Sustainable Resource Management, State University of New York College of Environmental Science and Forestry, Syracuse, USA
e-mail: niloufarmx@gmail.com

C. Godoy-Rangel (✉)
School of Architecture, Art and Design, Tecnologico de Monterrey, Ciudad de México, Mexico
e-mail: cgodoyr@tec.mx

1 Introduction

Urban areas concentrate more than 50% of worldwide population and about 70% of annual emissions due to high rate of energy consumption. This rate is predicted to present a sharp growth by 2050, when 70% of human will move to cities (Empowering Cities for a Net Zero Future, OECD 2021), resulting in immense energy and carbon emission increase. To mitigate these emissions and with the aim to limit global warming to below 2 °C (Salim et al. 2017), alliances between different sectors related to the construction sector have been making commitments in the sense of limiting energy consumption and generating clean energy. Here the concept of Regenerative Design becomes important, since it seeks to support climate change, changing the angle from designing less bad to better.

In regenerative concept, instead of typical one-way energy flow, focus is on a balanced partnership between nature and human innovations, seeking to reverse the causes of the climate crisis. In this context, urban district plays a key role in managing and reducing global emissions to achieve climate neutrality, which is why Positive Energy District (PED) is presented as energy efficient and flexible interconnected area of buildings, and additionally produce extra level of energy than demand, by using renewable sources in the district. PED as an example of regenerative cities, has multi directional renewable energy sources and incorporates nature as passive system to support human environment (Hofman and Valderrama 2021). But this concept still has a challenge ahead, the generation of a legislative framework and both national and international incentives that promote it as a priority.

The establishment of laws accompanied by incentive funds has as its aim to break down barriers and transition energy policies to national directives and standards that benefit citizens by providing clean energy and regenerating the environment. Instruments that an effective policy carries, include tax schemes, incentive plans, investigating institutions, monitoring and information access, certification and standards, training and education. However, such policies can only be effective if there are cultural and social changes that promote them in the climatic and specific context of a nation. This is evidenced by the speed with which they are promoted in different regions of the world, for example, in Europe and other developed countries they have consolidated progress since 1980, while in Latin America the promotion of these concepts has been slower since 2000 (Renewables Global Status Report—REN21 2023).

1.1 How Change Can Happen

Regenerative cities which promote energy efficiency and the generation of clean energy from renewable sources, as mentioned before, require a political vision and a clear legislative framework. Regulations regarding to energy efficiency can reduce energy demand extensively, for instance, due to tightening of European regulations, new constructed units consume 50% lower level of total energy for space heating, which derives with minimal citizens, as return of investment comes from reduction in energy cost in long term. Impact of policies on global emission, illustrates that with the current policies applied in some countries, emission level by 2100 can be kept around 2.5–2.9 °C, which is above to the Paris agreement limit (Boyd et al. 2015), however if these policies do not be implemented, global emission level will rise up to 4.8 °C.

Different sectors of society such as local and regional government bodies, as well as citizens are required to collaborate. The idea is to go through different stages of actions, including provision of cost optimum incentive plans, a comprehensive policy framework for energy efficient envelop, as well as clean energy provision and smart control system. Additionally, performance requirements of energy efficient buildings need to be defined based on individual components like residential building or commercial in terms of annual consumption per meter square. Furthermore, Renewable Energy (RE) policies framework require micro-regional planning with regards to climate and available resources. Also, other levels of support need to be provided, including access to data, which are transparent and robust, as well as financial and educational support.

But where to start? knowing the impact of cities and buildings, understanding the need to have positive cities and districts from the energy point of view, knowing the positive impact that has occurred in different regions of the world that have led policies in this regard, what should be the next steps for regions like Latin America? Mapping this region from the legislation and incentives points of view for energy in buildings will allow us to understand where the regions are right now and what could be the best way forward. In the following sections of this chapter, the Latin American economic and environmental characteristics, as well as its efforts in issues of energy efficiency in buildings and promotion of policies for the generation of energy from renewable sources will be discussed. In addition, the concept of regenerative district and its impact on urban environment will be explored. Finally, opportunities to implement PED strategies in Latin America will be identified.

1.2 Full Lifecycle Energy Policies

It is common for energy policies to limit themselves to reducing the energy consumption of buildings during their operation, but this leaves aside the environmental impact associated with obtaining and transforming raw materials (embedded energy) as well as the energy required to close the life cycle of the building (demolition energy). This limitation highlights the importance of considering the full life cycle of energy-related materials and processes. To develop effective energy policies, decision makers should consider incorporating methodologies such as Life Cycle Assessment (LCA) to comprehensively include carbon emissions associated with the entire construction process.

From the conception of the idea of constructing a building a series of requirements will be produced, transported, and processed to finally have a building in operation. Studies have confirmed that the materials most used in building construction, wood, aluminum, glass, concrete and ceramics, have a high embodied energy, with concrete being the largest contributor with more than 60% (Asif et al. 2007). Of the total energy use in buildings, about 10% is embedded energy in the materials, and although in comparison with the operational energy it does not seem to be too much, the reality is that this value has been increasing as the buildings are more efficient, reaching more than 45% in high performance housing (Koezjakov et al. 2018).

In addition to the above, the production and transport of materials related to energy generation, such as solar panels, wind turbines or batteries, have a significant environmental impact. And among the different possibilities there are options with greater and lesser impact, for example, the GHG emission for each kWh of energy produced taking into account the complete cycle of a photovoltaic system is up to twelve times more than an offshore wind system (Amponsah et al. 2014). This should help align policy development towards regenerative design.

Finally, promoting the recycling of materials can help reduce waste and limit the environmental impact of buildings. For this it is essential to move beyond "green washing" solutions that do not really consider the environmental impact of energy-related materials and processes. This is why policy makers need to take a holistic approach when setting energy policies. It must consider each phase of the building and the environmental impact associated with the energy related to it. By conducting life cycle assessments, policy makers can determine the embodied energy and associated carbon emissions of different energy sources and technologies. These measures will help develop efficient and sustainable energy policies that contemplate the environmental impact of energy-related materials and processes.

2 Background of Latin America

There are 18 countries in Latin America under three regions, which under North America, include only Mexico, and under Central America, there are Costa Rica, El Salvador, Guatemala, Honduras, Nicaragua, and Panama. Finally, under South America, there are Argentina, Bolivia, Brazil, Chile, Colombia, Ecuador, French Guiana, Paraguay, Peru, Uruguay, and Venezuela. This area is covered by 8.42% of the global population (Densidad de Población—East Asia and Pacific, Data 2022), which is around 667,590,193 people. The population density is 32 people per km^2, well below other regions of the world such as Europe, with 112 people per km, or East Asia with 96 people per km2 (Fig. 1). This population is expected to continue to grow, an indication is the sustained increase in urbanization where only in 2022 it increased to 82.5% (Santiago et al. 2020).

The increase in population also means an increase in the need for resources, as well as built environmental spaces to live. The Latin American region is known for its large reserve of natural resources, but also for its few policies in managing these.

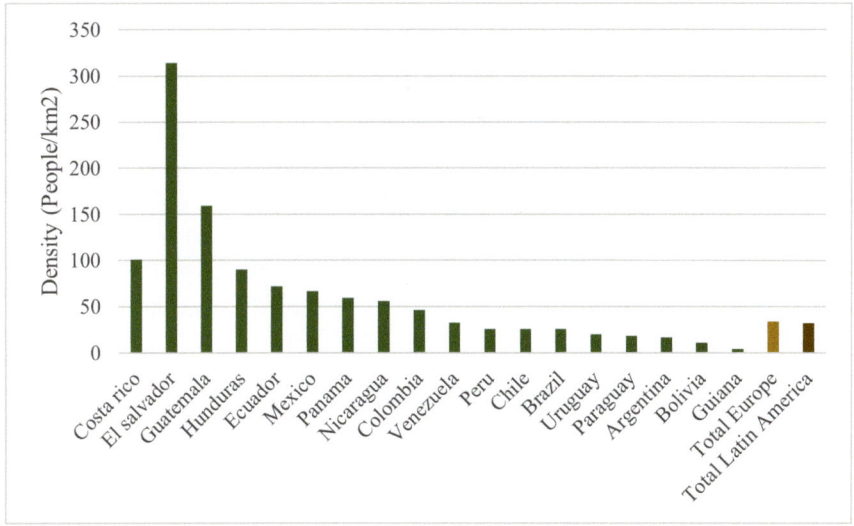

Fig. 1 Density of different countries in LA and comparison of its total to Europe

In addition, the very differences of the region, from the climatic and economic point of view, mean that some countries are more advantaged than others. Identifying commonalities and working based on them will be a strategy that could be highly effective when seeking to improve the behaviour of cities towards regenerative urban areas.

2.1 Economic Overview of LA Zone

Latin America and the Caribbean represents 8% of the world Gross Domestic Product (GDP) according to the Economic Commission for Latin America and the Caribbean (ECLAC). Brazil and Mexico lead the list and stand out for their strength and overall economic growth. Additionally, Brazil as a member of the BRICS, which are developing countries with high economic potential, plays a vital role worldwide (Mourato Pinto 2017). Argentina, Colombia, and Chile continue the list, the latter being the one that has climbed the most positions in recent years. Figure 2 illustrates portion of each member of LA in annual GDP for 2021 according to World Bank, Brazil and Mexico hold more than half (61%) of total GDP, which can directly impact energy budget.

After the impact of the COVID 19 pandemic in 2020–2021, the region begins to rebound at the end of 2022 with a 3.4% increase in its GDP (Policymakers Need Steady Hand as Storm Clouds Gather Over Global Economy 2022). This result will directly impact the supply and consumption of energy (Belke et al. 2011). Hence,

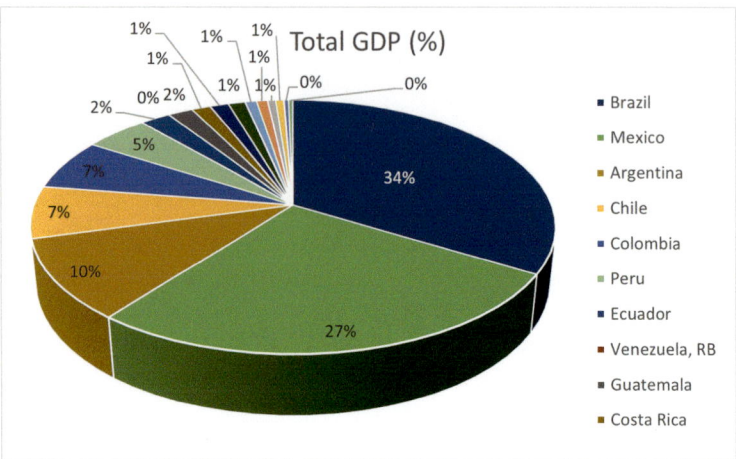

Fig. 2 GDP of different countries in LA

prior to policy and incentive planning, it is vital to consider the economic background and role of it in renewable energy supply and energy efficiency of buildings. Higher levels of urbanization and economic growth, results in larger portion of energy consumption, consequently energy efficiency policies and renewable energy incentives can provide emission reduction and energy security to the countries (Selvakkumaran and Limmeechokchai 2013). Thus, to achieve regenerative cities and buildings, an economy that encourages renewable resources and assess energy saving of buildings envelope, is demanded. Studies about LA has been highly neglected, whereas they could be a great lab for achieving environmental goals in line with economic targets as these countries adopted climate policies (Jakob 2017).

2.2 Climatic and Environmental Synopsis of LA

Buildings energy demand is highly related to climate characteristic of its location, including ambient temperature, humidity, global solar radiation, wind intensity and direction. The climate in Latin America ranges from desert areas to Antarctica, passing through tropical forests and coastal areas as presented in Fig. 3. The Koppen system subdivides the region into almost twenty climate types, where the tropical climate (A) is the one that most prevails. This type of climate can be subdivided into equatorial (Af), monsoonal (Am) and savannah (Aw), where in none of the cases the average temperature of the month is less than 18 °C. Dry climates range from desert (Bw) to semiarid (Bs), which can be hot (h) with a mean annual temperature greater than 18 °C or cold (k) with a mean annual temperature of less than 18 °C. These stand out mainly in the south of the region on the Pacific coast. Variations of temperate climates (C) prevail in the south of the region towards the Atlantic coast.

Its average temperatures range from −3 °C in the coldest months and exceed 10 °C in the warmest months (El Mapa de Los Climas de América Del Sur—Mapas de El Orden Mundial—EOM 2023).

Large areas of different countries share the same climatic characteristics, and the same visible consequences of climate change. As the rainfall patterns are shifting, temperature is rising and extreme weather appears. The average temperature increase in the region during the last decade was around 0.2 °C, while in 2021 the temperature increase was maximum with 0.59 °C, especially in the central zone and Mexico, which caused an increase in the water level in coastal areas. On the other hand, critical hazard in these areas been highly reminded through El Niño Southern Oscillation (ENSO) (Cai et al. 2020). For instance, ENSO brings drought in some areas of Amazon, while causing floods in west and south side of LA, resulting in socioeconomic impacts. An example of this is the drought in Chile, the situation lasted

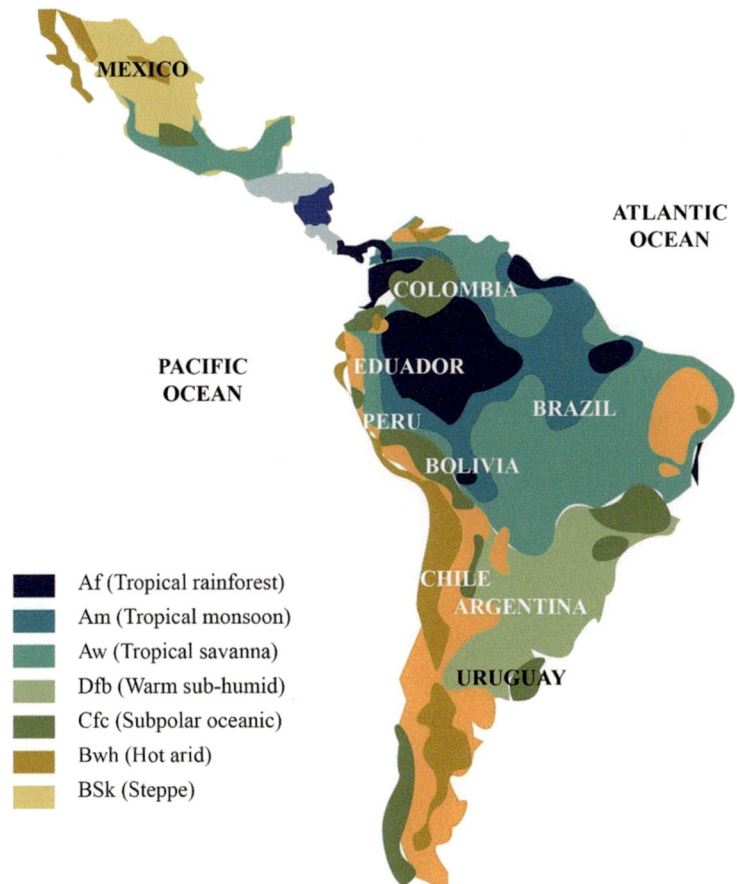

Fig. 3 Climatic map of Latin America according to Koppen

13 years and caused a drop in the agricultural industry, such as the reduction in soybean and corn production that caused a 2.6% drop in the total market in 2021. What is more, the rate of deforestation in Brazil it increased from 2020 by 22%, reaching its highest rate to date.

Finally, both positive and negative climate similarities create an opportunity for collaborative research and investment in policies that impact the energy behavior of cities. This is because building systems, including passive or active, are influenced by the climatic condition. In addition, cultural and socioeconomic factors also respond to the characteristics of the climate, which is why it is considered a good starting point to establish policies, regulations, and incentives.

2.3 Energy Efficiency Regulations in Latin America

The Latin American history has focused more on the differences than on the similarities, this is reflected in the generation of public policies around energy efficiency in buildings. Table 1 shows the individual efforts, standards, and codes unique to each country. The majority are mandatory regulations focused on the building envelope, lighting, construction materials and thermal comfort. The focus of these regulations is to regulate the energy consumption of the building, as is the case of NOM 008 ENER 2001 in Mexico, which limits the heat gain of commercial buildings through the envelope. Some present economic incentives, such as the case of Argentina with the promotion of green roofs. Colombia stands out in this type of incentives, since it has a rating system, codes and prizes for the promotion of green buildings.

To prosecute these political actions, many of the Latin American countries have adopted strategies such as NAMA (Nationally Appropriate Mitigation Action) (Nationally Appropriate Mitigation Actions (NAMAs)|UNFCCC 2020). This concept introduced in 2007 at the Conference of the Parties (COP) establishes a means for developing countries to generate mitigation actions. In addition, in recent years Latin America has generated a series of meetings that allow the region to consolidate, including the Latin American Summit on Climate Change held in Argentina in 2021 or the Climate Summit of Latin American Youth (Finalizó La Cumbre Climática de La Juventud de Latinoamérica: Las 20 Propuestas Para La Acción—Infobae 2022), held within the framework of COP27. In all these efforts, one of the most discussed aspects is the promotion of low emission cities (Das et al. 2020).

2.3.1 Process of Effective Policies Application

The initial stage in setting new policies in each country is institutionally, which means an institution that manages policy design, in this case energy efficiency in buildings. This body designs, operate and manages revised laws and respond to government ministry. Next Planning department develops diagnosis of opportunities

Table 1 Energy efficiency (EE) policies and financial incentives in different LA countries (CAC: command and control, IE: Economic incentives, TS: Technical standard)

Country	Name	Types of policies	Description of EE	References
Mexico	DOF 28/11/2008	CAC, IE	Renewable energy production promotion	Zepeda-Gil and Montiel-Castro (2023), Mexicana (2022), Romero-Pérez et al. (2017), Secretaría de Energía and Comité Consultivo Nacional de Normalización para la Preservación y Uso Racional de los Recursos Energéticos (2011)
	NOM-007ENER-2004	Standard	Efficient public lighting	
	NOM-008ENER-2001	Standard	Commercial envelope efficiency	
	NOM-009ENER-1995	Standard	Industrial building efficiency	
	NOM-018ENER-2011	Standard	Insulation materials	
	NOM-020ENER-2011	Standard	Residential building efficiency	
	NOM-023ENER-2010	Standard	HVAC efficiency	
	NOM-024ENER-2012	Standard	Glass efficiency	
	NMX-AA-164SCFI Code of sustainable housing	IE-NAMA	Sustainable building minimal standard	
	Article 293 + 294 of financial code of CDMX Zapopan council	CAC	Incentive for sustainable design	
		IE	Fiscal incentive for sustainable construction 100% discount on property tax to LEED project > 51 employees	
	NADF-008AMBT-2005	CAC	companies must have PV	
Ecuador	Decree No. 1048	IE	Energy efficiency and renewable energy Energy Audit	Moya et al. (2016)
	ISO 50001:2012	Standard		

(continued)

Table 1 (continued)

Country	Name	Types of policies	Description of EE	References
Brazil	Law 10.295 No.4.059 No. 7746/2012 No. 7.746 No. 3.5745 Qualiverde Property tax	CAC CAC CAC CAC IE IE TAX	Maximum energy level Energy rating system Public green building Low impact materials Incentive sustainable construction Incentives for Green building Tax reduction by 2% if Qualiverde	Marques and Chvatal (2013), ABNT (2003), Rupp and Ghisi (2019), Augusto Xavier Solange Goulart Renata De Vecchi Professor Roberto Lamberts (2022)
Panama	Law 69 Politica Nacional de climatico	CAC CAC + IE	Sustainable design strategies promotion Guideline for climate change strategies	
Peru	Law No. 28245 Law 27345 Law 28611 No.018-2006-VIVIENDA Decree No. 015-2012 D.S. No. 0032013-VIVIENDA	CAC CAC CAC CAC IE TS	National climate change strategies Energy efficiency as priority Constructions obligation to environmental evaluation Sustainable urban development Residential building energy efficiency Green housing rights	Avellaneda and Melanie (2018)

(continued)

Table 1 (continued)

Country	Name	Types of policies	Description of EE	References
Guatemala	Decree No. 522003	CAC	Renewable energy incentive	David et al. (2012)
Argentina	Law 13059/03 Law 4428 Law 449 Law 123 IRAM 11900 IRAM 62404 IRAM 1739 IRAM 11.601	CAC IE CAC CAC Standard Standard Standard Standard	Thermal comfort Green roof Prior evaluation study Extent evaluation study Efficiency calculation Lighting Insulating materials Thermal transmittance	Eficiencia EnergéticaIArgentina.Gob.Ar (2020), Filippín (2005), Norma (1996), NORMA ARGENTINA * Corresponde a La Revisión de La Edición de Noviembre de (1990). IRAM 11604* Segunda Edición, 1160
Chile	Ley 20.402 Article 162 NCh2677:2002 NCh3081:2007 Nch3184:2010 NCh3048/1:2007 NCh3049/1:2007 NCh3149/1:2008	CAC CAC Standard Standard Standard Standard Standard Standard	Efficiency agency of AChEE Mandatory building EE strategies Public lighting efficiency HVAC rating PV labelling Building sustainability indicators Building material sustainability EE in buildings	Urbanismo (2020, n.d.), Zepeda-Gil and Natarajan (2020), Ossio et al. (2012)

(continued)

Table 1 (continued)

Country	Name	Types of policies	Description of EE	References
Colombia	Ley 697/3683 Project 119 Hipoteca Verde Agr. No.186 National savings fund stamps	IE IE IE IE IE	EE as priority Incentives to green building Grants for green technologies Green building code Green building rating system	Rodríguez et al. (2019)
Costa Rica	Decreto No. 25584-96 INTE/ISO 15392 INTE/ISO 21929 ISO/TR 21932	CAC Standard Standard Standard	Energy efficiency minimum level Sustainable construction Building energy efficiency labeling Sustainable construction	Costa Rica (2020)

and challenges to achieve the goal and provides quantitative goals. Finally, to promote energy efficiency development, financial motivation for investment is necessary to boost the market. Hence incentive plans are required followed by policies to promote efficiency in buildings. In Latin America many countries promoted economic support in terms of decree and bills in their law, including Argentina, Brazil, Colombia, Dominican Republic, Ecuador, Guatemala, Panama, Peru, Uruguay, and Venezuela. On the other hand, Mexico, Chile, Costa Rica and El Salvador did not include any financial investment in their law (Ravillard et al. 2019).

2.3.2 Certification and Endorsement

To ensure effective implementation of policies and standards that been published by government for energy efficiency in buildings certification, endorsement, and enforcement regulations can describe how building designer and engineers had followed the ratings system based on the given policies and how much is the annual energy consumption rate. However, in some countries these endorsements are related to labeling of building materials, and efficiency of systems, while in others it refers to emission saving as in Guatemala, or in Ecuador, to people who apply efficiency. Energy labeling in Brazil, Chile, Argentina, and Panama illustrates the energy consumption of each product, which been proven in laboratory. In Mexico such certifications are managed by National Secretariat of Energy (CONUEE), and recognize projects with excellence in energy efficiency, if they meet the regulations requirement. Accreditation in Brazil and Uruguay is through bodies that deliver information and accredit organizations.

2.3.3 Education and Training

Training and informing society with new policies and their importance, shape individuals to contribute to a sustainable future. There are four general stages into training new laws and regulations, including workshops, demonstration program and campaigns, active participation of communities, consultancies, and access to information for all. Countries that performed training workshop for energy efficiency include Argentina, Chile, Colombia, Ecuador, and Uruguay.

Community participation and access information for all been applied in Brazil, Colombia, Costa Rica, and Panama. Finally, Argentina, Costa Rica, Ecuador, and Venezuela included energy efficiency in building in school studies.

2.4 RE Policies and Incentives in Latin America

Energy efficiency actions in buildings must be accompanied by the promotion of energy generation from renewable sources, much more when the objective would be to achieve goals such as regenerative cities. Hence, the Latin American situation in this sense becomes completely necessary. Table 2 illustrates latest policies and financial supports by government of five Latin American countries, to reduce emissions by using RE sources of energy. Mexico provided highest financial commitment in different areas including, solar energy and hydroelectric, as well as taxing those who contribute to emission by using fossil fuel. Also support Small and Medium-sized Enterprises (SME) to install RE technologies. Chile provided fund for efficiency in rural homes and efficient heating systems, in addition to loan for RE technologies used for heating and cooling. On the other hand, Brazil highly funded Wind power and biofuel energy across different states. Argentina allocated around 18 million for wind, solar and hydroelectric power generation, and 1.6 M for solar water heater.

Additionally, a series of mega renewable energy generation projects are projected for the next few years in the region. Among them, the Pampas Project in Argentina stands out, where it is expected to generate 2,000 MW onshore wind power, 500 MW photovoltaic and 2.2 million tons of green hydrogen per year by 2030, the H2 Magallanes Project in Chile which will have a capacity of 800,000 T/year of green hydrogen and 10,000 MW of onshore wind power with full operation by 2028 and the Ventos do Mar Portiguar Wind Complex, which by 2029 expects to generate 2,484 MW of offshore wind power (Los 8 Principales Proyectos de Energía Renovable de América Latina—BNamericas 2023).

Although there has been a big push and a further boom in renewable energy generation projects is expected in the region, the same cannot necessarily be said for ensuring the well-being of communities surrounding the projects. This is expressed in "Communities affected by hydroelectric dams in Latin America: A human rights perspective". Where he sheds light on the social and environmental impacts of hydroelectric dams on communities in Latin America (Lehmann and Tittor 2023). The project fails to involve and compensate affected communities in the decision-making process of hydroelectric projects in Brazil, such as the Madeira River Hydroelectric Complex, which has caused negative social, economic and environmental impacts, including limiting the passage of migratory fish, loss of income for local communities, and lack of transparency and control of information by the companies causing and monitoring the impact.

Several routes for improvement can be applied to remedy these shortcomings, and these criteria should be included in energy legislation. Some alternatives to consider are: (1) include the affected communities in the decision-making process of these projects and provide them with adequate compensation and mitigation initiatives; (2) companies should ensure transparency in reporting and monitoring the impact of their projects; (3) a comprehensive environmental and social impact assessment should be conducted prior to the implementation of any hydropower project to understand the potential impacts and design appropriate mitigation measures.

Table 2 Renewable energy policies and incentives in Latin America

Countries	Name of policy	Objective	Method of policies	Value committed (USD)
Argentina	National renewable cluster	Creation of equipment, services and technology in renewable services based on wind, solar and small hydroelectric generation	Budget or off-budget transfer (DT)	18 Million
	Development program solar thermal industry	Promote domestic production of solar water heaters through technical assistance, financing, and product certification	Budget or off-budget transfer (DT)	1.6 Million
Brazil	Creation of the national aviation biokerosene program	Assessment program for biofuels energy	Uncategorized	0
	Wind farm in ParaÃba	Generation of 48.5 MW clean energy generation during water crisis	Uncategorized	0
	Wind farms in CaiÃ§ara do Rio do Vento and Ruy Barbosa	Wind power generation of 121.8 MW of for 242,000 homes	Loan (Hybrid)	81 Million
	Authorization to import material for biofuels	The National Energy Policy Council in the public auction	New or extended regulation (IT)	0
	Brazilian National Development Bank (BNDES) energy efficiency fund	The National Program for Electrical Energy Conservation (Procel) to support energy efficiency projects through guarantees in different sectors of the economy	Budget or off-budget transfer (DT)	6 Million
	Financing for wind farms of Santa Luzia	Contribute to the strengthening of the wind turbine sector	Loan (Hybrid)	252 Million

(continued)

Table 2 (continued)

Countries	Name of policy	Objective	Method of policies	Value committed (USD)
Chile	Change your heat	Homes change to cheaper and cleaner heating methods to drop rate by 20 to 30%	New or extended regulation (IT)	0
	Investment in local energy efficiency	Implement innovative sustainable energy initiatives	New or extended regulation (IT)	4.79 Million
	Loan for renewable energy investment projects	Support eco-friendly investments	loan (Hybrid)	17 Million
	National strategy for green hydrogen	Produce the cheapest green hydrogen in 2030, being top hydrogen exports in 2040, with 5 GW of energy by 2025	New or extended regulation (IT)	50 Million
	Energy efficiency subsidy	This subsidy focuses on 11,000 vulnerable families and middle income sectors that do not have a budget to carry out work of this type in their homes, to follow established requirements	Other hybrid support measures	0
	Energy efficiency in hospitals	Reduction of GHG 2020–2030 carbon budget	Government procurement (DT)	457 Million
	Heating and cooling national strategy	Target for 80% of energy supply to be RE by 2050, reducing GHG in the heating and cooling sectors by 65%	New or extended regulation (IT)	0
Colombia	Eco-petrol investments in renewable energy	Reduction of leaks and venting, to decrease between 1.8 and 2.0 million tons of CO2e in 2022, incorporating 300 MW of renewable generation by 2022	Budget or off-budget transfer (DT)	50 Million

(continued)

Table 2 (continued)

Countries	Name of policy	Objective	Method of policies	Value committed (USD)
	Public funding solar panels in La Guajira	Provide electricity access to 4000 families in the Uribia, La Guajira	Budget or off-budget transfer (DT)	20 Million
	Law 2099 of 2021 (the energy transition law) with incentives for non-conventional renewable energy projects including blue and green hydrogen	Promoting incentives to energy production from non-conventional renewable energy sources	Tax or royalty or government fee break (GRF)	0
Mexico	Program to promote energy transition and sustainability for SMEs	By promoting the installation of renewable energy technologies, to improve the profitability of MSMEs, city decarbonization	Budget or off-budget transfer (DT)	0
	Investment in a photovoltaic power plant in the roof of the Central Market in Mexico City	The installation of the photovoltaic plant aims to reduce CO2 emissions, serve users in the Central Market, and illuminate surrounding areas	Budget or off-budget transfer (DT)	19 Million
	Integrated plan for the modernization of hydroelectric plants	Generate 1,860 GWh, avoid costs of new plants, strengthening the national electric system	Loan (Hybrid)	1 Billion
	Urban development with sustainable housing, natural resources 2020–2024	Policy aims to promote a sustainable, as well as housing with solid waste and wastewater management that reduces greenhouse effect emissions	Uncategorized	0

(continued)

Table 2 (continued)

Countries	Name of policy	Objective	Method of policies	Value committed (USD)
	First sustainable sovereign bonds linked to the UN sustainable development goals	First country in the world to issue a sustainable sovereign bond linked to the sustainable development goals building	Uncategorized	916 Million
	Emissions tax	To support the economy by taxpayers who are contributing in emission to environment and the negative effects generated	Increased fossil fuel taxation	0

3 Regenerative Positive Energy District and Its Impact on Urban Environment

Positive Energy District are designed to accelerate the solution to climate crisis regardless of country, as climate and planet is a global partnership. Generally, this concept contributes in reversing the carbon footprint of cities (positive carbon), using 100% renewable electricity, while regulations and incentives are tools to support their application. This system works at an energyefficient and flexible location that produces zero level of emissions and produces and supply surplus of energy using green sources, which require interactions between buildings. As Fig. 4 presents, while energy efficient building only consumes less and net-zero energy building produces as much as it consumes, PEB can supply energy more than its demand.

Positive energy district or building is a form of regenerative design that include energy flexible urban group of connected buildings that consume less energy than they produce. There are three functions to PED, firstly energy production must rely on clean energy sources (RES function), such as wind power, solar power, etc. Secondly, every building envelope and system is required to be efficient (EE function), which has local climatic conditions and available materials. Lastly, optimal flexibility between energy systems, that require smart systems to achieve energy balance through the urban network. For instance, smart control during the heat peak of the day, coordinate time and energy balance by using solar power, as it predicts consumption level increase due to space cooling, and solar energy radiation increase as well, so will the solar energy production. So, PED includes energy efficient buildings that produce zero emissions by using renewable sources with surplus production of energy stored to be managed intelligently through an urban network, in line with comfort and sustainable economic.

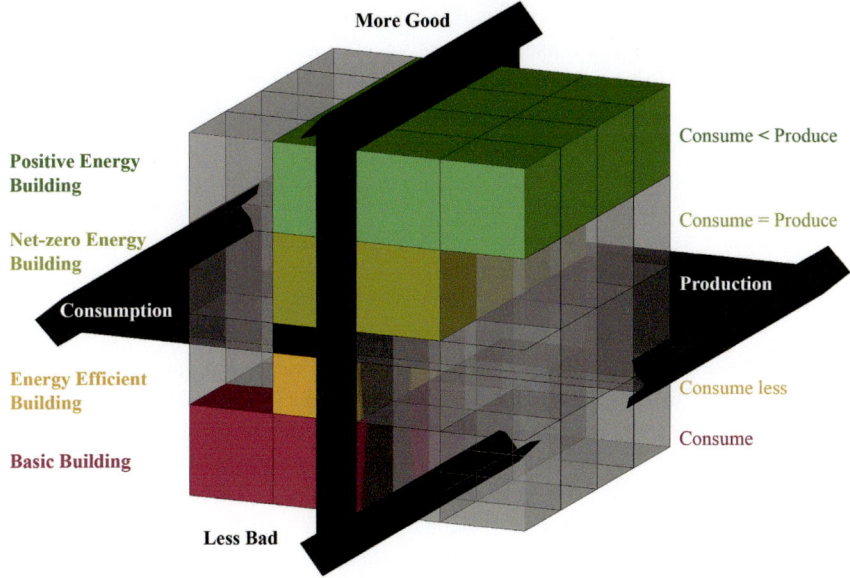

Fig. 4 PED concept diagram based on energy behaviour

Positive energy building at district level evaluates building forms and urban layout, as well as renewable energy production as a group with regards to site and its climatic situation. The energy cost of such systems will be more effective as they are on-site and accessible. The value of neighborhoods will increase, as homes will transform renewable energy into product, as a household income hence rental will also increase. Other benefits of PED are related to human comfort and health, cleansing the air, and design adaptation to harmonize with microclimate of each district, simply an architecture that can heal. Further impact of PED application is in career opportunity creation, new companies that are established for production of new solutions.

Citizens of PED need to be active and prosumer, in order to undergo the energy transition, hence role of government and citizens will change, and energy poverty reduces (van Wees et al. 2021). Energy communities support social changes by incorporating activities and Innovation Ateliers. Also, urban density has been found critical in solar energy potential, on one hand it can decrease mobility energy consumption (Madlener and Sunak 2011), while on the other hand it decreases solar potential (Kanters et al. 2014). Moreover, the form of cities influences the energy produced by wind and solar power, hence geometry have relative dependency with microclimatic conditions that need to be studied. In total, regenerative district level design, mitigate Urban Heat Islands (UHI) effect, be drop in consumption level, as well as renewable energy supply on site. UHI occurs by sparing organic land into urban density, as it will morph the reflectivity and absorption rate of surfaces from natural environment to human infrastructure, resulting in energy consumption upsurge.

Based on benefits and good results, PED is an opportunity to transform cities into low-impact spaces. In this sense, a legislative and incentive framework is essential for the adoption of this type of strategy. In Latin America, it was observed that great efforts have been made to establish efficiency measures in buildings in addition to promoting the generation of renewable energy. For this, the possibilities in this region must be analyzed.

4 Possibilities of Using PED in Latin America

Based on the differences and similarities, the possibilities of implementing Positive Energy Districts in Latin America are analyzed. Doing a count, the countries that applied the concept of zero energy in LA include Argentina, Brazil (Guenther and Vittori 2013), Chile, Colombia, Ecuador, El Salvador, Mexico, Peru, Uruguay, and Panama. In addition, different authors analyzed the capacities of residential buildings to achieve self-sufficiency in Cuba, Ecuador, Mexico and Peru (Hoque and Iqbal 2015). On the other hand, different climatic zones have been studied with various renewable energy productions and an energy supply higher than the demand has been achieved in some cases. All this suggests that there are opportunities for the application of positive energy buildings and districts in the region.

Building and energy codes are the first steps in delineating effective tools for achieving PED in different building types and climates. In this sense, it has been found that the countries of Latin America adopted the transition policies towards cities of positive and regenerative energy through (i) institutions, (ii) incentive funds, (iii) certifications and (iv) training systems. It is noteworthy that one of the institutions that initiated the promotion of policies regarding renewable energy was the Mercosur community, providing free trade between Latin American countries and supporting energy trade.

Other organizations are also strengthening the capacities of cities for a carbon neutral future. The Emerging and Sustainable Cities Initiative (ICES) (Emerging and Sustainable Cities Program|IDB 2022, as one of the Latin American initiatives towards regenerative cities, aims to achieve zero carbon emissions in the region. CityAdapt is another movement that tends to connect nature with urban cities by providing nature-based products (CityAdapt JEM—CityAdapt 2023.). As LA faces critical impacts from climate change, such as droughts and hurricanes, this organization is trying to develop solutions tailored to the countries' socio-economic backgrounds. Finally, the Euroclima + organization is an association between Europe and Latin America (EU-LAC), to work on solutions to climate change with a financing of 16.5 million euros.

Regarding the generation of renewable energy, it can be said that this aspect is much more advanced. The leadership of the region in this sense, in addition to the goals acquired for the coming years, more than 70% of energy coming from renewable sources, generate a good scenario for the future (América et al. 2022). In addition, efforts to reduce the gap in access to electricity between rural and urban areas are part of the positive aspects that can be integrated into the generation of Positive Energy Districts. The challenge is focused on the integration of policies in a common objective, both individually for each country and regionally. To achieve regenerative objectives, it will be essential to align energy efficiency policies with local energy generation, this is still a pending point for Latin America.

A good practice is to identify successful cases in other regions. In this sense, Europe, as a pioneer in the regenerative concept and PED, can generate some hints and paint a realistic goal. There are two main directives that European legislation has included in its way, on the one hand the energy performance of buildings (EPBD) and on the other the renewable energy directive (RED). All EU members must incorporate these directives into their national legislation and prepare financial and training support. However, as energy efficiency and renewable production indicators are very sensitive to climate, these values may differ (Santiago et al. 2020). Therefore, it can be said that the PED in all of Latin America could start with the generation of interconnected institutions as well as developing a common objective framework but adaptable to different climates and cultures. To achieve these transitions, two key actions are identified: on the one hand, seek interconnected national financing where several countries work together to reduce their emissions and take climate measures. On the other hand, promote collaboration in research and innovations, since they share a similar context.

4.1 A Successful Example in the Region

The Chilean Ministry of Energy has established a course of action called "Local Energy Management through the Commune Energy program". This seeks to promote sustainable energy management at the municipal level. The "Comuna Energética" program has three objectives (Home—Comuna Energética 2023): promote the participation of communities and local actors in the energy development of their territory, promote a local low-carbon energy market for the implementation of actions and strengthen the powers of municipalities associated with local energy management.

The program has a complete action path, which goes from the energy diagnosis and the definition of an action plan for the municipalities to the evaluation and generation of an energy commune seal. Among the benefits are the promotion of citizen participation, the promotion of the energy market by accessing different investment funds and the strengthening of skills associated with energy management. In addition, municipal cooperation, and the public image of municipalities in their climate action are promoted.

This project has allowed local governments in Chile to align their energy sustainability actions with other local and national plans. As of November 2022, 90 communities have joined the program, 59 local energy strategies have been implemented, and 50 projects have been implemented. To further strengthen the implementation of such initiatives, it is important to integrate transversal aspects of improvement from the cases analyzed, such as the "Comuna Energética" program in Chile, with the intention of strengthening future actions or strategies around the regulatory framework of each country. This can be achieved by promoting certifications and standards that support sustainable energy management at the local level, and by ensuring strategic financing and management among actors in different sectors to overcome the limitations of each sector. By doing so, local governments in Latin America can align their energy sustainability actions with other local and national plans and promote citizen participation in the transition towards regenerative cities.

5 Conclusion

Pathway to regenerative cities using PED concept can be developed, to not only benefit the environment with emission reduction, but also to grow the economy of societies and improve the air quality. One of the influential routes in reducing emissions is through energy efficient buildings, which reduce the energy consumption rate by occupants. Moreover, adoption of renewable energy sources to supply building demands. Embracing policies that contain guidelines for improvement in building systems performance, and implementation of energy labels, appliances, and lighting standards for existing and new construction, play a leading role in achieving this goal. Monitoring the building standards is another key parameter in application of policies and providing results.

However, based on the existing framework, sectoral policy packages are required to be designed that are sensitive to all circumstances. Presented work provided a review on existing policies and funding regarding the pillar of PED, which are regarding energy efficiency and use of renewable energy in buildings. Insights for Latin American policy makers are delivered to support the transition of countries into carbon neutral urban areas with long-run law formation. LA could follow EU, as a leading institution, on planning regional a holistic goal for efficiency of building material, active systems and design, as well as directive for renewable energy integration in cities.

Critical factors that result in in-efficient application of PEB in Latin America are regulatory, economic, and technological barriers. Hence, future challenges for government and decision makers are to overcome the obstacles of territorial regulations and financial support within a LA community. Furthermore, monitoring the construction conditions and electricity consumption pattern needs to be carried out with details. Also, experimental data on energy supply and demand supports the realistic practice in PEB. Overall, the study showed that with intensive regulatory and financial support throughout the whole of Latin America, PEB can be achieved throughout the future cities, which results in lower emission level and healing the planet.

In terms of Latin America's areas of opportunity in the regulatory framework, it is important to highlight the need for stronger enforcement and monitoring of building standards and regulations.

This includes developing clear and concise guidelines for building performance and energy efficiency, as well as implementing energy labels and standards for appliances and lighting in new and existing buildings. Additionally, it is important to recognize the role of certifications in promoting sustainable building practices, such as LEED or EDGE, and to encourage their adoption and integration into local regulations.

Regarding financing and recommended management, it is important to promote innovative financing mechanisms that encourage private sector investment in sustainable building practices, such as green bonds and energy performance contracts. It is also important to involve strategic actors, such as local governments, developers, and financial institutions, in the planning and implementation of sustainable building policies and practices.

Finally, addressing the limitations of each sector is critical for the successful implementation of sustainable building practices in Latin America. To this end, the first step could be to address the lack of skilled labor and training in sustainable building practices by integrating universities, higher education and state educational programs. In addition, promoting public–private financing programs that seek the rapid adoption of innovative technologies applicable to buildings. Finally, there is an urgent need to provide incentives and support for the adoption of renewable energy sources in buildings. Overall, there is a need for a comprehensive and collaborative approach to addressing the challenges and opportunities for sustainable building practices in Latin America.

References

ABNT N (2003) 15220-2. Desempenho Térmico de Edificações-Parte 2: Métodos de cálculo da transmitância térmica, da capacidade térmica, do atraso térmico e do fator de calor solar de elementos e componentes de edificações. Janeiro, Brasil

América DE, El LY, Los C, Del D (2022) BARÓMETRO DE LA ENERGÍA

Amponsah NY, Troldborg M, Kington B, Aalders I, Hough RL (2014) Greenhouse gas emissions from renewable energy sources: a review of lifecycle considerations. Renew Sustain Energy Rev 39. https://doi.org/10.1016/j.rser.2014.07.087

Asif M, Muneer T, Kelley R (2007) Life cycle assessment: a case study of a dwelling home in Scotland. Build Environ 42(3). https://doi.org/10.1016/j.buildenv.2005.11.023

Augusto Xavier Solange Goulart Renata De Vecchi Professor Roberto Lamberts A (2022) UNIVERSIDADE FEDERAL DE SANTA CATARINA CENTRO TECNOLÓGICO-DEPARTAMENTO DE ENGENHARIA CIVIL CONFORTO E STRESS TÉRMICO. www.labeee.ufsc.br. Accessed 8 Aug 2023

Avellaneda BC, Melanie W (2018) Confort trmico en las edificaciones de las aldeas infantiles de la provincia de Huancayo segn la Norma EM-110 del R.N.E. 2017. Universidad Peruana Los Andes. http://repositorio.upla.edu.pe/handle/20.500.12848/1857

Belke A, Dobnik F, Dreger C (2011) Energy consumption and economic growth: new insights into the cointegration relationship. Energy Econ 33(5). https://doi.org/10.1016/j.eneco.2011.02.005

Boyd R, Stern N, Ward B (2015) What will global annual emissions of greenhouse gases be in 2030, and will they be consistent with avoiding global warming of more than 2 °C? Centre for Climate Change Economics and Policy, May 2015

Cai W, McPhaden, MJ, Grimm AM, Rodrigues RR, Taschetto AS, Garreaud RD, Dewitte B, Poveda G, Ham YG, Santoso A, Ng B, Anderson W, Wang G, Geng T, Jo HS, Marengo JA, Alves LM (2020) Climate impacts of the El Niño southern oscillation on South America. Nat Rev Earth Environ 1(4). https://doi.org/10.1038/s43017-020-0040-3

CityAdapt JEM—EN—CityAdapt (2023). https://cityadapt.com/en/jem/. Accessed 10 Aug 2023

Costa Rica AL (2020) Creación del Servicio de Certificación de la Discapacidad (SECDIS). Decreto N° 40727-MPMTSS del 7 de diciembre de 2017 (publicado en "La Gaceta" No. 232 del 7 de diciembre de 2017). Consejo Nacional de Personas Con Discapacidad. http://aplicaciones.bibliolatino.com/handle/bibliolatino/204

Das P, Mathuria P, Bhakar R, Mathur J, Kanudia A, Singh A (2020) Flexibility requirement for largescale renewable energy integration in Indian power system: technology, policy and modeling options. Energ Strat Rev 29:100482. https://doi.org/10.1016/j.esr.2020.100482

David S, Morataya A, Heo E (2012) Electricity sector: design to ensure energy security

Densidad de población (personas por kilómetro)—East Asia & Pacific|Data (2022). https://datos.bancomundial.org/indicador/EN.POP.DNST?locations=Z4. Accessed 8 Aug 2023

Eficiencia Energética|Argentina.gob.ar (2020). https://www.argentina.gob.ar/economia/energia/eficiencia-energetica. Accessed 8 Aug 2023

El mapa de los climas de América del Sur—Mapas de El Orden Mundial—EOM (2022). https://elordenmundial.com/mapas-y-graficos/mapa-climas-america-del-sur/. Accessed 8 Aug 2023

Emerging and Sustainable Cities Program|IDB (2022). https://www.iadb.org/en/urban-development-and-housing/emerging-and-sustainable-cities-program. Accessed 8 Aug 2023

Empowering Cities for a Net Zero Future: Unlocking resilient, smart, sustainable urban energy systems|en|OECD (2021). https://www.oecd.org/publications/empowering-cities-for-a-net-zero-future7a222c8b-en.htm. Accessed 8 Aug 2023

Filippín C (2005) Energy use of buildings in Central Argentina. J Therm Envel Build Sci 29(1) https://doi.org/10.1177/1744259105051798

Finalizó la Cumbre Climática de la Juventud de Latinoamérica: las 20 propuestas para la acción—Infobae (2022). https://www.infobae.com/america/medio-ambiente/2022/09/17/finalizo-la-cumbreclimatica-de-la-juventud-de-latinoamerica-las-20-propuestas-para-la-accion/. Accessed 8 Aug 2023

Guenther R, Vittori G (2013) Sustainable healthcare architecture. Wiley. https://www.wiley.com/en-es/Sustainable+Healthcare+Architecture%2C+2nd+Edition-p-9781118416112. Accessed 10 Aug 2023

Hofman AA, Valderrama P (2021) Long run economic growth performance in Latin America 1820 2016. J Econ Surv 35(3). https://doi.org/10.1111/joes.12383

Home—Comuna Energética (2022). https://www.comunaenergetica.cl/. Accessed 8 Aug 2023

Hoque S, Iqbal N (2015) Building to net zero in the developing world. Buildings 5(1):56–68. https://doi.org/10.3390/BUILDINGS5010056

Norma IRAM (1996) 11605: Acondicionamiento térmico de edificios. Condiciones de habitabilidad en edificios. Valores máximos de transmitancia térmica en cerramientos opacos. Instituto Argentino de racionalización de materiales. Buenos Aires

Jakob M (2017) Ecuador's climate targets: a credible entry point to a low-carbon economy? Energy Sustain Dev 39:91–100. https://doi.org/10.1016/j.esd.2017.04.005

Kanters J, Wall M, Kjellsson E (2014) The solar map as a knowledge base for solar energy use. Energy Procedia 48:1597–1606. https://doi.org/10.1016/j.egypro.2014.02.180

Koezjakov A, Urge-Vorsatz D, Crijns-Graus W, van den Broek M (2018) The relationship between operational energy demand and embodied energy in Dutch residential buildings. Energy Build 165:233–245. https://doi.org/10.1016/j.enbuild.2018.01.036

Lehmann R, Tittor A (2023) Contested renewable energy projects in Latin America: bridging frameworks of justice to understand 'triple inequalities of decarbonisation policies. J Environ Policy Plan 25(2). https://doi.org/10.1080/1523908X.2021.2000381

Los 8 principales proyectos de energía renovable de América Latina—BNamericas (2022). https://www.bnamericas.com/es/noticias/los-8-principales-proyectos-de-energia-renovable-de-americ alatina. Accessed 9 Aug 2023

Madlener R, Sunak Y (2011) Impacts of urbanization on urban structures and energy demand: what can we learn for urban energy planning and urbanization management? Sustain Cities Soc 1(1). https://doi.org/10.1016/j.scs.2010.08.006

Marques THT, Chvatal KMS (2013) A review of the Brazilian NBR 15575 standard: applying the simulation and simplified methods for evaluating a social house thermal performance. Simul Ser 45(8)

Mexicana N (2022) Ventanas: CLASIFICACIONES Y ESPECIFICACIONES

Mourato Pinto J (2017) The BRICS and the future of global order. Europe-Asia Stud 69(2). https://doi.org/10.1080/09668136.2017.1296641

Moya D, Torres R, Stegen S (2016) Analysis of the Ecuadorian energy audit practices: a review of energy efficiency promotion. Renew Sustain Energy Rev 62:289–296. https://doi.org/10.1016/J.RSER.2016.04.052

Nationally Appropriate Mitigation Actions (NAMAs)|UNFCCC (2020). https://unfccc.int/topics/mitigation/workstreams/nationally-appropriate-mitigation-actions. Accessed 8 Aug 2023

NORMA ARGENTINA * Corresponde a la revisión de la edición de Noviembre de 1990. IRAM 11604* Segunda edición (1160)

Ossio F, De Herde A, Veas L (2012) Exigencias europeas para infiltraciones de aire: Lecciones para Chile. Rev De La Construcción 11(1):54–63. https://doi.org/10.4067/S0718-915X2012000100006

Policymakers Need Steady Hand as Storm Clouds Gather Over Global Economy (2022). https://www.imf.org/en/Blogs/Articles/2022/10/11/policymakers-need-steady-hand-as-storm-clouds-gatherover-global-economy. Accessed 8 Aug 2023

Ravillard P, Carvajal F, Lopez D, Chueca JE, Hallack M (2019) Towards greater energy efficiency in Latin America and the Caribbean: progress and policies. https://doi.org/10.18235/0002070

Renewables Global Status Report—REN21 (2023). https://www.ren21.net/reports/global-status-report/. Accessed 8 Aug 2023

Rodríguez CM, Medina JM, Pinzón A, García A (2019) A post-occupancy strategy to improve thermal comfort in social housing in a tropical highland climate: a case study in Bogotá, Colombia. Informes de La Construccion 71(555). https://doi.org/10.3989/ic.61006

Romero-Pérez CK, Rodríguez-Muñoz NA, Alpuche-Cruz MG, Martín-Domínguez IR (2017) Preliminary study of the condition of social housing in the city of Durango, México. Energy Procedia 134:29–39. https://doi.org/10.1016/j.egypro.2017.09.594

Rupp RF, Ghisi E (2019) Avaliação de modelos preditivos de conforto térmico em escritórios no clima subtropical brasileiro. Ambiente Construído 19(2). https://doi.org/10.1590/s1678-86212019000200310

Salim R, Rafiq S, Shafiei S (2017) Urbanization, energy consumption, and pollutant emission in Asian developing economies: an empirical analysis. In: ADBI working paper series

Santiago R, Fuinhas JA, Marques AC (2020) The impact of globalization and economic freedom on economic growth: the case of the Latin America and Caribbean countries. Econ Change Restruct 53(1). https://doi.org/10.1007/s10644-018-9239-4

Secretaría de Energía & Comité Consultivo Nacional de Normalización para la Preservación y Uso Racional de los Recursos Energéticos (2011) NOM 020 ENER2011 Eficiencia energética en Edificaciones. Envolvente de edificios para uso habitacional

Selvakkumaran S, Limmeechokchai B (2013) Energy security and co-benefits of energy efficiency improvement in three Asian countries. Renew Sustain Energy Rev 20. https://doi.org/10.1016/j.rser.2012.12.004

Urbanismo GS-RCdeAy (2020) Los materiales de construcción, su ausencia en la Certificación de edificio sustentable (CES), Chile. Rau.Cujae.Edu.Cu. https://rau.cujae.edu.cu/index.php/revistaau/article/download/583/549. Accessed 10 Aug 2023 van Wees M, Revilla BP, Fitzgerald H, Ahlers D, Romero N, Alpagut B, Kort J, Tjahja C, Kaiser G, Blessing V, Patricio L, Smit S (2021) Energy citizenship in new energy concepts. Environ Sci Proc 11(1):27. https://doi.org/10.3390/environsciproc2021011027

Zepeda-Gil C, Montiel-Castro AJ (2023) Cold housing in Central Mexico: environmental dissatisfaction and underheating lowers self-perceived health in Central Mexico. Buildings 13(3):814. https://doi.org/10.3390/BUILDINGS13030814

Zepeda-Gil C, Natarajan S (2020) A review of "green building" regulations, laws, and standards in Latin America. Buildings 10(10):188. https://doi.org/10.3390/BUILDINGS10100188

Urban-Bioclimatic Aptitude as Nature-Based Solution

Marisol Ugalde-Monzalvo

Abstract Nature-based solutions are inspired, supported, or integrated with nature; they restore natural flows, are adaptive, efficient, systemic, and incremental. They increase the resilience and sustainability of a city. Bioclimatic applies natural adaptation strategies so that the building and the built environment itself collect, accumulate, distribute, or dissipate energy. These can be applied at different scales and in both new projects and existing built environment. The urban bioclimatic aptitude of a city describes its current state, its limitations, and its potential. The methodology for constructing an Urban-Bioclimatic Aptitude Index consists of four steps: (1) description of the urban-bioclimatic system, (2) description of qualitative and quantitative variables, (3) generation of urban-bioclimatic variables, and (4) construction of the aptitude index. Urban Bioclimatic Aptitude is the measure of the capacity of the city to meet the climate needs and depends on the characteristics of the built environment as a new nature.

Keyword Urban sustainability · Urban bioclimatic · Aptitude

1 Introduction

Sustainability proposes that the provision of environmental, economic and social goods and services for humans be performed so that it does not reduce over time the quantity and quality of goods and services that natural, economic and social systems can provide (Schipper et al. 1992). Although sustainability criteria are paradigmatic in different disciplines, a substantial amount of their implementation and monitoring will occur in cities and communities (Kawakubo et al. 2018), including those related to energy use in buildings.

Nature-based solutions are those that are inspired by, supported by, or integrated with nature. They are responses that restore natural flows. They are adaptive, efficient, systemic, and incremental solutions that generate benefits across the three axes

M. Ugalde-Monzalvo (✉)
School of Architecture, Art and Design, Tecnologico de Monterrey, Estado de México, Mexico
e-mail: marisol.ugalde@tec.mx

of sustainable development: (a) environmental (ecosystem restoration, soil remediation, flood protection, and thermal regulation); (b) social (civic empowerment, sense of place, social capital through community support, environmental education, habitability, recreation, and social cohesion); and (c) economic (green jobs, food, and capital through community support). Nature-based solutions increase the resilience and sustainability of a city. This research proposes a model to develop an Urbanbioclimatic index Aptitude as a Nature-based solution. Bioclimatic architecture is focus on the design and construction of buildings considering the climatic conditions of the region or country in which it is being built. The thermo-physical properties of covered and construction materials in contemporary cities and the urban geometrical characteristics affect the microclimatic conditions inside urban centers (Dimoudi et al. 2014).

Climate parameters such as temperature, daylight, wind or humidity are duly analyzed to provide optimal indoor conditions and considerable energy savings, a thorough location context analysis takes into account the building orientation, taking maximum advantage of the sun, wind and earth energy using as a key element of sustainable and bioclimatic architecture is the ability to draw conclusions from vernacular architecture (Widera 2014).

Bioclimatic architecture promotes the relationship between the environment, people, and habitable spaces to reduce the environmental impact generated by energy consumption for space heating and cooling. Natural acclimatization strategies are applied so that the building itself collects, accumulates, distributes, or dissipates energy through appropriate solar orientation, use of thermal mass, proper window placement, installation of breezeways and solar shading, and choice of colors and textures. These strategies, among others, can be applied at different scales and in both new projects and existing built spaces.

The strategies applied to architecture can be extended to the city: Some countries have established the systematic use of design and construction standards that ensure a significant and cumulative reduction in energy demand for interior fittings like Energy Saving Standard of 1999 in Japan (Kuma et al. 2007), or the European Community issued Directive 2002/91 on Building Energy Certification (Valdiserri et al. 2016). Systematize the application of bioclimatic strategies presents the challenge of generating specific solutions in a wide range of climatic environments, at different scales, at different urban conditions and with varied temporal extents (in the new, in what exists and in what is going to be built) to achieve the objective of obtain adaptive, particularized, efficient, and aggregated results to improve the city.

Systematizing the application of bioclimatic strategies presents the challenge of generating specific solutions in a wide range of climatic environments, at different scales, in different urban conditions, and with different temporal extents (in the new, in what exists, and in what is to be built) to achieve the goal of obtaining adaptive, particularized, efficient, and aggregated results to improve the city.

Nature-based solutions applied to restore ecological flows in cities that increase their resilience and sustainability can be easier introduced in new constructions and new urban developments: environmental efficiency, social balance and economic viability are elements that are frequently considered in these processes. However, the existing city and its restructuring present a greater challenge when implementing sustainable strategies. Furthermore, as innovative technologies and energy efficiency measures for buildings are well known, the main issue is to identify those that will prove to be the more effective and reliable in the long term (Asadi et al. 2012). Strategies are usually focused on solving problems with a large social dimension that require an immediate solution, making it difficult and even preventing the introduction of attention to long-term problems.

The city is a complex and energetically demanding system; an energetically poorly planned city leads to an energetically poorly solved city. But cities are also built resources, and their renovation and adaptation can be seen as a solution based on this new nature that is the existing city, which has a large content of primary energy and plays the role of both supplier and consumer of energy. This rehabilitation would reduce both the need for infrastructure and the consumption of territory.

The challenge is to apply nature-based solutions to redesign and renovate the existing city. Create urban diversity, richness and complexity through physical renewal and preservation of established communities. Some elements can be changed quickly (such as bus routes), others such as the specific conditions of buildings can be changed gradually (such as green areas or bioclimatic urban conditions), while natural conditions cannot be changed. The redesign must consider not only the social, economic, and environmental goals, but also the structure and specific characteristics of each city to identify the goals that can be achieved.

Figure 1 shows the differences between a city in which there is no renovation (systems in degradation), a renovated city (systems in maintenance), and a renovated system with a sustainable approach (solution Based on the aptitude of the system), considering that:

- The system: Refers to the built environment (city, neighborhood, building).
- Investment (or renovation): Refers to material or energy added to the final system.
- Suitability: Is the ability of the system to meet the needs of its inhabitants.
- Final Suitability: The suitability of the system after renovation.
- Entropy (or Degradation): The material and energy lost by the system over time.

A city "without renovation" is a system in the process of entropy and degradation. A city "with renovation" is characterized by the addition of material or energetic information to a built system with the purpose of preserving or recovering the services that the environment provides to its inhabitants, that is, only a "maintenance" investment is made. And in the city with "sustainable renovation", this investment should have the purpose of increasing the aptitude of the environment to reduce the energy consumption of its inhabitants in the long term. Nature-based solutions used in sustainable redevelopment add information to the system to improve and maintain it, so that the rebuilt cities, rehabilitated buildings, and revitalized neighborhoods could ensure the sustainability of this redevelopment if they are based on technical solutions that consider social trends, integration with existing policies, and impacts on the environment and economic development.

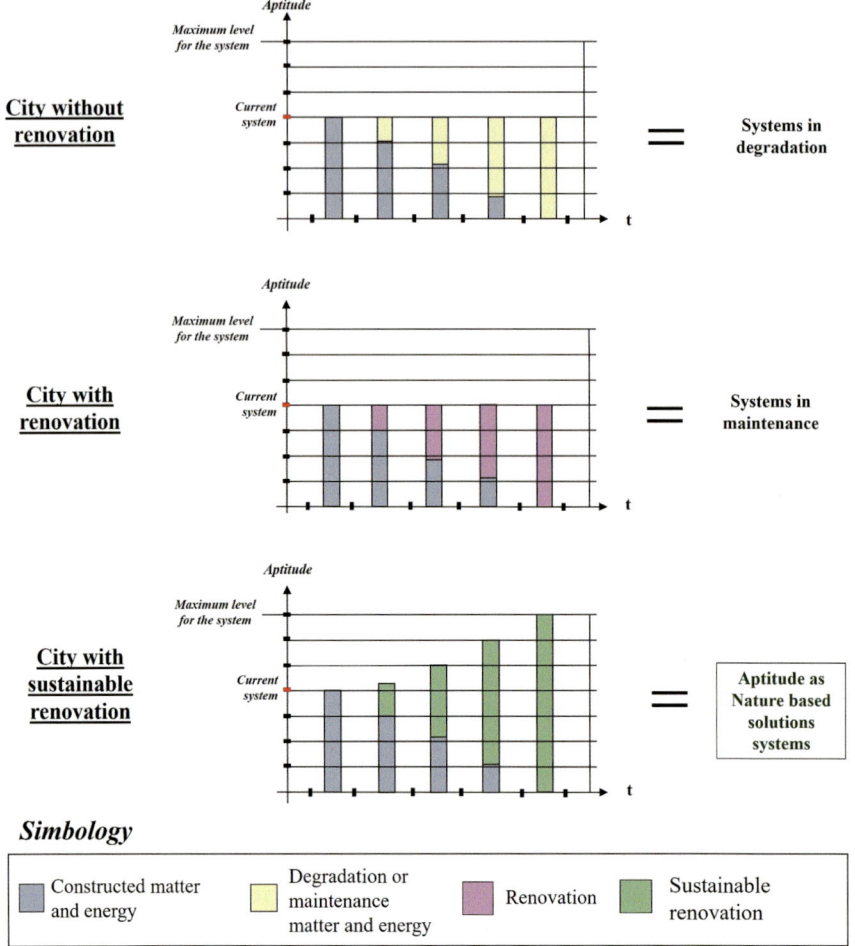

Fig. 1 Systems without renovation, with renovation and with sustainable renovation (own elaboration)

In Design with Nature, Ian L. McHarg developed an ecological planning theory and method for analyzing and evaluating biophysical and sociocultural landscape characteristics to determine appropriate land uses. His suitability analysis method links planning and design to ecology and evolutionary biology with concepts such as energy, order and disorder, adaptation, health, creativity, and human agency, as well as his parallel constructs: syntropic fitness-health and entropic mismatches-morbidity and death (Bryant and Turner 2019). The foundation of these concepts is respect for the environment based on its analysis, i.e., ecology as the basis of design, which opens the possibility of coexistence of artificial systems with natural ones. As well as the integration of human ecology (ethology, ethnography, and anthropology), which includes the behavior of organisms that inhabit the environment as an adaptive strategy. Ian McHarg's notion of planning stems from two fundamental characteristics of natural processes: creativity and aptitude, which derive from Darwinian notions of how organisms adapt and survive (Cohen 2019) These concepts coincide with Henderson's description in his book, "The fitness of the environment", where he states that a fit environment can be defined as one in which the greatest consumer needs are met by the same environment, which means that the organism has less adaptive work to do "The fitness of the environment" is one part of a reciprocal, the fitness of the organism being the other. And the fitness of the environment is the result of characteristics which constitute a series of maxima so numerous, so varied, so nearly complete among all the things involved in the problem, that together they certainly form the greatest possible fitness (Henderson 2016). Figure 2 shows the adaptative work to the environment and the aptitude of the system.

In this work, 'aptitude' refers to:

(1) The measure of the capacity of a system to meet the needs of its inl habitants.
(2) Aptitude depends on the characteristics of the environment and the needs of the inhabitants.
(3) If these characteristics or these needs change, the aptitude of the environment changes, it can increase if the characteristics are improved or decrease if the demands increase.
(4) The economy defines that "Resources are scarce and needs are unlimited", so we can say that, considering current growth trends, the aptitude of a Built System cannot increase indefinitely, in other words, aptitude has limits.

Fig. 2 Adaptative work and aptitude (own elaboration)

(5) On the contrary, the inability of the system to meet the needs of the inhabitants can grow indefinitely if the favorable characteristics of the system are degraded or lost and the demands increase, that is, the "inadequacy" of the system can tend to infinity.
(6) If there is no system renewal, its condition of "unfitness" will grow.
(7) Each system has an aptitude range with an upper or maximum limit and a lower limit of infinity.

The growth of new urbanized spaces and the renewal of existing ones should be based on their ability to determine the increase in organized information, the increase in energy efficiency and the reduction in resource consumption. Because the city is a system that, like a living organism, consumes and exchanges energy with its environment. Their survival depends on their ability to adapt to changes in the environment and evolve before let the new order destroys it. Thus, the value of the built environment is found both in the historic centers and in the small houses, in the old neighborhoods, in the characteristic neighborhoods, on the outskirts of the city and the new subdivisions, which together have generated the current cities. Thus, the greater the sustainable renewal, the greater the aptitude, the less energy consumption, the less pollution, the same or better quality of life, the greater integration into what exists and the greater economic reactivation.

Buildings and cities have different energy consumptions because they have different aptitudes to cover the basic needs of the organisms that inhabits them. If we know the needs and value the aptitude of the environment, we will know:

(a) The adaptation work that must be done (by the system or by the organism).
(b) The "unnecessary" or "extra" energy consumption of the organism and,
(c) If the aptitude of the environment is at the limit or above its capacity. This means, *Solutions Based on the capacity of Nature*.

1.1 Related Work

Different approaches have been found in literature review for the systematization of bioclimatic strategies. These can be organized into three types (Fig. 3): (1) those that link the city and climate and propose generalizations of bioclimatic strategies by climate type (approaches based on generalizations); (2) those that link human needs with climate through acclimatization and comfort indices (referential approaches); and (3) those that link the effects of climate in cities with human needs based on limitations, goals, and potentials (approaches based on optimization). These approximations are shown in the Fig. 3.

The first approach, generalization, involves analysis ranging from the general to the particular and includes studies of the differences between macroclimate and microclimate (Clarke and Bach 1971; Gubenskiĭ 2009; Bianchi et al. 2019), and the microclimate design potential (Yuan et al. 2017). Related to this approach is the phenomenon of the urban heat island, which is defined as the differences in climatic

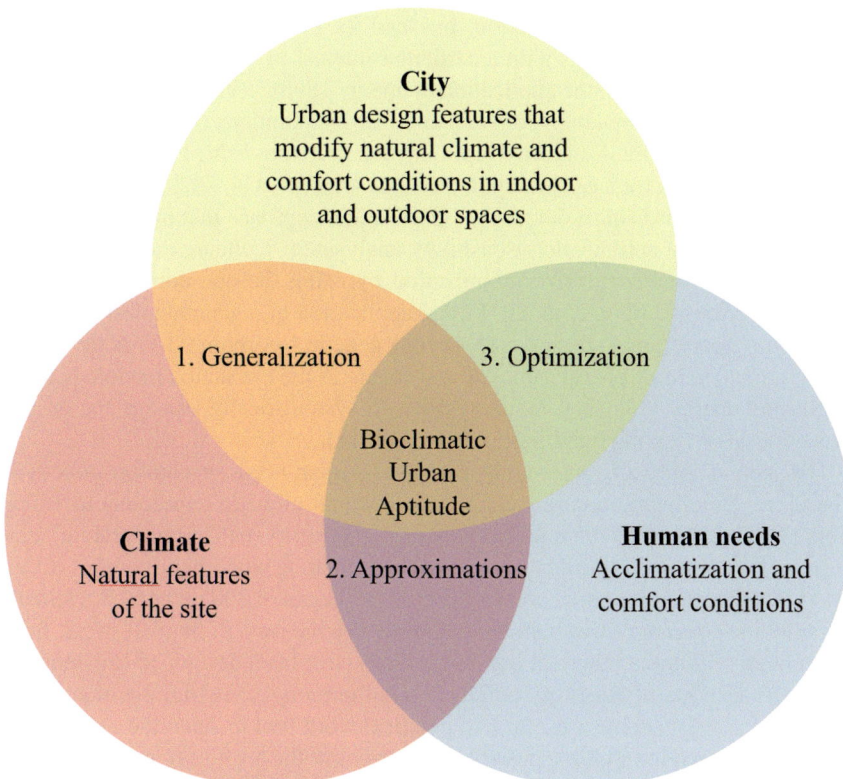

Fig. 3 Approaches to systematize the application of bioclimatic strategies in cities: generalization, approximations and optimization (own elaboration)

features between urban areas and their natural and rural counterparts (Toy and Yilmaz 2010; Mirzaei 2015; Akbari et al. 2016; Nardino and Laruccia 2019) and the manner in which different factors are involved in this phenomenon, such as solar radiation (Saá et al. 2012; Katafygiotou and Serghides 2015); air temperature, relative humidity and wind speed (Patania et al. 2010; Andrade et al. 2011; Gobo et al. 2018), urban trees (Rantzoudi and Georgi 2017), vegetation cover (Rantzoudi and Georgi 2017; Notaro et al. 2012), the effects of different surfaces (Karakounos et al. 2018; Back et al. 2021) and the total effect of an urban area on bioclimatic comfort (Goldberg et al. 2013). Other potential contributors that have been studied are anthropogenic factors like the relationship between the urban thermal environment and city size or road density variables as indicators of urbanization (Román et al. 2017; Kodis et al. 2018), biozone proposals that have been undertaken by state action plans and regulations (Cárdenas-Jirón and Morales-Salinas 2019) and urban morphology as an influential factor on energy performance and the livability of a city (Salat and Vialan 2010).

The third approach, optimization, has had its most important application in retrofitting existing buildings with a particular interest in the potential costs, benefits and impacts involved: the application of energy-saving techniques in an existing building using both simulation software and real interventions (Boukli Hacene and Chabane Sari 2020); multi-objective optimization problems using Pareto to select the best design solution for a case study in Morocco (Ameur et al. 2020); the evaluation of optimal solutions using a developed optimization approach that includes a model for multi-objective optimization; sensitivity analysis for reducing the dimension of input variables, multi-objective optimization by using the non-dominated sorting genetic algorithm II (Gou et al. 2018); the application of a generate design system that applies goal-oriented design combining a genetic algorithm (GA) as the search engine with a building energy simulation software as the evaluation module (Caldas 2008); and energy efficiency rating systems to systematically manage the energy consumption of existing buildings (Koo et al. 2014).

The second approach, approximations, uses human needs or comfort indices as a baseline. Bioclimatic comfort defines the optimal climatic conditions in which people feel healthy and dynamic, bioclimatic components such as temperature, relative humidity, and wind speeds are important in evaluating bioclimatic comfort (Cetin 2015). The interest in thermal comfort conditions began in 1905 when J.S. Haldane proposed the concept of wet-bulb temperatures as a measure of thermal stress, from which nearly 40 human thermal comfort indices have been developed (Epstein and Moran 2006). One of these key parameters is thermohygrometric: the bioclimatic index, which has become a useful and practical work tool in scientific research by obviating the need for multiple possible evaluations of the positive or negative potential that a region's climate has on human health. The mathematic formula by W.J. Kyle is based on the measured values of the air temperature (°C) and relative air humidity (%) (Grigore et al. 2020). The atmospheric environment has a few meteorological parameters (temperature, humidity, air movement and radiation) that determine the bioclimatic conditions of someone out of doors, and there is another relationship with buildings and urban design. Some empirical indices use only one of these parameters, while others use two or more (Clarke and Bach 1971). Human discomfort indices, account for environmental factors but are not universally applicable under a broad range of human metabolic and work exposure environments (Mistry 2020); they are employed in mathematical optimization methods to support the design of buildings in which thermal comfort is an objective function (Carlucci et al. 2014). The physiological equivalent temperature is a universal index that integrates the thermo-physiology of the human body (gender, height, activity, clothing resistance for heat transfer, shortwave albedo and long-wave emissivity of the surface) and multiple relevant meteorological parameters such as air temperature, relative humid and wind velocity (Ren et al. 2019). In complex urban environments, wind speed and mean radiant temperature are highly volatile and can thus only be estimated by micro-scale models. One easy way to apply a model for determining thermal indices in urban environments is the Sky Helios model, which is designed to resolve—temporally and spatially and using a small amount of computation time—the sky view factor sunshine duration, global radiation, wind speed, wind direction, considering reflections and thermal indices (Fröhlich and Matzarakis 2018). For the study of the bioclimatic comfort conditions of Burdur city center, measurements were taken at 20 different points over the city in the same time zone, the physiological equivalent temperature (PET) index which includes the human energy balance and multiple linear regression models for concluding that urban high building density areas have higher PET values in most of the day than in other terrestrial areas; between these areas and open areas was a difference of 0.4–1.2 °C, between parks 2.2–3 °C, between locations 1.4–2.2 °C, and between rural areas 1.7–2.5 °C per day (Cetin et al. 2019).

The third approach, optimization, has had its most important application in retrofitting existing buildings with a particular interest in the potential costs, benefits and impacts involved: the application of energy-saving techniques in an existing building using both simulation software and real interventions (Boukli Hacene and Chabane Sari 2020); multi-objective optimization problems using Pareto to select the best design solution for a case study in Morocco (Ameur et al. 2020); the evaluation of optimal solutions using a developed optimization approach that includes a model for multi-objective optimization; sensitivity analysis for reducing the dimension of input variables, multi-objective optimization by using the non-dominated sorting genetic algorithm II (Gou et al. 2018); the application of a generate design system that applies goal-oriented design combining a genetic algorithm (GA) as the search engine with a building energy simulation software as the evaluation module (Caldas 2008); and energy efficiency rating systems to systematically manage the energy consumption of existing buildings (Koo et al. 2014).

2 Methodology

This research proposes a model to evaluate urban bioclimatic aptitude based on nature that integrates the climate, human comfort needs and urban design features that modify natural climate in indoor and outdoor spaces. The importance of climate in urban planning is highlighted in the study of Kahramanmaras city: the influence of Climate comfort based caused by geographical conditions is analyzed in human activities, living standards, settlement forms and it was determined that people moved and developed a different way of life due to the temporal-spatial variation of climate comfort, situation that should be reflected in climate comfort-based planning (Cetin 2020).

The viability and effectiveness of bioclimatic architecture differs in each city or geographic space: The bioclimatic aptitude depends on natural factors and the climate (approaches based on generalizations), human needs (referential approaches) and on limitations and objectives generated by the built environment and urban structure (approaches based on optimization).

Bioclimatic design in cities generally relies on the optimization of the interface between the city's built environment and the outdoors (Salat and Bourdic 2013). The macroclimate, microclimate and each building make up thermal subsystems that can be defined as the set of interactions that occur between the physical, climatic and design variables that determine both the thermal comfort of the inhabitants and the energy consumption required for heating or cooling. The closer the subsystems are to comfort conditions, the lower the energy consumption required.

A bioclimatic aptitude index of a city to describe its current state, its limitations, and its potential. The methodology for constructing this index has three steps: (1) Description of the urban-bioclimatic system, (2) the integration of the qualitative and quantitative aspects of the system for defining urban-bioclimatic variables and (3) the construction of aptitude index based on the maximum and minimum values of each variable.

2.1 Description of Urban-Bioclimatic System

Urban-Bioclimatic system can be defined as the set of interactions that occur between the physical and climatic variables of a city, the characteristics of its urban design and the characteristics of its buildings that determine the thermal comfort of the inhabitants and energy consumption required for heating or cooling (the built environment and city as new nature). The Fig. 4 shows the elements that make up the urban-bioclimatic system.

Although there are processes, tools, and methods for the qualitative and quantitative evaluation of a contained (architectural) space, qualitative analysis predominates for (urban) open spaces. The aptitude index includes an assessment of the urban variables at a quantitative level based on a classification of the variables used in the calculation of heat gain or loss.

The thermic conditions of a building depend on the magnitude of heat gain and loss at a given time. The building will tend to warm up when heat gains exceed losses, and it will cool when the converse is true. According to the equation of thermic balance, a building achieves thermic equilibrium when the sum of heat gain and loss is equal to zero, a neutral point that is expressed by the following equation:

$$Q_c + Q_r \pm Q_v \pm Q_i + Q_a + Q_m = 0 \quad (1)$$
Thermic balance

These values refer to the primary sources of heat gain and/or loss of a building: (a) conduction (Qc), (b) radiation (Qr), (c) convection (Qv), (d) infiltration (Qi), (e) devices (Qa) and (f) metabolism (Qm). The basic calculus variables for infiltration correspond to the ambit of architectural characteristics and for this reason are not considered in the urban aspect. Heat by devices and heat by metabolism depend on the project's characteristics (like the number of people in the building or their activities) and are therefore not regarded as controllable variables in architectural or urban design.

Conduction: Heat transfer by conduction through walls, roofs and windows of the building constitutes an energy propagation process (heat) by means of direct contact

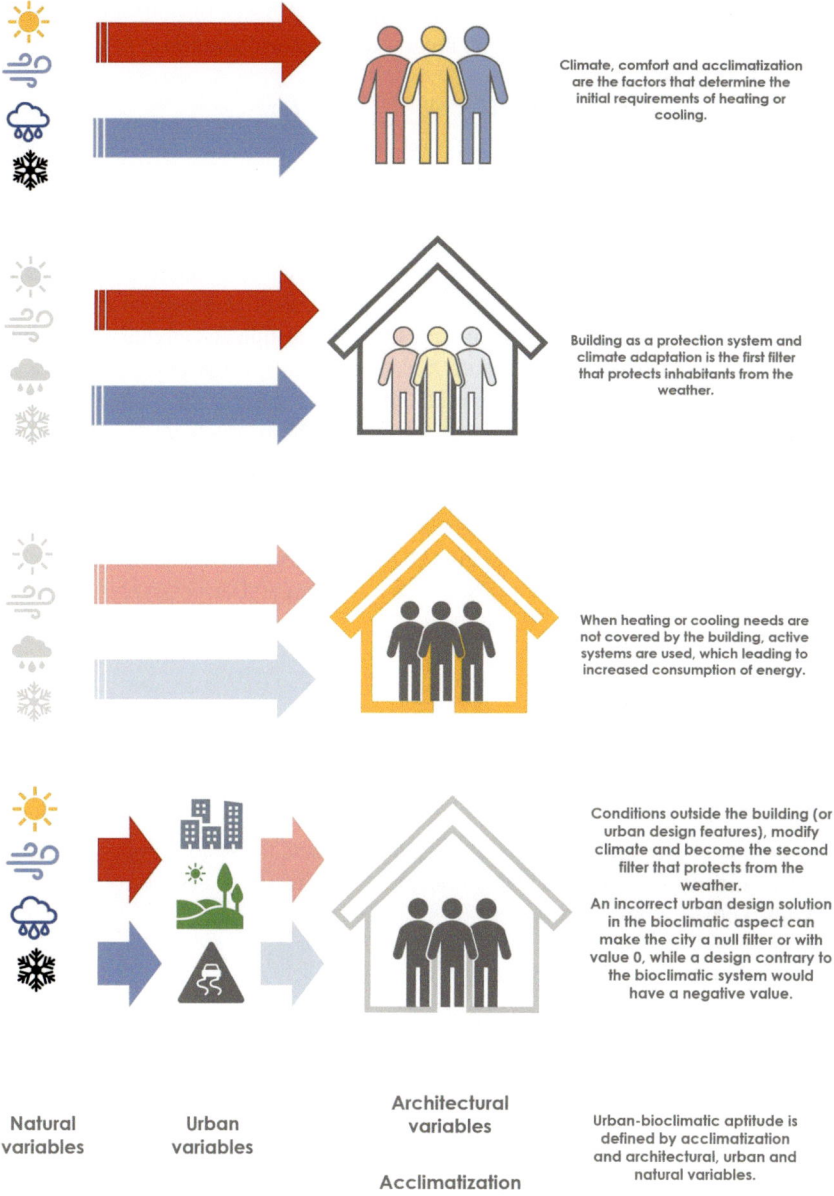

Fig. 4 Urban-bioclimatic aptitude system (own elaboration)

between bodies at different temperatures. Heat transfer takes place from the region with the most elevated temperature to area(s) of lower temperature.

$$QC = U \times A_s \times (T_{sa} - T_i)$$
Heat transfer by conduction
(2)

Radiation: Unlike conduction, in which energy transport passes through material means, heat spreads by radiation even in a void. Radiation penetrating a material causes changes that are expressed in the form of primary flux attenuation, whether by energy absorptivity or redistribution in a different direction from the original direction of propagation. The intensity, spectral composition and angular distribution of this emission depend on the nature and temperature of the body analyzed. By emitting radiation, all bodies lose internal energy and cool so the temperature may stay constant; this must be compensated for by taking the energy from the surrounding bodies and environment. The gain or loss of heat by direct radiation is presented only in translucent materials (glass) and is the amount that passes through a window and comes from direct solar radiation; part of it increases the temperature of the inner air and is immediately added to the instantaneous charge, while the rest is stored in furniture, carpeting and other fixtures, where it remains and successively contributes to increasing the interior air temperature:

$$QR = A_v \times t \times S \times H_t$$
Heat transfer by radiation
(3)

Figure 4 presents a diagram of the variables and elements that make up, in turn, the variables of the Eq. 3.

Ventilation: Heat gain and loss by ventilation refers to the air that flows into and out of windows or vents built for that purpose. The heat flow caused by ventilation depends on the following variables:

$$QV_t = Q_{vs} + Q_{vt}$$
Heat transfer by ventilation
(4)

2.2 Description of Qualitative and Quantitative Variables.

Description of the qualitative aspect of the system: (a) The variables that describe a quantitative relationship between the outdoor climate and a building's interior climate, are identified in the calculating heat gains/losses and they are called basic calculus variables (VBC); (b) these variables can be classified into two groups: those that can be modified with architectural design and those that cannot: The first are called controllable basic calculus variables (VBCc) and the second non-controllable basic calculus variables (VBCnc); and (c) and there is a fourth type of variables, those urban characteristics that are an intermediaries between the building envelope and the natural climate. These are called in this study, Urban-bioclimatic variables (UBV) and modify some controllable basic calculus variables (VBCc).

Basic Calculus Variables (VBC) generation: Once the calculation of heat gains and losses through conduction, convection and radiation is reviewed, the results are physical, climatic and design variables used for (see Table 1).

These basic calculus variables are classified into two groups: (a) variables controllable by design (VBCc), which are those that can be modified by architectural or urban design, and (b) variables uncontrollable by design (VBCnc), which depend on the physical conditions of the site and cannot be altered by design (Table 2).

Table 1 Basic calculus variables (VBC) (own elaboration)

Variable	Classification	Unit
1. Material thermal conductivity coefficient	Kn	W/m^2°C
2. Exterior temperature	Te	°C
3. Interior Temperature	Ti	°C
4. Wind Convection coefficient	He	W/m^2°C
5. Inner air convection coefficient	Hi	W/m^2°C
6. Solar radiation	Ht	W/m^2
7. Wind speed	W	m/s
8. Air density	P	Kg/m^3
9. Air specific heat	Cpa	Kj/Kg°C
10. Vapor humidity	HVAD	Kj/Kg
11. Exterior air specific humidity	Wo	Gr/Kg
12. Inner air specific humidity	Wi	Gr/Kg
13. Exposed surface area	AS	M2
14. Material thickness	On	M
15. Roof inclination	SLP	Degrees
16. Suface absortance coefficient	AS	Factor
17. Surface emittance coefficient	E	Factor
18. Window area with direct solar radiation	Av	M2
19. Glass transmittance	T	Factor
20. Shadow coefficient	S	Factor
21. Direct opennings area	AA	M3
22. Oblique opennings area	AB	M2
23. Effective oppening factor	Cva	Factor
24. Number of changes per hour	Cvb	Factor
25. Cloud reflection temperature	Tsky	°K
26. Exterior reflection temperature	Tsli	°K

Table 2 Variables uncontrollable by design (own elaboration)

Way of heat gain and loss	Variable
Conduction	*Exterior Temperature (Te)
	*Ambient Temperature (Tamb)
	*Inner Temperature (Ti)
	*Wind convection coefficient (he)
	*Inner air convection coefficient (hi)
	*Stefan Botzman Constant (S)
Radiation	Solar radiation (Ht)
Ventilation	*Wind speed (W)
	*Air density (P)
	*Air specific heat (Cpa)
	*Vapor humidity (Hvap)
	*Ambient Temperature (Tamb)
	*Inner Temperature (Ti)
	*Absolute ambient humidity (Wo)
	*Inner absolute humidity
Infiltration	*Air density (P)
	*Air specific heat (Cpa)
	*Vapor humidity (Hvap)
	*Ambient Temperature (Tamb)
	*Inner Temperature (Ti)
	*Absolute ambient humidity (Wo)
	*Inner absolute humidity (Wi)
Devices	*Number of devices
	*Heat emitted (Qa)
Metabolism	*Number of people that carry out an activity (N)
	*Activity Heat (Qp)
	*Activities (n)
Total	100%
	Heat gain/loss

2.3 Urban-Bioclimatic Variables Generation

Urban-bioclimatic variables (UBV) generation: From VBCc, those that can be modified modifiable by urban design, i.e. were separated, that is, the characteristics of an urban settlement that have a direct relationship with the microclimate and that can modify it, such as orientation of streets, topography, urban trace, street width and building height, occupation index, settlement age, property shape, predominant colours and textures, vegetation, paving and bodies of water (Table 3).

Table 3 Urban-bioclimatic variables generation (own elaboration)

Form of heat gain/loss	Design-controllable variables	Urban bioclimatic variable	Architectural variable
Conduction	Surface area exposed to the exterior	Orientation of the property and streets	Shape of the building
	Heat transfer coefficient (U)	Topography	Volumetry
	Heat transfer coefficient (k)	Urban layout	Choice of optional materials
	Thickness of the material	Density	Optional materials
	Roof pitch	Street width/Building height	Insulating material
	Absorption coefficient of the surface	Open space index	Change thicknesses
	Surface emittance coefficient	Age of settlement	Increase material layers
		Site shape	Change roof pitch
		Textures and colors	Material thickness
		Tree cover	Alternative materials
		Pavement material	Material color
		Water bodies	
Radiation	Area of window or glass (domes) that receives direct solar radiation	Orientation of the property and streets	Area of translucent material
	Transmittance of the glass or translucent material	Topography	Geometry of translucent material
	Shading coefficient	Urban layout	Layout of translucent material
		Density	Material with different optical characteristics
		Street width/Building height	Implementation/removal of shading elements (architectural or natural)
		Open space index	
		Age of settlement	
		Shape of property	
		Wooded	

(continued)

Table 3 (continued)

Form of heat gain/loss	Design-controllable variables	Urban bioclimatic variable	Architectural variable
Convection	Area of direct openings	Orientation of the property and streets	Ventilation area
	Area of oblique openings	Topography	Arrangement of vents on facade
	Opening effectiveness factor	Urban layout	Arrangement of vents with respect to the floor or roof
		Density	
		Street width/Building height	
		Open space index	
		Age of settlement	
		Tree cover	
		Water bodies	
Infiltration	Number of air changes per hour		Increase/decrease thermal insulation in ceilings, door and window frames
	Volume of the room	–	Vary room volume
Devides			
Metabolism			
Total	100%		
	Heat gains/losses		

Once the urban bioclimatic variables have been defined, the existing qualitative information can be used to order the nominal variables and convert them into ordinals. Thus, the specific conditions of the UBV´s are hierarchized, selected, or rejected based on an evaluation expressed by scores, preference values or intensities according to the positive or negative criteria of each type of climate.

2.4 Aptitude Index

Description of the quantitative aspect of the system. To determine the size of the VBC's, first, the extreme values for a built urban-bioclimatic system (city) are identified and the extreme values of heat gains or losses (Q) are calculated for each of the basic calculation variables. Then, maximum, and minimum values for Q are obtained for the maximum and minimum values of each basic calculus variables. With these values, the relevance of each basic calculus variable for the total heat gain or loss can be determined if the difference between the maximum and minimum values of the basic calculus variable is distributed between the proportions of the Q values obtained.

$$i = \frac{VBCM - VBCm}{Qm/QM} \quad (4)$$

Aptitude index (own elaboration)

Once the proportional values of the basic calculus variables have been obtained and it has been determined which ones are related to the urbanbioclimatic variables, the values obtained are applied to the qualitative assessment to obtain an approximation of the importance or relevance of the urban bioclimatic variable in the built urban-bioclimatic system.

Formulas for heat gain and loss were developed on a calculation sheet. The maximum (QTM) and minimum (QTm) values were obtained for extreme analysis conditions. To accomplish that, the following analysis was carried out:

(a) For each of the basic calculus variables (Qvbn), a maximum possible value for the specific study case was used, while the rest of the values were managed with minimum values. This is how the value of QT was obtained for a maximum value of Qvbn.
(b) For each of the basic calculus variables (Qvbn), a minimum possible value for the specific study case and more explicitly discussed in Sect. 2 was used, while the rest of the values were managed with maximum values. This is how the value of QT was obtained for a minimum value of Qvbn.

A graph corresponding to the values of the previously described was made; it represents the extreme values of Qvbn and its relationship with the extreme values of QTR (relative). The graph allows for the assessment of each basic calculus variable regarding the total gain or loss for the study case.

The proportion of each basic calculus variable was obtained as follows: the difference between the extreme heat gain and loss values (QTm/QTM) was obtained to find its value without units (QTA). The difference between the extreme values of each basic calculus variable (Qvbn MAX − Qvbn MIN) was obtained. This represents the absolute value of Qvbn. The proportion of each absolute Qvbn regarding the value of QTA was obtained as follows: basic variable n is equal to a maximum X% of the total absolute heat gain or loss (QTA).

Both the uncontrollable basic calculation variables (weather conditions) and the design controllable variables can be measured, calculated or predicted. In order to know the extreme conditions for both weather and design variables, the extreme values for each basic calculus variable have been identified. For example, if we consider that radiation is a function of four basic variables, we can evaluate each one by taking the minimum value of the first and the maximum value of the other three. This is how we obtain the values of heat gain or loss (Q) shown in red in Figs. 4 and 5. If the analyzed variable has a minimum Q value, the first graph is obtained; if the maximum Q value is found, the second graph is applied.

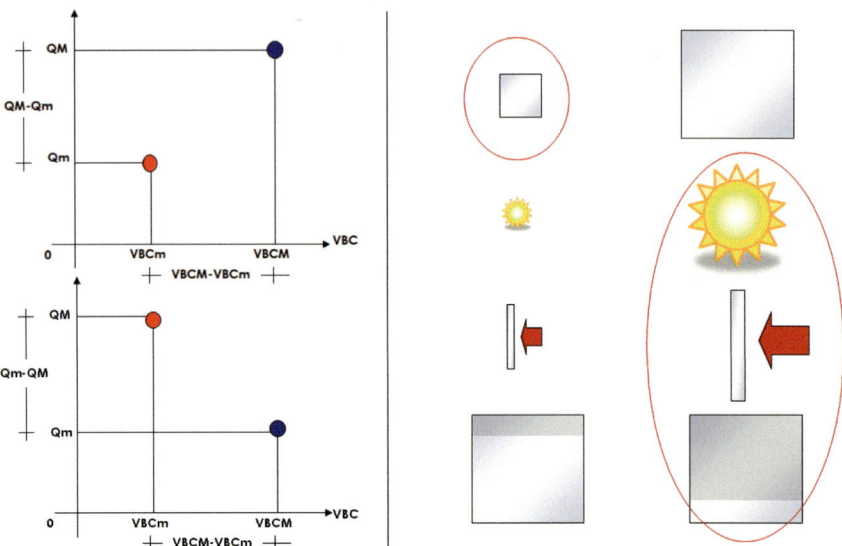

Fig. 5 Minimum value of a basic calculus variable (own elaboration)

On the other hand, the maximum value of the same valuable is considered along with the minimum values of the other three, the heat gain and loss values in blue in Figs. 4 and 5 are determined. If the analyzed variable has a minimum Q value, the first graph will be obtained; if the maximum Q value is found, the second graph will apply (Fig. 6).

This will identify whether the relationship is positive or negative and the maximums and minimums values obtained for Q for maximum and minimum values of each basic calculus variable. These values enable the determination of the proficiency rate (i) for each basic calculus variable.

Once these values are obtained, the following deductions may be made (see Table 4):

- If the basic calculus variables are too similar, we find ourselves in conditions of mild weather, because the difference between their maximum and minimum values will tend to zero.
- If the basic calculus variables are far apart from one another, there is extreme weather, and the difference between their maximum and minimum values will be greater than zero.
- If the maximum and minimum Q values are very similar, the basic calculus variable analyzed has little influence overheat gain or loss, and the relation between maximum and minimum Q will tend to one.
- If the maximum and minimum Q values are far apart from one another, the basic calculus variable analyzed has significant influence overheat gain or loss, and the relation between maximum and minimum Q will tend to zero.

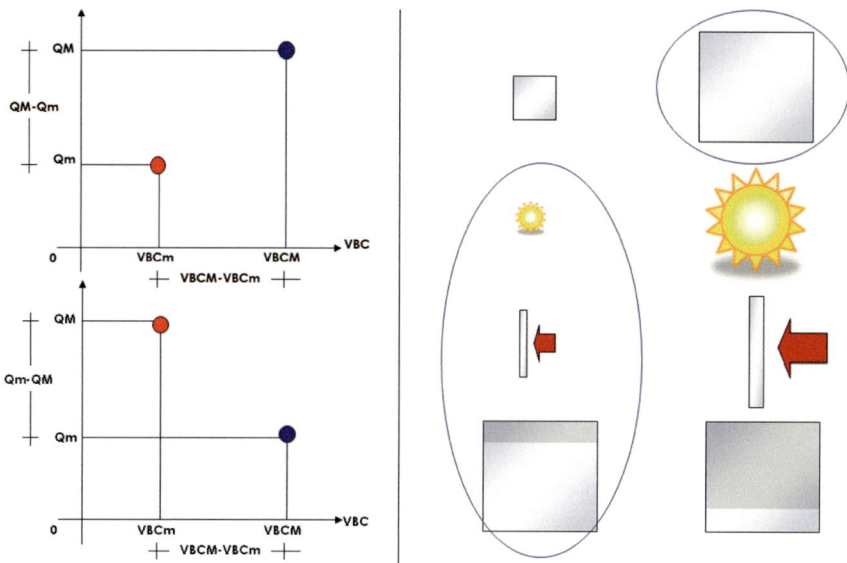

Fig. 6 Maximum value of a basic calculus variable (own elaboration)

Table 4 Proficiency rate (own elaboration)

Proficiency rate	$i = \frac{VBCM - VBCm}{Qm/QM}$		
		Temperate, moderate conditions => VBCM-VVCm ≈ 0	Extreme climates => VBCM-VVCm > 0
QM Qm => QM/Qm ≈ 1		0/0 => i ≈ 0	N/1 => i ≈ N
QM = Qm => QM/Qm ≈ 1		0/0 => i = ind	N/0 => i = ∞

There are four possible outcomes regarding the relations between these four possible situations.

- In mild weather with little variation in heat gain or loss, the significance of the basic variable analyzed will tend to zero; in other words, it will have a very low value.
- In mild weather with a great deal of variation in heat gain or loss, the significance of the basic variable analyzed will tend to zero; in other words, it will have an undetermined value.
- In extreme weather with little variation in heat gain or loss, the significance of the basic variable analyzed will have an intermediate value.

- In extreme weather with a great deal of variation in heat gain or loss, the significance of the basic variable analyzed will tend to infinity; in other words, it will have a high value.

3 Case Study

Pachuca, Mexico, is located between 20°01' and 20°12' north latitude and 98°41' and 98°52' west longitude, at an altitude between 2400 and 3000 m. It is situated 100 km northeast of Mexico City at the end of the southern slope of the Real del Monte and Pachuca Mountain range. It is the capital city of the Mexican state of Hidalgo and has a metropolitan population of approximately 314,000.

Its climate is temperate subhumid with summer rains, low moisture (43.0%), semi-cold subhumid with summer rains, high humidity (40%), with a temperature range of 10–16 °C and a precipitation range of 400–900 mm.

Comfort conditions between the tropics are defined as 22–28 °C and 30–70% humidity. Outside the tropics, the range is 14–20 °C and 18–26 °C. For Pachuca, the comfort temperature range based on the Szokolay formula is as follows:

$$Tn = (17.6 + 0.31Te) \pm 2.5\ °C$$
Comfort temperature range
(5)

23.9 to 18.9 in January, 24.2 to 19.2 in February, 24.8 to 19.8 in March, 25.3 to 20.3 in April, 25.4 to 20.4 in May, 15.04 to 20 in June, 24.8 to 19.8 in July, 24.8 to 19.8 in August, 24.7 to 19.7 in September, 24.5 to 19.5 in October, 24.3 to 19.3 in November and 24.06 to 19.06 in December. Therefore, the annual comfort temperature lies in the 25.40 °C–18.90 °C range.

For the application of the urban-bioclimatic aptitude index in the case study:

1. Climatic data of the last 10 years of the national meteorological system were obtained and averaged to determine the climatic conditions (variables that cannot be modified with the design). Tables 5, 6, 7 and 8.
2. A cartographic and visual review of the urban characteristics by neighborhood (urban-bioclimatic variables) was carried out.
3. Quantitative calculations were made in a spreadsheet to determine the aptitude per colony and the marginal contributions by type of heat gain or loss.
4. Thematic maps were generated in ArcGIS with fitness ranges. The lower the values, the more apt the area to be renovated in terms of bioclimatic-urban and better energy yields by type of intervention in conduction, convection or radiation strategies in winter and summer.

Table 5 Average temperatures, relative humidity, radiation, and wind velocity at Pachuca. Own elaboration based on data from Servicio Metereológico Nacional

Hour	January	February	March	April	May	June	July	August	September	October	November	December
1:00	5.3	6.8	8.3	10	11.4	12.5	11.9	11.5	11.4	9.2	7.2	6.1
2:00	4.2	5.7	7.3	9.1	10.6	11.7	11.2	10.8	10.8	8.5	6.3	5
3:00	3.2	4.8	6.4	8.2	9.8	11.1	10.6	10.2	10.2	7.8	5.6	4.1
4:00	2.5	4.1	5.7	7.5	9.2	10.6	10.2	9.7	9.7	7.3	5	3.4
5:00	1.8	3.4	5.1	7	8.7	10.2	9.8	9.3	9.4	6.8	4.5	2.8
6:00	1.3	2.9	4.6	6.5	8.3	9.9	9.4	9	9.1	6.5	4.1	2.3
7:00	1.9	2.5	4.2	6.2	8	9.6	9.2	8.7	8.8	6.2	3.7	1.9
8:00	−0.4	1.2	4.4	6.3	8.1	9.7	9.3	8.8	8.9	6.3	3.9	0.7
9:00	2.8	4.4	8.9	10.6	11.9	12.9	12.3	11.9	11.8	9.7	7.7	3.7
10:00	8	9.5	14.4	15.8	16.5	16.7	16	15.7	15.3	13.8	12.4	8.6
11:00	13.2	14.8	19.4	20.6	20.7	20.3	19.4	19.2	18.5	17.5	16.6	13.6
12:00	17.6	19.1	23.1	24.1	23.8	22.9	21.9	21.8	20.8	20.3	19.8	17.7
13:00	20.5	22	25.3	26.2	25.7	24.4	23.4	23.3	22.2	22.2	21.7	20.5
14:00	21.9	23.4	26.1	26.9	26.3	24.9	23.9	23.8	22.7	22.5	22.3	21.8
15:00	22.1	23.6	25.8	26.6	26	24.7	23.7	23.6	22.5	22.3	22	22
16:00	21.3	22.8	24.6	25.4	25	23.8	22.9	22.8	21.8	21.4	21	21.2
17:00	19.8	21.3	22.8	23.8	23.6	22.6	21.7	21.6	20.6	20.1	19.5	19.8
18:00	17.9	19.4	20.8	21.8	21.8	21.2	20.3	20.1	19.3	18.5	17.8	18
19:00	15.8	17.3	18.6	19.8	20	19.7	18.9	18.6	18	16.9	15.9	16.1
20:00	13.7	15.2	16.5	17.8	16	18.2	17.4	17.2	16.6	15.3	14.1	14
21:00	11.7	13.2	14.4	15.9	14.4	16.8	16.1	15.8	15.3	13.8	12.4	12.1

(continued)

Table 5 (continued)

Hour	January	February	March	April	May	June	July	August	September	October	November	December
22:00	9.8	11.3	12.6	14.1	12.9	15.5	14.8	14.5	14.1	12.4	10.8	10.3
23:00	8.1	9.6	11	12.6	11.6	14.3	13.7	13.4	13.1	11.2	9.4	8.7
0:00	6.6	8.1	9.5	11.2	10.4	13.3	12.8	12.4	12.2	10.1	8.2	7.3

Table 6 Average relative humidity percentage at Pachuca. Own elaboration based on data from Servicio Metereológico Nacional

Hour	January	February	March	April	May	June	July	August	September	October	November	December
1:00	76	70	66	68	76	85	85	85	86	85	83	78
2:00	79	73	69	71	78	87	88	88	88	87	86	81
3:00	82	75	71	73	80	89	90	90	90	89	88	84
4:00	84	77	72	74	82	91	91	91	91	91	90	86
5:00	86	79	74	76	83	92	93	93	93	92	92	88
6:00	87	80	75	77	85	94	94	94	94	94	93	89
7:00	88	81	76	78	85	95	95	95	95	94	94	90
8:00	92	84	75	77	85	94	94	94	94	94	94	94
9:00	83	76	65	67	74	83	84	84	84	83	82	85
10:00	69	63	52	54	61	70	72	71	71	70	67	71
11:00	54	50	41	43	49	58	60	60	60	58	53	56
12:00	42	39	32	34	40	49	52	51	51	49	44	44
13:00	34	31	27	29	35	44	47	46	46	43	38	36
14:00	30	28	35	27	33	42	45	45	45	41	36	32
15:00	29	27	26	28	34	43	46	45	45	42	37	32
16:00	32	29	29	31	37	46	49	48	48	45	40	34
17:00	36	33	33	35	41	50	53	52	52	49	44	38
18:00	41	38	37	40	46	55	57	57	57	54	50	43
19:00	47	43	43	45	51	60	62	62	62	60	56	49
20:00	53	49	47	57	56	65	67	67	67	65	61	55
21:00	58	54	52	62	61	70	72	71	71	70	67	60
22:00	64	58	56	67	66	74	76	76	76	74	72	66

(continued)

Table 6 (continued)

Hour	January	February	March	April	May	June	July	August	September	October	November	December
23:00	68	63	60	71	69	78	79	79	79	78	76	70
24:00	73	67	63	75	73	82	83	83	83	82	80	75

Table 7 Average radiation (w/m2) at Pachuca. Own elaboration based on data from Servicio Metereológico Nacional

Hour	January	February	March	April	May	June	July	August	September	October	November	December
0:00	−4	−4	−4	−4	−4	−4	−4	−4	−4	−4	−4	−4
1:00	−4	−4	−4	−4	−4	−4	−4	−4	−4	−4	−4	−4
2:00	−4	−4	−4	−4	−4	−4	−4	−4	−4	−4	−4	−4
3:00	−4	−4	−4	−4	−4	−4	−4	−4	−4	−4	−4	−4
4:00	−4	−4	−4	−4	−4	−4	−4	−4	−4	−4	−4	−4
5:00	−4	−4	−4	−4	−4	−4	−4	−4	−4	−4	−4	−4
6:00	−4	−2	−4	−4	−4	−4	−4	−4	−4	−4	−4	−4
7:00	−4	−2	16	22	121.24	155.15	46.25	25	18	14	5	−4
8:00	100.28	127.32	132.15	198.91	271.28	290.11	237.36	180.58	13	24	5	−4
9:00	272.22	184.11	272.86	292.86	440.6	499.25	537.18	219.53	127.6	142.15	278.72	148.25
10:00	421.24	484.89	428.18	321.27	480.79	527.53	602.18	389.14	319.64	279.65	301.33	345.24
11:00	639.75	678.22	523.16	427.45	524.78	648.77	677.18	528.94	489.56	419.45	389.45	456.66
12:00	692.98	649.34	637.78	633.88	765.18	788.16	725.78	648.75	604.65	609.16	529.19	580.29
13:00	762.38	675.22	681.61	622.32	675.29	688.9	625.6	578.93	518.43	588.65	578.56	723.3
14:00	678.27	485.72	483.34	422.3	529.19	539.2	489.2	514.25	434.56	389.14	289.88	356.66
15:00	529.98	327.18	239.14	318.8	423.19	438.14	415.11	389.54	312.12	302.14	201.33	233.45
16:00	106.18	128.16	139.12	128.8	320.13	389.21	318.15	289.54	245.34	177.55	141.22	188.23
17:00	72.12	183.36	151.34	127.8	229.1	289.14	245.55	189.34	32.28	34.4	19.43	18.3
18:00	11.32	71.89	118.6	121.5	124.45	120.94	94.47	64.34	−3	−3	−3	−3
19:00	−3	−3	87.89	14.31	−3	−3	−3	−3	−3	−3	−3	−3
20:00	−3	−3	−3	−3	−3	−3	−3	−3	−3	−3	−3	−3
21:00	−3	−3	−3	−3	−3	−3	−3	−3	−3	−3	−3	−3

(continued)

Table 7 (continued)

Hour	January	February	March	April	May	June	July	August	September	October	November	December
22:00	−3	−3	−3	−3	−3	−3	−3	−3	−3	−3	−3	−3
23:00	−3	−3	−3	−3	−3	−3	−3	−3	−3	−3	−3	−3
0:00	−3	−3	−3	−3	−3	−3	−3	−3	−3	−3	−3	−3
Promedy	154.4475	157.9764	154.7268	144.368	194.3688	213.14	198.7204	158.8752	122.6472	117.2516	107.6044	119.7352

Table 8 Average wind direction and velocity (m/s) at Pachuca. Own elaboration based on data from Servicio Metereológico Nacional

Hour	January		February		March		April		May		June	
	Direction	Velocity	Direction	Velocity	Direction	Velocity	Direction	Velocity	Direction	Velocity	Direction	Velocity
0:00	–	0	–	0	–	0	NO	5.7	NO	4.5	NO	4.6
1:00	–	0	–	0	SO	4.4	NO	2.2	–	0	NO	3.9
2:00	–	0	–	0	NE	3.4	–	0	NO	5.4	NO	3.3
3:00	–	0	–	0	–	0	–	0	–	0	NO	2.9
4:00	–	0	–	0	–	0	–	0	–	0	NO	2.6
5:00	–	0	–	0	–	0	–	0	–	0	NE	3.3
6:00	–	0	–	0	NE	5.1	–	0	–	0	NE	3.9
7:00	NO	5.1	NE	5.6	SE	6.3	NE	6	NE	4.4	NE	5
8:00	NO	6.3	NE	6.8	SE	7.5	NE	7.2	NE	5.6	NE	6.3
9:00	NO	9.7	SE	8.2	NE	9	SE	8.8	SE	6.8	NE	7.5
10:00	SE	9	SE	9.7	SE	10.7	NE	10.4	NE	8.2	NE	9
11:00	NO	10.7	SE	11.4	NE	12.4	NE	11.9	NE	9.7	NE	10.4
12:00	SE	12.2	S	12.9	NE	13.9	NE	13.5	NE	11.3	NE	11.7
13:00	NO	13.5	SE	14.2	NE	15.4	NE	14.9	SO	12.6	NE	12.9
14:00	NO	14.3	NE	15.1	NE	16.4	NE	15.8	SE	13.9	NE	13.8
15:00	NO	15	S	15.8	NE	17.2	NE	16.7	SE	14.9	NE	14.4
16:00	SO	13.5	NE	14.2	NE	15.4	NE	14.9	SO	15.6	NE	12.9
17:00	SO	12.2	NE	12.9	NE	13.9	NE	13.5	SO	13.9	N	11.7
18:00	NO	10.7	SO	11.4	NE	12.4	NE	11.9	SO	12.6	NE	10.4
19:00	SO	9	NE	9.7	NE	10.7	NE	10.4	S	11.3	N	9
20:00	–	0	–	0	NE	9	NE	8.8	SO	9.7	NE	7.5
21:00	–	0	–	0	NE	7.5	NE	7.2	NE	8.2	NE	6.3
22:00	–	0	–	0	NE	6.5	NE	6.3	SO	6.8	NE	5.3
23:00	–	0	–	0	NE	5.8	N	5.8	SO	5.8	–	4.6

(continued)

Table 8 (continued)

Hour	January		February		March		April		May		June	
	Direction	Velocity	Direction	Velocity	Direction	Velocity	Direction	Velocity	Direction	Velocity	Direction	Velocity
Hours	July		August		September		October		November		December	
	Direction	Velocity	Direction	Velocity	Direction	Velocity	Direction	Velocity	Direction	Velocity	Direction	Velocity
0:00	NO	4.2	NO	4	NE	4.3	NE	4.5	NE	5.1	–	0
1:00	NO	3.6	NE	3.5	NE	3.8	NE	6.4	–	0	–	0
2:00	SE	3.1	NO	2.9	NE	3.2	–	0	NE	4	–	0
3:00	SE	2.6	–	0	NE	2.8	–	0	NE	3.6	–	0
4:00	SE	2.4	O	2.2	NO	2.5	–	0	NE	3.3	–	0
5:00	SE	3.1	–	0	NO	3.2	–	0	NE	4	–	0
6:00	SO	3.6	–	0	N	3.8	–	0	NE	4.6	–	0
7:00	SO	4.6	NE	4.4	NO	4.7	–	0	NE	5.6	–	0
8:00	S	5.7	–	0	NE	6	–	0	NO	6.9	–	0
9:00	NE	6.9	SE	6.8	NE	7.2	NE	7.9	NO	8.5	SE	10.6
10:00	SE	8.3	SE	8.2	NE	8.8	NO	9.4	NO	10.1	NE	12.1
11:00	NE	9.7	SE	9.6	NE	10.1	NE	11	NO	11.7	NE	13.5
12:00	NE	11	SE	10.8	NE	11.4	NE	12.4	NO	13.1	NE	14.7
13:00	NE	12.1	SE	11.9	NE	12.6	N	13.6	NO	14.3	NE	15.7
14:00	NE	12.9	SE	12.8	NE	13.5	N	14.6	NO	15.3	NE	16.3
15:00	NE	13.5	SE	13.3	NE	14	NE	15.1	NE	15.8	NE	14.7
16:00	NE	12.1	S	11.9	N	12.6	NE	13.6	NE	14.3	NE	13.5
17:00	NE	11	NE	10.8	NO	11.4	NE	12.4	NE	13.1	NE	12.1
18:00	NE	9.7	NO	9.6	NO	10.1	NE	11	NE	11.7	NE	10.5
19:00	NE	8.3	NO	8.2	NE	8.8	NE	9.4	NE	10.1	NE	8.9
20:00	SO	6.9	NO	6.8	NE	7.2	NE	7.9	NE	8.5	SO	7.4
21:00	O	5.7	NO	5.6	NE	6	NE	6.5	NE	6.9	O	6.3
22:00	O	4.9	NO	4.7	NE	5	NE	5.4	NE	5.8	NO	0
23:00	SO	4.2	N	4	NE	4.3	NE	4.7	NE	5.1	–	

4 Results

4.1 Qualitative and Quantitative Analysis

Once the values of heat gain or loss were obtained, the amount of participation of each basic calculus variable could be determined. Physical characterization per neighborhood was performed and the aptitude index was applied to each of the colonies in a spreadsheet. A matrix was generated with the values for conduction, convection, and radiation.

Figures 7, 8 and 9 show the marginal contributions of the basic calculation variables for conduction, convection and radiation and represent the contribution that the urban bioclimatic system of the city has in the losses or gains of heat inside the building.

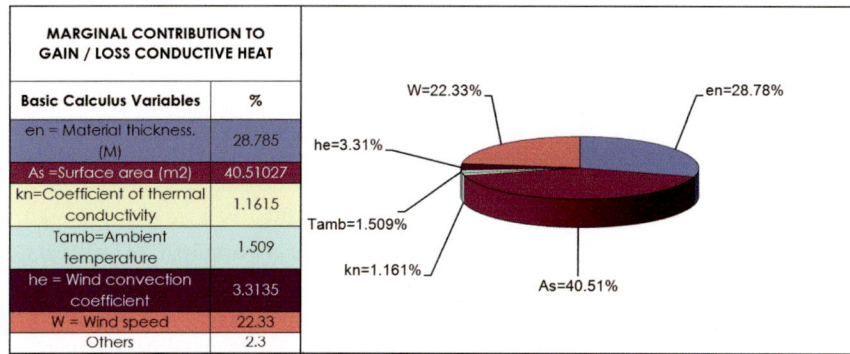

Fig. 7 Marginal contribution of conduction to heat gain or loss in Pachuca (own elaboration)

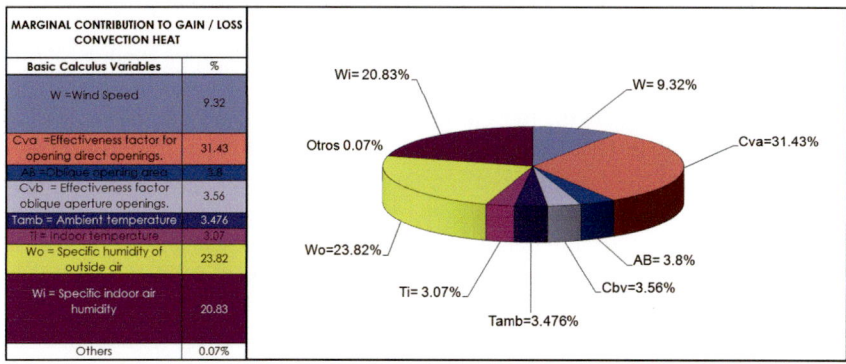

Fig. 8 Marginal contribution of convection to heat gain or loss in Pachuca (own elaboration)

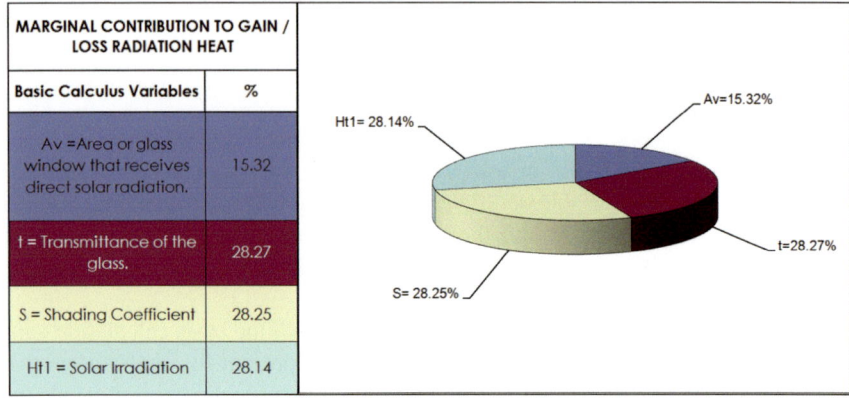

Fig. 9 Marginal contribution of radiation to heat gain or loss in Pachuca (own elaboration)

(a) The Basic Design variable with the highest percentage of participation in the Conduction Heat Gain/Loss is the Surface Area, followed by the material thickness.
(b) The Basic Design variable with the highest percentage of participation in the Convection Heat Gain/Loss is the Opening Effectiveness Factor Direct, followed by the Specific Indoor Air Humidity.
(c) Radiation heat gains/losses, the basic design variables "Glass transmittance", "Shading coefficient" and "solar radiation" have a participation percentage practically the same, followed by the Area of the window.
(d) The largest variations in the Gain/Loss Values for Driving are presented in Conduction by Solar Radiation, followed by the Coefficient of Air Convection interior, that is, they are the most sensitive variables of heat by conduction.
(e) The greatest variations of the Convection Gain/Loss Values are presented in Convection is the Specific Heat of the Air, followed by the Density of the air, that is, are the most sensitive variables of convective heat.
(f) The greatest variations of the Values of Gains/Loss by Radiation are presented in radiation is the Transmittance of the glass, followed by the shading coefficient, that is that is, they are the most sensitive variables of heat by radiation.
(g) Integrating Urban-Bioclimatic Variables to the aptitude index, the most relevant variables in urban-bioclimatic design are density, the building height-width ratio street and open space index, followed by orientation and topography.
(h) Density, building height-street width ratio and the open space index, the orientation and topography are all variables related to "access to the sun" and "access to the wind".
(i) The most repeated urban-bioclimatic variables (mode) by Conduction are density, Width of the street/Height of the building and Trees ratio; by convection are the density, the street width/building height ratio, the open space index, and the water bodies; and by radiation they are the orientation, the topography, the density, the relation street Width/Building Height and Open Space Index.

(j) The urban-bioclimatic variables with the most repetitions (mode) are density and the relationship street Width/Building Height.

4.2 Territorial Analysis

The city was characterized with the twelve urban-bioclimatic variables (UVB) both in summer and winter by district and they were evaluated with the results of the marginal contributions by convection, convection, and radiation. Figures 10, 11, 12, 13, 14 and 15 present the classification method of standard deviation in five classes. The lower the value, the greater the urban bioclimatic aptitude of the district.

(a) The predominant orientation of the lots is northwest-southeast with an approximation of 32%, without However, this data is not notorious since the geographical distribution is random with a certain predominance of north–south and east–west orientations in the northern area, the oldest in the city and with a greater slope and certain dominance of the Northwest–Southeast orientation in the valley. The northwest–southeast orientation is distributed in the same way in any area of the city. Combining these data with the age of the settlement, it is observed that in the original settlement, the southern slope is preferred over the northern slope. Radiation is favored solar for being located halfway up the slope, open to the south.

(b) Regarding the topography, 59% of the surface of the urban area has a slope of maximum terrain of 10, 26% corresponds to properties with slopes between 10

Fig. 10 Bioclimatic aptitude map in Pachuca in summer by conduction (own elaboration)

Fig. 11 Bioclimatic aptitude map in Pachuca in summer by convection (own elaboration)

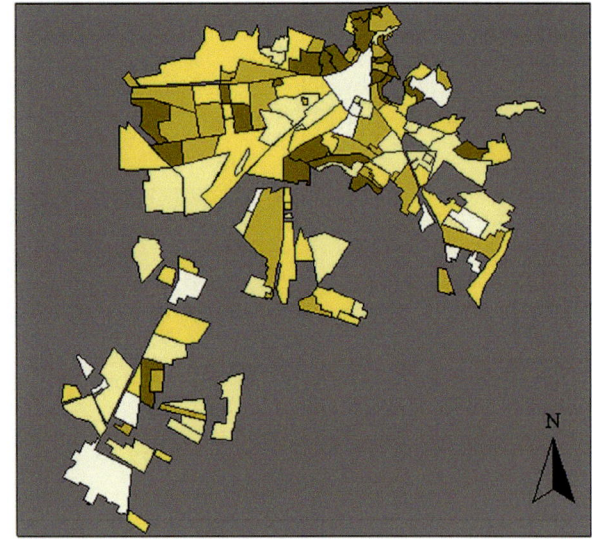

Fig. 12 Bioclimatic aptitude map in Pachuca in summer by radiation (own elaboration)

and 25% and the remaining 14% to land with more than 25% slope. The use of this characteristic will depend on the orientation conditions of the property and streets. For him analyzed urban settlement, the determining factor of the form of the city was, in its origin, closely related to its economic activity and

Urban-Bioclimatic Aptitude as Nature-Based Solution

Fig. 13 Bioclimatic aptitude map in Pachuca in winter by conduction (own elaboration)

Fig. 14 Bioclimatic aptitude map in Pachuca in winter by convection (own elaboration)

topography as can be observed in graphs; At present, its growth is governed by the proximity to Mexico City. However, from the review of the maps generated, it is observing some characteristics that can be related to adaptations of the design and the urban form to the environment: The growth of the city indica a

Fig. 15 Bioclimatic aptitude map in Pachuca in winter by radiation (own elaboration)

preference for the gentle slope over the plain as well as a position sheltered from cold prevailing winds, places with watercourses are preferred stable surface, or with ease for the exploitation of groundwater. These factors prevail over the fact of building on firm ground and with capacity important, which in the case of the city of Pachuca, is an important factor in the nearby are-as to mine shots.

(c) The dominant urban layout is the grid with 54% of the total area with blocks closed rectangular, very compact, and well-defined alignments, followed by the condominium with 37% and 29% of irregular traces, being clearly located in the oldest areas of the city. The urban layout of the grid standardizes the conditions of the properties and favors the application of design strategies for the use of solar energy.

(d) The city has an average density of 95 inhabitants/ha, areas with very low densities (less than 80 inhabitants/ha) occupying 70% of the urban sprawl and located. the high densities (greater than 160 inhabitants/ha) are located mostly in the up-per areas of the city and the older. Smaller batches may prevail here as well.

(e) The street width-height ratio of the predominant building (55%) in the city is the ratio 0.5 to 1, that is, a street width of half the height of the buildings and ex-tending into the areas with minor slopes, while the 1 to 1 relationship occurs in the oldest areas of the city, this being so due to the smaller dimension of the width of the streets, rather than the presence of taller buildings.

(f) The statistics of the index of open spaces indicate that an index of 0.75 prevails, that is, with free spaces of 25% of its total area, presenting this case for 60% of the city. The second place is occupied by the 0.50 index, which is geo-graphically located in the areas that surround the center of the city that correspond to colonies of more recent creation and that represent 23% of the urban sprawl, while the center of the city and other neighborhoods of the center and south of the city are located with an index of 0.90, that is, with properties with constructions that occupy almost the totality of its surface. These cases represent 11% of the urban stain.

(g) In relation to the age of the settlement, it was obtained that 20% of the urban sprawl is occupied by old neighborhoods, 36% by modern constructions and the remaining 42%, by contemporary developments, indicating a high rate of decentralization and growth of the city.

(h) In relation to the front-bottom ratio of the lots, the 1 to 1 ratio predominates with 54% of the urban sprawl occupying the central zone of the city. In this case, one can distinguish the colonies that have apple cores, and the majority that are made up of batches Complete blocks that occupy the block without central areas. The second place is held by the proportion front-bottom 1 to 2 with 42% and which corresponds to the most current real estate developments.

(i) In the textures of the city, smooth textures predominate with 91%, followed by mediums with 9%.

(j) The tree planting of the city is reduced, considering that it is limited to 4 parks and some green areas belonging to educational institutions. So, it can be considered that only one subdivision is immersed in a wooded area, occupying 0.016% of the urban sprawl and that only 0.12% of the city is on the edge of or close to a wooded area.

(k) The predominant pavement material in the city is concrete with 72%, followed by earth 12%, stone 0.04%, asphalt 0.04% and paving stones 0.02%

(l) There are no large bodies of water in the city, considering for the study only those areas close to the rivers that cross the city. Thus, it is considered that 0.16% of the urban area is in the vicinity of bodies of water.

5 Discussion

In the calculation of heat gains/losses, 26 Basic Calculation Variables were identified, including calculations for Conduction, Convection, Radiation, Infiltration, Devices and Metabolism. The basic variables of calculation by infiltration correspond to the field of architectural characteristics and therefore are not considered in the urban analysis. The basic variables of calculation by Devices and by Metabolism are dependent on the characteristics of the project (such as the number of occupants of the building or their activity) and therefore are not considered as controllable variables by architectural or urban design.

Within the variables for Conduction, Convection and Radiation, two types can be identified: Variables that can be controlled or modified by design (architectural or urban) and Variables that can be controlled by design that can be modified with design strategies. Of the 26 Basic Calculation Variables, 14 are variables that can be controlled by design and can be linked to 12 urban characteristics or qualities called

"Urban-Bioclimatic Variables" in this research. The Urban-Bioclimatic Variables have different and, sometimes, opposing qualitative values, according to the heat gain/loss criteria under which they are analyzed (Conduction, Convection or Radiation) and for the climate in which they are considered. This happens because, being an abstraction of reality, the Urban-Bioclimatic Variable in question was considered in a totally isolated manner and only considering the type of heat gain and the type of climate to carry out the qualitative assessment. This assessment is integrated back into a real context by combining it with the spatial aspect (Fig. 16).

The Basic Calculation Variables are numerical values that can correspond to the physical properties of the materials, to the design characteristics of the building (such as dimensions, location, etc.), to the physical conditions of the place or, to coefficients that correspond to any of the above. These numerical values can be calculated and thus find a maximum value and a minimum value for each of the Basic Calculation Variables for a specific place. Consequently, the fitness index obtained in the exercise is applicable only to the city under study since some of the maximum and minimum numerical values depend on the specific physical conditions of the place.

The aptitude index for another city can be obtained by changing the maximum and minimum values depending on the place. The Basic Design variable with the highest percentage of participation in Heat Gains/Losses by Conduction is the Surface Area, followed by the material thickness. The Basic Design variable with the highest percentage of participation in Heat Gains/Losses by Convection is the Opening Effectiveness Factor Direct, followed by the Specific Humidity of the Indoor Air.

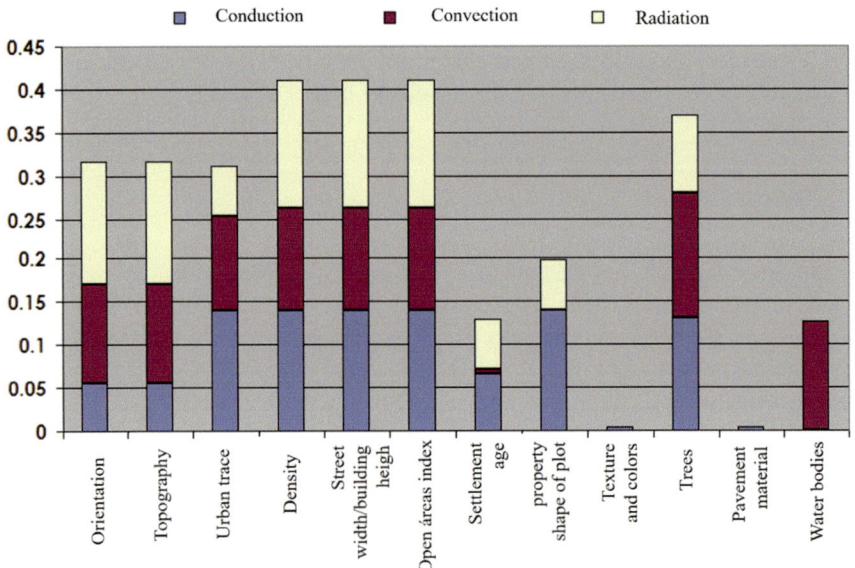

Fig. 16 Accumulated percentage of the participation of the urban-bioclimatic variables by conduction, convection and radiation (own elaboration)

Heat Gains/Losses by Radiation, the basic design variables "Glass Transmittance", "Shading Coefficient" and "Solar Radiation" have a practically equal participation percentage, followed by the Window Area. The greatest variations of the Driving Gains/Losses Values are presented in Conduction by Solar Radiation, followed by the Convection Coefficient of the Interior Air, that is, they are the most sensitive variables of heat by conduction.

The greatest variations of the Values of Gains/Losses by Convection are presented in Convection it is the Specific Heat of the Air, followed by the Density of the air, that is, they are the most sensitive variables of heat by convection. The greatest variations of the Values of gains/Losses due to radiation are presented in radiation it is the Transmittance of the glass, followed by the shading coefficient, that is, they are the most sensitive variables of heat by radiation.

When integrating the Urban-Bioclimatic Variables to the aptitude index, the most relevant variables in the urban-bioclimatic design is density: the height of the building-width ratio of the street and the index of open spaces, followed by the orientation and the topography. Density, building height-to-street width ratio, and open space ratio, orientation and topography are all variables related to "access to the sun" and "access to the wind". The most repeated urban-bioclimatic variables (mode) by Conduction are Density, the Width of the street/Height of the building ratio and Trees; by convection are the density, street width/building height ratio, open space index, and water bodies; and by radiation are the orientation, the topography, the density, the relation Width of the street/Height of the building and the index of open spaces.

The urban-bioclimatic variables with more repetitions (mode) are the density and the relation Width of the street/Height of the building (Fig. 17).

From the spatial analysis it was found that the predominant orientation of the lots is Northwest–Southeast with an approximation of 32%, however this data is not noticeable since the geographical distribution is random with a certain predominance of the North–South and East–East orientations. West in the north, the oldest in the city and with the steepest slope and a certain dominance of the Northwest–Southeast orientation in the valley. The Northwest–Southeast orientation is distributed equally in any area of the city. Combining these data with the age of the settlement, it is observed that in the original settlement the south slope is preferred over the north slope. Solar radiation is favored because it is located halfway up the slope, open to the south.

Regarding the topography, 59% of the surface of the urban area has a maximum land slope of 10, 26% corresponds to properties with slopes between 10 and 25% and the remaining 14% to land with more than 25% slope. The use of this characteristic will depend on the orientation conditions of the property and streets. However, from the review of the generated maps, some characteristics are observed that can be related to adjustments of the design and the urban form to the middle. The growth of the city denotes a preference for the gentle slope over the plain as well as a position sheltered from the prevailing cold winds, places with stable surface watercourses are preferred, or with ease for the exploitation of groundwater. These factors prevail

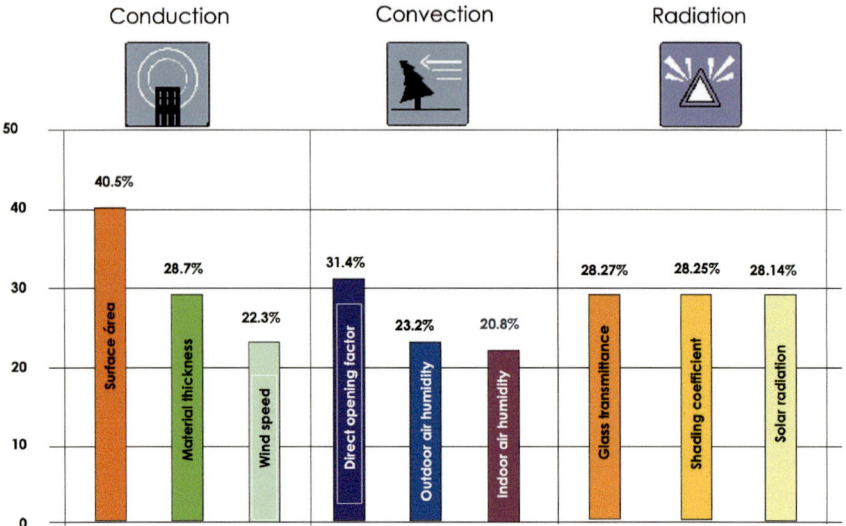

Fig. 17 Main urban-bioclimatic variables (own elaboration)

over the fact of building on firm land with bearing capacity, which in the case of the city of Pachuca, is an important factor in nearby areas.

The dominant urban layout is the grid with 54% of the total area with closed rectangular blocks, very compact and well-defined alignments, followed by the condominium with 37 and 29% of irregular layout, clearly located in the oldest areas of the city. The urban grid layout standardizes the conditions of the properties and favors the application of design strategies for the use of solar energy (Fig. 18).

6 Conclusions

Urban-Bioclimatic aptitude is a Nature-based solution or a Solution Based on the capacity of nature. That integrate different approaches to the systematize of bioclimatic strategies: (1) those that link the city with climate and propose generalizations of bioclimatic strategies by climate type; (2) those that link human needs with the climate through acclimatization and comfort indices (referential approaches); (3) those that link the effects of the climate in cities with human needs based on limitations, goals, and potentials (approaches based on optimization). This methodology makes it possible to integrate the physical variables of the site with the urban variables that modify it and that make cities a new nature to adapt.

Urban-bioclimatic aptitude is the measure of the capacity of a system (city, neighborhood) to meet the needs of its inhabitants and capacity of nature and the existing

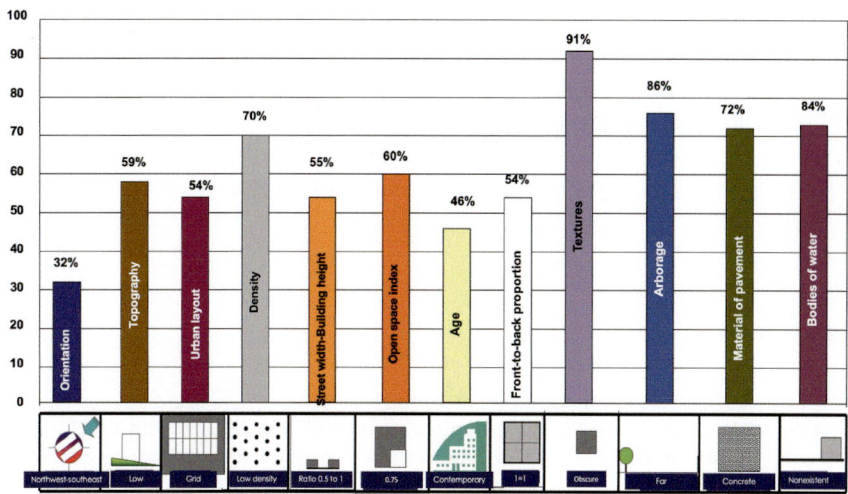

Fig. 18 Predominant urban-bioclimatic features in Pachuca (own elaboration)

city. It depends on the characteristics of the environment and the needs of the inhabitants. As these characteristics or these needs change, the aptitude of the environment changes, it can increase if features are improved or decrease if the needs increase.

The aptitude of a built system cannot increase indefinitely, development has limits. On the contrary, the inability of the system to cover the needs of the inhabitants can grow indefinitely if the favorable characteristics of the system are degraded or lost and the demands increase, that is, the "ineptness" of the system can tend to infinity. If there is no system renewal, the condition of "unfitness" will grow. Each system has an aptitude range.

The Basic Calculus Variables are numerical values that correspond to the physical properties of materials, the design features of a building, the physical conditions of the site, or coefficients that correspond to any of the above classifications. These numerical values can be calculated and then a maximum and minimum value can be determined for each basic calculation variable. Consequently, the index obtained in the exercise is applicable only to the city analyzed, since some of the maximum and minimum values depend on the physical conditions of a specific place. Indices for another location can be obtained by changing the maximum and minimum values to those of the location to be analyzed. A range of urban-bioclimatic aptitude can be defined by city or by climate type, allowing comparisons within a given city or between different cities with the same climate.

The evaluation of the Urban Bioclimatic aptitude of a city would allow the assessment of energy consumption related to heating/cooling systems in different areas of the city; developing strategies at different scales that promote and facilitate the sustainable development of a locality and application of nature-based solutions considering the constructed city as a new nature; the identification of optimal strategies at the right time and place; the implementation of bioclimatic systems by zones at the urban level; the generation and apply aptitude indexes for different cities and the definition an aptitude index by types of weather; develop a sustainability indicator based on urban renewal indices and the monitoring of infrastructure, operation, evaluation and improvement of applied bioclimatic systems and/or ecotechnologies; as well as the application of public policies for urban renewal of the city based on bioclimatic strategies. Urban-bioclimatic aptitude is a solution based on the capacity of Nature.

References

Akbari H et al (2016) Local climate change and urban heat island mitigation techniques - the state of the art. J Civ Eng Manag 22(1):1–16. https://doi.org/10.3846/13923730.2015.1111934

Ameur M, Kharbouch Y, Mimet A (2020) Optimization of passive design features for a naturally ventilated residential building according to the bioclimatic architecture concept and considering the northern Morocco climate. Build Simul 13(3):270–285. https://doi.org/10.1007/s12273-019-0593-6

Andrade H, Alcoforado MJ, Oliveira S (2011) Perception of temperature and wind by users of public outdoor spaces: relationships with weather parameters and personal characteristics. Int J Biometeorol 55(5):665–680. https://doi.org/10.1007/s00484-010-0379-0

Asadi E, Da Silva MG, Antunes CH, Dias L (2012) Multi-objective optimization for building retrofit strategies: a model and an application. Energy Build. 44(1):81–87. https://doi.org/10.1016/j.enbuild.2011.10.016

Back Y, Bach PM, Jasper-Tönnies A, Rauch W, Kleidorfer M (2021) A rapid fine-scale approach to modelling urban bioclimatic conditions. Sci Total Environ 756:143732. https://doi.org/10.1016/j.scitotenv.2020.143732

Bianchi C, Thomas J, Smith AD (2019) Impact of microclimate and macroclimate on building energy consumption. ASHRAE Trans 125

Boukli Hacene MA, Chabane Sari NE (2020) Energy efficient design optimization of a bioclimatic house. Indoor Built Environ. 29(2):270–285. https://doi.org/10.1177/1420326X19856668

Bryant MM, Turner JS (2019) From thermodynamics to creativity: McHarg's ecological planning theory and its implications for resilience planning and adaptive design. Socio-Ecol Pract Res 1(3–4):325–337. https://doi.org/10.1007/s42532-019-00027-1

Caldas L (2008) Generation of energy-efficient architecture solutions applying GENE_ARCH: an evolution-based generative design system. Adv Eng Inf 22(1):59–70. https://doi.org/10.1016/j.aei.2007.08.012

Cárdenas-Jirón LA, Morales-Salinas L (2019) Urbanismo bioclimático en Chile: Propuesta de biozonas para la planificación urbana y ambiental. Eure 45(136):135–162. https://doi.org/10.4067/S0250-71612019000300135

Carlucci S, Pagliano L, Sangalli A (2014) Statistical analysis of the ranking capability of long-term thermal discomfort indices and their adoption in optimization processes to support building design. Build Environ 75:114–131. https://doi.org/10.1016/j.buildenv.2013.12.017

Cetin M (2015) Determining the bioclimatic comfort in Kastamonu City. Environ Monit Assess 187(10):1–10. https://doi.org/10.1007/s10661-015-4861-3

Cetin M (2020) Climate comfort depending on different altitudes and land use in the urban areas in Kahramanmaras City. Air Qual Atmos Heal 13(8):991–999. https://doi.org/10.1007/s11869-020-00858-y

Cetin M, Adiguzel F, Gungor S, Kaya E, Sancar MC (2019) Evaluation of thermal climatic region areas in terms of building density in urban management and planning for Burdur, Turkey. Air Qual Atmos Heal 12(9):1103–1112. https://doi.org/10.1007/s11869-019-00727-3

Clarke JF, Bach W (1971) Comparison of the comfort conditions in different urban and suburban microenvironments. Int J Biometeorol 15(1):41–54. https://doi.org/10.1007/BF01804717

Cohen WJ (2019) The legacy of design with nature: from practice to education. Socio-Ecol Pract Res 1(3–4):339–345. https://doi.org/10.1007/s42532-019-00026-2

Dimoudi A, Zoras S, Kantzioura A, Stogiannou X, Kosmopoulos P, Pallas C (2014) Use of cool materials and other bioclimatic interventions in outdoor places in order to mitigate the urban heat island in a medium size city in Greece. Sustain Cities Soc 13:89–96. https://doi.org/10.1016/j.scs.2014.04.003

Epstein Y, Moran DS (2006) Thermal comfort and the heat stress indices. Ind Health 44(3):388–398. https://doi.org/10.2486/indhealth.44.388

Fröhlich D, Matzarakis A (2018) Spatial estimation of thermal indices in urban areas-basics of the skyhelios model. Atmosphere (Basel) 9(6):1–14. https://doi.org/10.3390/atmos9060209

Gobo JPA, Faria MR, Galvani E, Goncalves FLT, Monteiro LM (2018) Empirical model of human thermal comfort in subtropical climates: a first approach to the Brazilian Subtropical Index (BSI). Atmosphere (Basel) 9(10):391. https://doi.org/10.3390/atmos9100391

Goldberg V, Kurbjuhn C, Bernhofer C (2013) How relevant is urban planning for the thermal comfort of pedestrians? Numerical case studies in two districts of the City of Dresden (Saxony/Germany). Meteorol. Zeitschrift 22(6):739–751. https://doi.org/10.1127/0941-2948/2013/0463

Gou S, Nik VM, Scartezzini JL, Zhao Q, Li Z (2018) Passive design optimization of newly-built residential buildings in Shanghai for improving indoor thermal comfort while reducing building energy demand. Energy Build 169:484–506. https://doi.org/10.1016/j.enbuild.2017.09.095

Grigore E, Constantin (Oprea) DM, Bogan E, Cristea M-A, Tatu F (2020) The thermo-hygrometric index on the territory of the Southern Dobrogea plateau – a component of the balneoclimateric treatment. Present Environ Sustain Dev 14(1):89-98. https://doi.org/10.15551/pesd2020141007

Gubenskiĭ ID (2009) Physical factors of the urban dwelling environment: ecological and hygienic aspects. Gig Sanit 5:11–15

Henderson LJ (2016) The fitness of the environment, an inquiry into the biological significance of the properties of matter. Am Nat 47(554):105–115. The American Society and C. Press, The Universit (1913)

Karakounos I, Dimoudi A, Zoras S (2018) The influence of bioclimatic urban redevelopment on outdoor thermal comfort. Energy Build 158:1266–1274. https://doi.org/10.1016/j.enbuild.2017.11.035

Katafygiotou MC, Serghides DK (2015) Bioclimatic chart analysis in three climate zones in Cyprus. Indoor Built Environ 24(6):746–760. https://doi.org/10.1177/1420326X14526909

Kawakubo S, Murakami S, Ikaga T, Asami Y (2018) Sustainability assessment of cities: SDGs and GHG emissions. Build Res Inf 46(5):528–539. https://doi.org/10.1080/09613218.2017.1356120

Kodis M, Galante P, Sterling EJ, Blair ME (2018) Ecological niche modeling for a cultivated plant species: a case study on taro (Colocasia esculenta) in Hawaii. Ecol Appl 28(4):967–977. https://doi.org/10.1002/eap.1702

Koo C, Hong T, Lee M, Seon Park H (2014) Development of a new energy efficiency rating system for existing residential buildings. Energy Policy 68:218–231. https://doi.org/10.1016/j.enpol.2013.12.068

Kuma Y, Fukuda H, Ozaki A (2007) Performance evaluation of residences by dynamic simulation: heat load based on changing the location, plan and specification of residences. J Asian Archit Build Eng 6(1):183–188. https://doi.org/10.3130/jaabe.6.183

Mirzaei PA (2015) Recent challenges in modeling of urban heat island. Sustain Cities Soc 19:200–206. https://doi.org/10.1016/j.scs.2015.04.001

Mistry MN (2020) A high spatiotemporal resolution global gridded dataset of historical human discomfort indices. Atmosphere (Basel) 11(8):835. https://doi.org/10.3390/ATMOS11080835

Nardino M, Laruccia N (2019) Land use changes in a peri-urban area and consequences on the urban heat Island. Climate 7(11):133. https://doi.org/10.3390/cli7110133

Notaro M, Mauss A, Williams JW (2012) Projected vegetation changes for the American Southwest: combined dynamic modeling and bioclimatic-envelope approach. Ecol Appl 22(4):1365–1388. https://doi.org/10.1890/11-1269.1

Patania F, Gagliano A, Caponetto R, Nocera F, Galesi A (2010) A neural network model for the estimation of bioclimatic indexes. WIT Trans Ecol Environ 136:237–247. https://doi.org/10.2495/AIR100211

Rantzoudi EC, Georgi JN (2017) Correlation between the geometrical characteristics of streets and morphological features of trees for the formation of tree lines in the urban design of the city of Orestiada, Greece. Urban Ecosyst 20(5):1081–1093. https://doi.org/10.1007/s11252-017-0655-4

Ren Z, Fu Y, Du Y, Zhao H (2019) Spatiotemporal patterns of urban thermal environment and comfort across 180 cities in summer under China's rapid urbanization. PeerJ 2019(8):e7424. https://doi.org/10.7717/peerj.7424

Román E, Gómez G, de Luxán M (2017) Urban heat Island of Madrid and its influence over urban thermal comfort. In: Sustainable development and renovation in architecture, urbanism and engineering

Saá C, Míguez JL, Morán JC, Vilán JA, Lago ML, Comesaña R (2012) A study of the influence of solar radiation and humidity in a bioclimatic traditional Galician agricultural dry storage structure (horreo). Energy Build 55:109–117. https://doi.org/10.1016/j.enbuild.2012.05.020

Salat S, Vialan D (2010) Sustainable Mediterranean urban development affordable to all, a morphological approach

Salat S, Bourdic L (2013) Passive zones, bio-climatic design and scale hierarchic urban fabric. WIT Trans Ecol Environ 173:563–572. https://doi.org/10.2495/SDP130471

Schipper L, Meyers S, Howarth, RB, Steiner R (1992) Energy efficiency and human activity: past trends, future prospects. https://doi.org/10.1016/0959-3780(93)90032-g

Toy S, Yilmaz S (2010) Evaluation of urban-rural bioclimatic comfort differences over a ten-year period in the sample of Erzincan city reconstructed after a heavy earthquake. Atmosfera 23(4):387–402

Valdiserri P, Biserni C, Garai M (2016) Energy performance of a ventilation system for an apartment according to the Italian regulation. Int J Energy Environ Eng 7(3):353–359. https://doi.org/10.1007/s40095-014-0159-4

Widera B (2014) Bioclimatic architecture as an opportunity for developing countries. In: 30th international PLEA conference: sustainable habitat for developing societies: choosing the way forward - proceedings, vol 2

Yuan J, Emura K, Farnham C (2017) Is urban albedo or urban green covering more effective for urban microclimate improvement?: A simulation for Osaka. Sustain Cities Soc 32:78–86. https://doi.org/10.1016/j.scs.2017.03.021

Learning Systems and Experiences

Introduction

Llamas et al. present a retrospective on the experience at the Universidad del Medio Ambiente (UMA), which was founded in Mexico in 2009 with the purpose of 'Promoting a regenerative, sustainable and ethical future by preparing agents of change capable of promoting initiatives that transform socio-environmental systems'. The intention of the UMA was the flourishing of human cultures, in a way that cares for and replenishes planetary life systems. Since regeneration holds promise in this regard, many of its principles were adopted. Regeneration has been a constant source of inspiration and guidance for students, professors and alumni of the UMA. But it also poses challenges: it is a novel perspective, difficult to put into practice, has many interpretations and can be ambiguous. Being grounded in the regenerative perspective has been a positive experience for the UMA and its community, which has the ability to foster a new paradigm and a more holistic perspective in efforts to heal the planet and our relationships with it. Looking ahead, a new phase in the work lies ahead in which the UMA, as an institution, will seek to provide its community of students and faculty with more pointed definitions of sustainability, regeneration and their interrelationships, intended to bring clarity to the project design process and a firm basis from which to develop impact metrics. This is particularly important for the UMA's action research approach, which will maintain an open approach to different perspectives and understandings of complex ideas and challenges to explore different points of view and paradigms but will always be invited to reflect, investigate and be aware of the evidence that might transform them.

Terrado shares an innovative experiential learning initiative within the Argentinean educational field, whose main intention was to support and co-evolve educational initiatives by creating a safe, nurturing and learning collective space, the Evolutionary Learning Ecosystem (ELE). This initiative joined forces and ignited collective intelligence to enhance its impact in its territory, achieved by creating and preserving a shared holistic worldview and a space of trust. As the initiatives evolved and faced numerous challenges, a methodology for solving them emerged, the Community Response Circles (CRCs), which aimed to engage in a creative problem-solving process to address challenges that only through collective and collaborative work can truly transcend the current educational paradigm. The revelatory effect of ELE and the CRC was the psychological and social regeneration of the people and initiatives involved, as a new way of being and doing together. An important learning was how CRC became an innovative solution to the challenges of the initiative but fragmented the ecosystem by not including and opening financial concerns to all ELE participants, which led to a feeling of exclusion and mistrust. Instead of taking the challenge as an opportunity to grow together, to be genuine and transparent, the small leadership team made a wrong decision that cost the ecosystem its existence. But this creative process of the CRC was offered to another school's project of the Living Gratefully Foundation. The participants of this other project created 290 CRCs in public schools in the north and east of the country. All these CRCs gave birth to an innovative solution to their current problems. The most notable aspect of this growth was the modelling and space to engage holistically in collaboration with others, learning to live and engage together to create a more connected world.

Hueso-Kortekaas talks about how nature-based education (NBE) is growing worldwide. Although the idea is not new, having been practised in ancient times, it became an organised activity in the early 20th century as an initiative to improve children's health. Exist variety of approaches, from family-led playgroups to associations and entrepreneurs offering outdoor projects in both the formal and informal sectors, to forest school sessions run by existing schools at different curricular levels. From 2022 this topic is formally taught at the university level and the literature on the subject has grown enormously. However, this educational model has not been officially recognised and supported by the authorities, which brings financial, social, cultural and logistical challenges. To achieve this recognition is necessary to establish standard quality criteria, covering pedagogical as well as logistical and environmental aspects, and to expand the supply of vocational training. These key aspects can open the door to green, sustainable and fair access to nature-based education.

In recent years, and especially after the onset of the pandemic, NBE has been perceived as safe, healthy and effective and how families and professionals are increasingly demanding this model. Although still far from being a real option on an equal footing with other pedagogical alternatives, some of the older projects are now well established, with a clear identity, and genuinely regenerative in nature, as they promote a healthy, resilient and sustainable lifestyle for both children and the school community as a whole and impact positively on the local community. Training of professionals is improving, with a wide range of courses catering for those interested in the practical aspects as well as for those who aspire to managerial positions, entrepreneurs and researchers in the field. Both popular science books and academic works are being published on the subject, although there are still some gaps to be filled. This approach is likely to be extended to higher educational levels, such as primary and secondary schools, or even to higher education. Gradually, society is accepting, celebrating and demanding NBE as a regenerative approach to improving the quality of society and nature that will contribute to better people for a better planet.

Crowson et al. present their qualitative research conducted in community projects as the approach focused on shared processing of the emotional impact of the huge biodiversity loss, we are experiencing that can empower participants to become agents of change and the use of ecological literacy approaches and permaculture principles to engage in deeper conversations. The authors identify attitudinal shifts needed in our current paradigm, in particular the need to move from gratitude to reciprocity, exploring in depth the ramifications of interdependence and well-being in our work with others. When people feel connected and that they belong, they see the value of their actions as part of the system. There is a growing desire and demand for people moving into rural spaces to find resources to support this transition; at the same time, there is a lack of understanding of permaculture as a holistic framework to address this demand, that more training is needed so that it can have a greater impact, detecting the value of sharing tools for its applications in all areas of community design. More work is also needed on scientific indicators to have a more universally recognisable way of assessing impact.

It is especially important to the economic viability of such change projects, as although 'invisible values' such as ecosystem services, increased biodiversity, etc., are understood, they are not easily translated into economic benefits. Creating economically productive models of regenerative agriculture will be a key step for the growth of regenerative cultures. Connecting with love is an important key to this. There is a need for commitment to inner work to facilitate and hold space for others exploring a shift towards a regenerative culture and a need for mutual support in exploring radical tenderness, self-care as a basis for caring for others, and the usefulness of expanding the notion of agency and working with nature. Local impact can be amplified through sharing experiences and practices and discussing the evolution of their experiences with a promising future, with actions in terms of climate-adapted reforestation, syntropic agriculture, Miyawaki forests, design of local art and ecology projects around ecological literacy and water cycle restoration, with the achievement of some funding for capacity building, among other initiatives.

A University focused on Regeneration: What Have We Learned?

Federico Llamas, Victoria Haro, Andreea Dani, Ana Lucía Castaño, and Elizabeth García Rojas

Abstract Universidad del Medio Ambiente (UMA) was founded in Mexico in 2009 with the following purpose: *"Foster a regenerative, sustainable, and ethical future by preparing change agents capable of promoting initiatives that transform socio-environmental systems"*. The intention of our efforts is the flourishing of human cultures, in a way that cares for and replenishes planetary life systems. Because regeneration is especially promising in this regard, we have adopted many of its principles in our educational model and our seven faculties. Regeneration has been a constant source of inspiration and guidance for students, faculty, and alumni. But there are also challenges: it's a novel perspective that is difficult to practice, has many interpretations, and can be ambiguous. This chapter is divided in four sections: (1) the context in which UMA adopted the regeneration perspective; (2) UMA´s regenerative project design methodology; (3) description of a project developed with this methodology; (4) the two main conceptual challenges that seem to be generating confusion. By sharing how we have integrated regeneration into our programs and highlighting some of the questions we have encountered, we hope to help others in their implementation of this perspective.

F. Llamas (✉)
Chairman, Universidad del Medio Ambiente, Valle de Bravo, Mexico
e-mail: flv@umamexico.com.mx

V. Haro
Rector, Universidad del Medio Ambiente, Valle de Bravo, Mexico

A. Dani
Faculty of Sustainable Architecture, Universidad del Medio Ambiente, Valle de Bravo, Mexico

A. L. Castaño
Faculty of Agroecology, Universidad del Medio Ambiente, Valle de Bravo, Mexico

E. García Rojas
Faculty of Sustainable Architecture, Universidad del Medio Ambiente, Valle de Bravo, Mexico

1 Context

Universidad del Medio Ambiente (UMA) started with a dream of a place where people from different parts of Latin America could come together to learn, reflect, and act, on our local and global environmental challenges. From 2006 to 2009 a small team of educators, change agents, and entrepreneurs scouted the world for inspiration and guidance with the following question: *what educational model is best suited to address our socio-environmental challenge?* Coming from a new paradigm, we developed a disruptive learning process with elements like project-based learning, systemic thinking, co-design, action research, and personal development.

We were inspired by movements and concepts that were already talking about a worldview aligned with what is now being called regeneration: Biomimicry, Deep Ecology, Permaculture, Gift Culture, Ethical Banking, Systems Thinking, Developmental Work, U-theory, Collective Intelligence, Teal organizations, Integral Theory, and others.

In 2011 we discovered the work of Regenesis Group, an interdisciplinary ensemble of practitioners that have developed theoretical and methodological frameworks for regeneration. For UMA it was exciting to find a new terminology that integrated many elements already in our educational model and offered new frameworks and insights to enrich our project-based learning. The influence was such that we changed our mission statement to include the word regenerative, and we incorporated this perspective into our curricula. Nevertheless, in UMA's tradition, we also adopted an explorative and action-research philosophy toward this new idea, attentive to evidence of its usefulness and applicability.

One of the main challenges we faced when designing our educational model, was how to integrate different, but complementary frameworks in a project-based learning environment. Specifically, how to integrate systemic action research, socio-environmental purpose-driven initiatives, communities of practice, personal transformation, and a regenerative practice into our model. After 6 years of working with these elements with hundreds of students, we arrived at a design process for regenerative projects that is described in the following section.

2 UMA's Regenerative Project Design Process

UMA has a project-based learning process through communities of practice: students form groups in which they exchange experience and learnings from their projects. The Project Model we have developed at UMA uses Action Research (McNiff et al. 1996) as its core methodology and has three main characteristics regarding Purpose, Perspective, and Process: a socio-environmental purpose, a systemic perspective, and a regenerative process (Fig. 1):

Fig. 1 UMA's regenerative project design process

"The best way to understand a system is to try to change it"
Lewin (1947)

Action Research refers to designing actions to research questions and hypotheses: finding out by experimentation. In the UMA context, the action is the project, so they are formally researched and adapted according to findings and learnings.

UMA projects use a socio-environmental purpose as NorthStar: the main goal is to generate social and environmental benefits that are interrelated and helpful for sustainable development. Following the Regenerative Development framework (Mang and Reed 2012a, b), as well as Theory U (Scharmer 2009), the benefits sought represent the materialization of key potentials of the context or system that the project aims to change, helping to bring forth a better version of itself.

Working with potential is one of the main features of using a regenerative framework since developing potential implies actualizing some of the system's own capacity for renewal. Systems can be of any scale: your family, your organization, your neighborhood, or your city. The main question for choosing a NorthStar-purpose is: what key social and environmental potentials can we bring forth in this system?

To be able to identify these key socio-environmental potentials, as well as to design strategies that may be successful in bringing them to life, a systems perspective is needed. System mapping (Senge 2006) is essential to understand the system in a better way, as well as to better identify where and how the system may be intervened for potentials to materialize. A strategy is composed of several chosen interventions that synergize to actualize these potentials. In this way, both the map of the system and the strategy are hypotheses that are put to the test when the project is implemented. As Kurt Lewin -who first developed the action research methodology in 1947—famously said: the best way to understand a system is to try to change it.

Finally, if we aim to regenerate a system in the specific sense of bringing forth key qualities or potentials, exploring a regenerative process is essential and congruent with this purpose. This follows Mahatma Gandhi's idea that *the means are the ends* (Gandhi 1966): the process needs to mirror the goal. First, because the DNA of the goal is in the process: apples cannot be harvested if you plant pear seeds. But also, because the process provides a way to research the goal: if we explore regeneration in the process, we may find out how to better provide conditions for regeneration as the goal.

The UMA Project Model proposes the exploration of a regenerative process in five dimensions:

(a) Personal regeneration: How can team members use the project to develop personal key capacities and potentials? As a strategy, the Model invites each member to strengthen a specific source of personal enthusiasm within the project. This provides a fruitful way of finding out what renewing oneself might mean.
(b) Relational regeneration: How can the project deepen relationships and build trust between the humans involved so that their relationships reach a new, higher level? An effective way the Model invites for achieving this is by codesigning the project with the others involved: develop projects with others, instead of for others. This allows weaving diverse perspectives, which in turn permits the project to be appropriated by those involved in it. The weaving process provides excellent opportunities for relationships to evolve, as well as for the social potential of the project to emerge.
(c) Environmental regeneration: How can the project help to develop ecosystem capacities for self-renewal? This is intimately related to the environmental potential the project is trying to materialize. The key questions the Model invites are: What are the process conditions that permit this potential to arise? How can we measure them on the way?
(d) Learning regeneration: How can we renew our knowledge about the system we want to change as well as our learning capacities? What indicators and reflection processes may help us in this endeavor? The Model uses Action Research as a powerful method for regenerating our learning.
(e) Project regeneration: How can the project itself generate conditions for its own renewal? How can the project process continue without the project team or leaders? Codesign provides an effective means for appropriation and thus continuity; additionally, the Model proposes a Cyclical Theory of Change for the sustainability of the project. A Theory of Change (Taplin and Clark 2012) represents the connection between strategy and purpose: a chain of results that we believe will take us from the initial interventions in the system, to the actualization of the socio-environmental potentials: a theory of a path through which we can achieve the purpose. This path can be linear from strategy to purpose, or cyclical, providing feedback loops so that once the specific objective of the purpose is achieved, the original strategies are reinforced, so that an evolutionary process around the purpose may be continued and further objectives related to it may be achieved.

A regenerative process is something to be explored and researched. Each student begins this exploration from her or his developed capacities. Having different starting points and different conditions, each project achieves different levels and results regarding the regenerational aims. Our purpose is that with each project, the capacity for a regenerative process is strengthened and developed.

UMA's Regenerative Project Design Process has been used since 2016 in our 7 master's programs. Hundreds of projects have been designed with this framework. In the following section, we offer an example of one project from our Faculty of Architecture.

3 Example of a Regenerative Project from UMA´s Architecture Faculty

In September 2017, the traditional adobe-house town of Totolapan in the state of Morelos in Mexico, experienced an earthquake that damaged 27% of its 746 houses. UMA participated in the reconstruction process by providing technical accompaniment in the co-design and construction of the Totolapan Community Center promoted by "Creativo Totolapan", a local nonprofit organization. UMA's participation was complemented with the project of Elizabeth García Rojas, a student of UMA's Sustainable Architecture, Design and Construction Master's Degree. Elizabeth named her project: *"Potentialization of the dissemination of sustainable constructive knowledge in the building process of the Totolapan Community Center"*.

Following UMA's Project Model, Elizabeth facilitated the co-creation of a socio-environmental purpose based on potential, developed a systemic perspective and explored a regenerative process.

3.1 Socio-environmental Purpose

A participatory identification of socio-environmental potentials that could be materialized through the project, revealed the value of strengthening the traditional constructive culture characterized by adobe (raw earth), so as to enhance the identity of the town as well as the use of the region's rich soil (Fig. 2).

The project aimed to actualize this potential, establishing as purpose to build the Totolapan Community Center according to the local adobe-based constructive knowledge, in a way that achieved its effective dissemination and promoted the use of raw earth in substitution of industrialized construction materials, thereby fostering environmental benefits and enhancing social wellbeing by improving habitability.

The initial phase of the construction of the Community Center (which was the phase included in Elizabeth's project; Fig. 3), achieved the purpose successfully: the closing surveys showed that all the community members that participated in the project elevated significantly their level of knowledge regarding the use of earthquake-resistant adobe building system and were highly interested in disseminating this knowledge; the portion of participating youth interested in building their future houses with adobe increased by 50%; and the size of the target group of

Fig. 2 Liévanos house. García Rojas (2019)

community adults interested in using the Center's constructive techniques in their homes was more than tripled.

3.2 Systemic Perspective

A systemic perspective of the project was developed by approaching the place as a unique, multilayered network where complex interactions arise between the human and natural systems. Using the Assessment Scope Framework for Pattern Understanding developed by the Regenesis Group (Mang and Reed, 2012a), nine aspects of the place were studied: geology, hydrology, biology, settlements, economy, culture, education, psychology, and spirituality; identifying the main patterns and interactions between them.

This understanding was translated into a Story of Place (Mang and Reed, 2012b), which narrated how Totolapan sustains itself and what makes it unique; and also, into a text based on the VerSus methodology (Correia et al. 2014), which organized the ideas to represent how vernacular architecture embodies environmental, socioeconomic and sociocultural sustainability.

Fig. 3 Construction of adobe walls of Community Center. García Rojas (2019)

This systemic understanding of place allowed the design of an effective strategy to achieve the project's purpose: To build the Community Center with adobe and seismic-resistant techniques in a demonstrative way, through workshops where students, community volunteers, and construction workers developed collective capacities and bonds that strengthened both identity and social fabric; providing a space for exchange where the community could recognize the value of its ecosystem and constructive culture.

3.3 Regenerative Process

Throughout the project process, renewal was sought in the five dimensions addressed by the UMA Project Model:

a. Personal regeneration

The process invited Elizabeth to question and evolve the professional skills with which she arrived at Totolapan, since the codesign context required an attitude of humility for her to be able to present a technical proposal that was valued from the community's perspective. She also realized that to identify the community's needs, she had to apply active listening and observation beyond the specific dialogues and spaces planned in the co-design process. One key source of renewed enthusiasm was finding out how this moved her from *observing* to *living* the place, blurring the line between offering professional and personal support.

b. Relational regeneration

The codesign process between UMA and "Creativo Totolapan" cultivated valuable relationships of trust and development. This was noticeable for example, in how several of the organization's members became co-facilitators in UMA's workshops. They also ended up managing part of the project's funds. Friendships arose beyond the space of the project.

c. Environmental regeneration

The design of the Community Center included composting toilets that close the nutrient cycle and improve the soil, as well as an integrated water management system that contributes to a change in the local water culture, transforming the perception of scarcity into one of abundance and increasing ecosystem vitality. The use of raw earth as a reusable building material placed the architectural proposal under a sustainability paradigm. Evidence was found that old adobe houses used reinforcement elements made with the branches of an endemic plant called "tanalocote" (*Dodonaea viscosa*), which explained the very thick joints between the adobes and the presence of stones in the joints, giving a characteristic image to the town (Fig. 4). The interest in renovating this technique could lead the community to develop tanalocote plantations that could have an ecosystemic regeneration effect.

d. Learning regeneration

The construction of the Community Center was conceived as a learning process for participants to learn about seismic-resistant construction systems that use local and low-cost materials and eco-techniques. This allowed new construction capacities to be developed by community members, construction workers, and UMA students. Workshops offered great moments for knowledge sharing: UMA provided solutions to improve stability and resistance in adobe structures, while the artisans shared the adobe technology passed on from generation to generation (Fig. 5). The remembrance of the use of tanalocote as a reinforcement element also regenerated the local constructive knowledge in a valuable way.

e. Project regeneration

The project was conceptualized as having a cyclical Theory of Change, where the impact is sustained and evolved by the local organization "*Creativo Totolapan*", which, thanks to the acquired capacity in adobe production, daub construction, and earth plasters, is now able to continue building the Community Center through communal work, as well as to organize adobe workshops with students from regional schools.

Fig. 4 Adobe wall. García Rojas (2019)

Architectural projects can be vehicles for actualizing social and environmental potentials, as well as creating favorable conditions for regeneration. Although limited project time scales might show small direct effects on ecosystem regeneration, embracing the understanding of place as an ongoing process allows us to renew our knowledge and envision new possibilities for regeneration through architecture.

This is only one example of a project designed and implemented with the regenerative design process applied at UMA. Hundreds of others have been using this methodology, some of them getting only to the first initial phases of implementation and some of them developed further. UMA is at this moment conducting an impact assessment of these projects to assess the strengths and areas of opportunity of the model.

4 Conceptual Challenges Faced with the Regenerative Perspective

We are aware at UMA that developing projects with a socio-environmental purpose, a systemic perspective, and a regenerative process, is a complex task. We have researched and developed these ideas with our students and faculty for the last 12 years. Although a more systematic analysis of the benefits of this model is

Fig. 5 Workshop with construction workers. García Rojas (2019)

underway, we have found empirical evidence that projects and communities are enriched by the regenerative perspective when designing and implementing projects. It prompted UMA´s faculty and students to look beyond superficial or conventional solutions. They were challenged to look into nature, themselves, and their communities for the necessary wisdom. Students are naturally excited about contributing to the world in a positive manner, and not see themselves only as problem-fixers. We found regeneration to be an appealing, inspirational, and useful proposal.

Although we have encountered all these benefits from the regenerative perspective, we have also encountered challenges difficult to overcome. The main obstacles we have identified are:

1. There was confusion on what it meant to say a project was "regenerative". This ambiguity created confusion between students and faculty and made it almost impossible to measure if a project was in fact "regenerative". Without being able to evaluate a project's effects, our focus on Action Research turns subjective and ineffective.
2. We saw a tendency from some students and faculty to frown upon projects that were not considered "pure" in the regenerative understanding, generating unclear and unfounded judgments.
3. Students started naming all their ideas as "regenerative", without clear guidelines on what that meant. We started to see the risk of a green-washing phenomenon.

The architecture project in Totolapan presented above, can be an example of these ambiguities. Although it is a very successful project that was enriched by UMA´s model, it is still difficult to assess its regenerative qualities. Specifically in the section explaining the "Environmental Regeneration" dimension, it would be useful to have a more precise understanding of what it means by a regenerative or sustainable effect. For example, when stated that: "The use of raw earth as a reusable building material placed the architectural proposal under a sustainability paradigm", it would be crucial to have clear criteria as why this would be sustainable or regenerative.

We can see this happening very often with UMA´s project model. Upon reflection, we believe one of the reasons this could be happening, is that UMA´s model is focused mainly on processes but is somewhat ambiguous as to its desired results. For example, when speaking of the "socio-environmental purpose" the model states that "the main goal is to generate social and environmental benefits that are interrelated and helpful for sustainable development". It is unclear as to what these benefits could be, or how they are related to sustainability. We believe the main way we can address this challenge is to provide a clear understanding of what we mean by sustainability, regeneration, and their interrelatedness. In the next sections, we offer a deeper reflection on these conceptual challenges.

4.1 Understanding the Regeneration Perspective

As stated above, one of the main difficulties we have encountered at UMA is having a clear and common understanding of what it means to be "regenerative". Without this understanding, we generate unnecessary confusion in our design process, and it's difficult to apply Action Research methodologies and evaluate our progress. Following, we offer an analysis of the different ways to understand the regenerative approach and what questions are alive right now in our community.

According to Lyle (1994), the term regenerative was first used by Robert Rodale in 1989 to describe his work in agriculture beyond organic and sustainable, emphasizing the renewal of the complex life of soil in the absence of artificial chemicals (Chap. 1. Sect. 5, para. 2).

Now we find a vast array of definitions and descriptions as to what we can understand by regeneration. First, we will compare a very simple definition to a very complex one, to illustrate their usefulness in our socio-environmental efforts.

Thompson (2021), a market research firm that has studied the Regeneration concept for the last few years, offers us a simple and easy to understand definition: "Regeneration goes beyond sustainability and mitigating harm, to actively restoring and nurturing, creating conditions where ecosystems, economies and people can flourish." (p.7). It is very common to see regeneration defined as "beyond sustainability", which in our experience is not very useful. It generates confusion as to what is the value of a sustainability intent. We would have to clarify our understanding of sustainability and what we mean by "beyond" (more on this in the following section).

The second part of this definition emphasizes the common understanding that regeneration is seeking conditions where we can flourish. This fundamental element has proven useful because it inspired our faculty and students with the exciting proposal of working towards a better future and not only solving yesterday's problems.

Gibbons (2020), a researcher and regenerative practitioner from the U.S.A., who has been developing frameworks to evaluate regenerative projects, offers this very comprehensive explanation of regeneration:

> Regenerative sustainability sees humans and the rest of life as one autopoietic system in which developmental change processes manifest the unique essence and potential of each place or community. Regenerative sustainability's aspirational aim is to manifest thriving and flourishing living systems (i.e., complex adaptive systems) in the fully integrated individual-to-global system. It calls for humans to live in conscious alignment with living systems principles of wholeness, change, and relationship, as nature does. (p. 3)

Although several of the elements in this definition invite for deep relevant questions, it is a challenging definition that immediately faces us with complex questions: How do we know what is the essence or potential of a community? What is necessary to align our consciousness with living systems? How can we know if we are actually working with principles like wholeness, change, or relationship? This complexity is one of the great contributions of this approach, helping us to work with the real complexities that arise in systems, but at the same time can undermine the efforts of advancing and implementing projects.

A definition in a interesting middle ground of complexity is offered by Sustainable Brands, a community of brand innovators that explore socio-environmental challenges: "Regeneration means (1) restoring, renewing and/or healing systems we all depend on, while also (2) improving the ability of said systems to restore, renew and/or heal themselves more effectively" (Sustainable Brands 2021, p. 3). A very useful element of this definition is the emphasis on working on the systems' own capability to "restore, renew or heal themselves." This is a capacity that can be objectively observed, and it gives the system not only the capacity to sustain itself but also to evolve to new and better adapted stages. Furthermore, it does not necessarily imply for us to define an "essence" or "potential" as a starting point of a design process, but rather promote the healing and restoring process innate to the system and let itself realize its potential over time.

So, what could be a useful understanding of regeneration? From our experience at UMA, that is still a question to be explored. If we adopt a unique way of understanding this concept, it will give us clarity, but it will restrict an exciting concept that is still very young and has much potential for evolution. If we leave the concept to be interpreted in any way possible, it will lose its meaning. As mentioned above, we find especially useful the idea of being able to identify a regenerative system by its ability to self-organize, evolve, renew, and heal itself. These characteristics can give us evidence that the system is regenerating itself. Given these basic characteristics, the system would be able to find and materialize by itself its own essence and potential, not in a static and predetermined way, but in a continuously changing and self-emerging process. These intentions are already alive in the current UMA´s Project Design Model as stated in the "Environmental Regeneration" aspect of the model that states the following question: *"How can the project help to develop ecosystem capacities for self-renewal?"*.

Having this in mind, UMA is now offering the following definition of regeneration to its community: *A process that strengthens the capacities of a system to restore, heal or renew itself and thus increase its adaptive capacity and resilience.* This definition can provide a common starting point to align different efforts of a project and can enable the development of indicators to measure its progress and its purpose. We believe this approach can give clarity to design efforts as well as to the Action Research methodology that requires evaluating results and impacts.

4.2 Regeneration and Sustainability

As we mentioned, a second and crucial conceptual challenge with the Regenerative Perspective is the explicit proposal that it is something beyond sustainability (Thompson 2021). In our experience, students and faculty often interpret this to mean that regeneration should *replace* sustainability. The idea that sustainability should be replaced or left behind often comes from a misunderstanding of the term sustainability as something only focused on mitigating negative impacts on the environment. This can probably be traced back to the very first time Rodale (1989) spoke about regeneration as something that is more "appealing, inventive, and fruitful" than sustainability.

We have found that this point of view creates confusion and apathy in many people that have been working for decades or have been attracted to the environmental arena from the sustainability perspective. This could be acceptable if in fact the idea of sustainability had lost its usefulness, but we believe that it's a matter of what we mean by sustainability. It is true that the sustainability movement over the last decades has focused on reductionistic, end-of-the-pipe solutions. Until recently, proposals that had a more systemic approach were rare. As clearly reflected by studies like Planetary Boundaries from the Stockholm Resilience Center (Rockström et al. 2009), this approach has not given the desirable results and indeed has to be complemented with deeper and more systemic perspectives.

So, what is the most useful way of understanding of the term sustainability? Is it only efforts that mitigate negative impacts? Is it only reductionist and superficial in its essence? We believe sustainability offers a much broader and challenging question. In its most comprehensive approach, it faces us with the reality that human activities are impacting the whole earth's living systems. On this path, our quest for humans´ wellbeing, and maybe even our existence and that of millions of other species, are at risk. Under the term sustainability, there has been a myriad of proposals and approaches that have a much broader sense than only mitigating impact or only achieving balance. We find a more useful understanding of sustainability through the approach that Johan Rockström, one of the main researchers of the Planetary Boundaries Framework, offers in the open online course "Planetary Boundaries" (Rockström n.d.):

> We used to define sustainable development as the three pillars of development: social, economic, and ecological development -the modern thinking around sustainable development. This course will challenge that concept. We will be putting forward evidence to suggest that we have to redefine development, and we now have to think of the world in terms of providing wealth, development, livelihoods, and human prosperity, within the safe, resilient life support systems on Earth.

In alignment to the above understating, UMA is working with the following definition of sustainability: *Preserve the planetary conditions necessary to enable human well-being and the planet's biological diversity.* This approach to the sustainability challenge has a much broader perspective than only one that mitigates impacts and can be synergistic with concepts like regeneration. The complementary approaches between these two crucial terms can leave us better suited to find how we can collaborate and reinforce efforts across a great variety of actors, perspectives and contexts. Gibbons (2020) uses the term *Regenerative Sustainability*, which emphasizes the complementary aspects of these two terms:

> Over time, sustainability paradigms have evolved from meeting human needs throughout time to improving human wellbeing and the viability of ecological systems. Regenerative sustainability (RS), the next wave of sustainability, includes and transcends these goals, aiming for thriving living systems in which whole-system health and wellbeing increase continually. (p. 3)

Additional to the emphasis that Gibbons makes on recognizing a synergy between sustainability and regeneration, we find especially important the idea of "include and transcend" that she provides. This is a concept often used in the developmental process research area and sometimes overlooked when working with ideas that seem to come from a higher developmental stage like regeneration. Ken Wilber, who has worked for decades in understanding the different levels of consciousness, and one of the proponents of the "include and transcend" concept, explains it like this:

> when we are transforming into integral stages of thinking, views, and values, it's not just a "reaching up" that brings us into 2nd-tier, but also a "reaching down" to re-align and re-integrate all previous stages of development that remain alive within us… if we don't allow that descending self-love to fully take root, then we risk carrying all sorts of 1st-tier shadows, allergies, and addictions with us into Integral stages, or even sabotaging our own transformation into those stages. (2022)

With this perspective, we must pay special attention to the idea that we not only transcend the levels from conventional sustainability to regenerative sustainability, but we must include all previous levels. That helps us integrate from conventional sustainability to regenerative sustainability and adopt the most appropriate framework for the different contexts of the socio-environmental challenge.

We believe that promoting regeneration by undermining sustainability can be detrimental to our larger socio-environmental efforts for three reasons:

First, sustainability, as we just discussed, is a very complex term that gathers a vast array of worldviews, proposals, and efforts. If we say that it's no longer relevant, we are oversimplifying the term and undermining the work of millions of people who have strived for decades to address an existential, global threat.

Second, we need both sustainable and regenerative processes. These are not competing but complementary efforts. When we focus on sustainability, we have to ask ourselves what we have accomplished as a society so far and be sure we can sustain it. There is a growing narrative that everything in the world is wrong and should be replaced. This is losing sight of the wondrous advancements that human civilizations (south *and* north, east *and* west) have made in the last 12,000 years. This is clearly seen in the improvement of universal well-being indicators such as life expectancy, education, child mortality, and democracy (Roser 2020).

Third, a vast array of explorations, proposals, solutions, and narratives striving for a better future gives us a better chance to find complementary ways of arriving there. Diversity is one of the hallmarks of healthy living systems. If we reduce the whole socio-environmental movement to one specific way, such as regeneration, we will risk finding out too late it was not the right path. It is very possible that reductionistic, technological, and mitigating impact-focused solutions will be part of the mix that takes us to a better future. We need humility and respect for other efforts and initiatives.

Lyle (1994), the principal architect for the Lyle Center for Regenerative Studies, and presumably the first architect to apply the regenerative paradigm to the built environment, offered a way of understanding these two concepts in an elegant, complementary way:

> This means that, in order to be sustainable, the supply systems for energy and materials must be continually self-renewing, or regenerative, in their operation. That is, sustainability requires ongoing regeneration (Chap. 1. Sect. 5, para. 1)

5 Conclusions

Through this chapter we have described the origins of the educational model of UMA, how the regeneration perspective aligned with our intents and values, and how it was incorporated in UMA´s Project Design Model. An example of this application in a real project was described. Although UMA is still working in a systematic way of assessing the impacts of our model, we have already identified some advantages and challenges of incorporating regenerative processes and principles.

The regenerative perspective has been a positive experience for UMA and its community. We believe it has the capacity of driving a new wave of energy and a more integral perspective in our efforts to heal our planet and our relationships. We have pointed out a few challenges working with regeneration, such as its ambiguous meaning or dismissal of other perspectives.

Going forward we now see a new phase in our work where UMA as an institution will offer its community of students and faculty more punctual definitions of sustainability, regeneration, and its interrelations. This will have the intention of providing clarity to the design process of projects and a firm ground from which to develop impact metrics for our initiatives. This is especially important for UMA's Action Research approach which is highly dependent on measuring projects´ impacts. We speak of "offering" these definitions and reflections because UMA will maintain an open approach to the different perspectives and understandings of complex ideas and challenges. Members of our learning community are encouraged to explore different standpoints, but will always be invited to reflect, research and be aware of evidence that could transform them.

References

Cooperativa Las Cañadas (n.d.) Plan de manejo forestal. https://bosquedeniebla.com.mx/que-hacemos/manejo-forestal-sustentable/plan-de-manejo-forestal/
Correia M, Dipasquale L, Mecca S (2014) VERSUS: heritage for tomorrow: vernacular knowledge for sustainable architecture. Firenze University Press, Italia
Gandhi MK (1966) An autobiography or the story of my experiments with truth. Navajivan Publishing House
Gibbons L (2020) Regenerative-the new sustainable? ResearchGate 12(13):5483. https://doi.org/10.3390/su12135483
Lewin K (1947) Frontiers in group dynamics: concept, method and reality in social science; social equilibria and social change. Hum Relat 1(1):5–41. https://doi.org/10.1177/001872674700100103
Lyle JT (1994) Regenerative design for sustainable development. John Wiley & Sons Inc., New York, N.Y.
Mang P, Reed B (2012a) Regenerative development and design. In: Meyers RA (eds) Encyclopedia of sustainability science and technology. Springer, New York, NY. https://doi.org/10.1007/978-1-4419-0851-3_303
Mang P, Reed B (2012b) Designing from place: a regenerative framework and methodology. Build Res & Inf 40:1, 23–38
McNiff J, Lomax P, Whitehead J (1996) You and your action research project. Taylor & Francis
Regenesis Group (n.d.) The regenerative practitioner series. https://regenesisgroup.com/the-regenerative-practitioner-working-program-overview/
Regenesis Group (n.d.) The regenerative practitioner series. https://regenesisgroup.com/manifesto/
Rockström J, Steffen W, Noone K, Persson Å, Chapin FS, III, Lambin E, Lenton TM, Scheffer M, Folke C, Schellnhuber H, Nykvist B, De Wit CA, Hughes T, van der Leeuw S, Rodhe H, Sörlin S, Snyder PK, Costanza R, Svedin U, Falkenmark M, Karlberg L, Corell RW, Fabry VJ, Hansen J, Walker B, Liverman D, Richardson Crutzen KP, Foley J (2009) Planetary boundaries: exploring the safe operating space for humanity. Ecol Soc 14(2):32. http://www.ecologyandsociety.org/vol14/iss2/art32/
Rockström J, Module 1. The big picture. Lecture 1: welcome to the anthropocene. In: Planetary boundaries. edX. https://www.edx.org/course/planetary-boundaries?index=product&queryID=544bcc1fb2709cd0d3039e565128f2fc&position=1
Roser M (2020) The short history of global living conditions and why it matters that we know it. Our World in Data. https://ourworldindata.org/a-history-of-global-living-conditions-in-5-charts
Scharmer CO (2009) Theory U: learning from the future as it emerges. Berrett-Koehler Publishers
Senge PM (2006) The fifth discipline: the art and practice of the learning organization. Random House Books
Sustainable Brands (2021) The road to regeneration. California, U.S.A.
Taplin DH, Clark H (2012) Theory of change basics: a primer on theory of change
Thompson W (2021) Regeneration rising. United Kingdom
Wilber K, DeVos C (2022) The momentous leap to integral consciousness. https://integrallife.com/the-momentous-leap-to-integral-consciousness/

The Birth of an Evolutionary Learning Ecosystem

Delfina Terrado

Abstract This chapter aims to describe in a testimonial form an innovative experience within Argentina's educational field during the time of the world pandemic. The main intention was to support and co-evolve educational initiatives by creating a collective safe, nurturing and learning space in order to develop an interbeing awareness approach to problem-solving for individuals, communities and institutions. The Evolutionary Learning Ecosystem (ELE) was the term coined to host and potentiate through mutual interaction experiential learning. These initiatives decided to join forces and ignite collective intelligence to enhance their impact in their territory, rather than transforming education through each individual endeavor. The exploration of different practices was crucial to their growth which was achieved by creating and sustaining a shared holistic cosmovision and a trusting space. After two years of working together, the need for financial support to maintain the ELE sustainability over time became a burden to the project. As the initiatives evolved and faced many challenges, one methodology emerged to solve them, the Circles of Community Response (CRC). Their focus was and is, till today, to engage with a creative process of problem solving to face challenges within communities or institutions that only through collective and collaborative work can really transcend the current educational paradigm. The revealing effect of ELE and the CRC was the psychological and social regeneration of the people and initiatives that participated during such uncertain and complex times. This is a detailed account of a real time innovative initiative application that serves as an example to other regenerative educational projects.

Keywords Fragmentation · Crisis of perception · Social container · Regeneration · Education · Evolutionary learning ecosystem · Circles of community response

---

D. Terrado (✉)
Iniciativa Murmullo, Buenos Aires, Argentina
e-mail: terradodelfina@gmail.com
URL: https://iniciativamurmullo.com/

1 Introduction

During December 2019, in the oblivious times of a pre-pandemic world, one of the largest School Climate and Social-Emotional Learning Programs in the province of Buenos Aires was ending. A report written by a small group of researchers was unraveling an incipient phenomena. Teachers and principals in different communities around the provinces were coming alive with intrinsic motivation by engaging with their inner ecology (thinking, feeling, sensing and intuition) in their every day work, strengthening their relationships with others around them and their environment, and creating pedagogical practices that involved holistic participation of their students to their context, outer ecology. All of these qualities were emerging as the main themes in the research interviews to public schools. The report stated the absolute necessity of transforming the current educational system towards a fully embodied experiential learning that had a direct impact on the individuals, their self-organizing capacities in communities of practice and thus the positive commitment to their environment. The program was taken down by changing political parties in power at the beginning of 2020. The School Climate and Social-Emotional Education Program for 2000 public schools had an interesting effect on the current pedagogical and organizational approaches of the provincial education system. However, it did not last enough to create significant and scalable change in the higher structures of the system.

The year 2020 uncovered the underlying structural challenges within the educational system that persist until today. Most of the current problems-symptoms in education originate from the development of a fragmented perception of the world, based on the human illusion that we are separate from Nature, others (humans and other beings) and the world we collectively create. The crisis face is first and foremost one of mind, perception and values. Scharmer (2018) mentions this phenomena as the gap between the collective unconscious and the collective actions, a deep divide between human and nature, the Self and others and the Self and Self. Bohm (1998), explains how the preponderant paradigm in which humanity is immersed to understand its reality only leads to a fragmented experience. This insistence that the human being recognizes itself separate from others and the environment has brought countless negative effects. The educational system of most countries in Latin America responds to this fragmentation, it is visible in the design of their infrastructure, the curriculum, the pedagogy in practice, the processes schools follow and so on.

The purpose of education in the following years to come is to integrate and heal in many dimensions this approach to life, restoring the relationship of man with nature, body and mind, feeling with knowledge, object and subject, and so on. Thich Nhat Hanh, the late Buddhist monk, invites humanity to transcend the prevailing ideas of separation and individuality by expressing the term of **interbeing** to show how interconnected all forms of life are. To reconnect humanity with the ability to see the interrelatedness that underlies the world we inhabit. The complexity of the current context with its variables requires an integrated and systemic view that allows to perceive its totality, to perceive wholeness. This leads into changing the substance and processes of education contained in a curriculum and its expression in the community. To return the locality and the emergence of learning in everyday interactions, comprehending that children and adults can be instructed by birds, rivers, mountains, seasons and the ecology of their places. To acknowledge the nestedness in alive networks, to honor the intrinsic mutuality so thus it is possible to consciously participate in creating an integrated future for all.

The explicit integration of other human capabilities in the government social-emotional learning program showed in research a phenomenon called **Regenerative Education**. The program was working with what was most alive in public schools, with exploring purpose and with creating community, it engaged with emotions. It was opening the doors to fill classes and meetings centered in human capabilities. It was igniting the capabilities to connect with one another, to live and thrive together. Maturana and Varela (1984) wrote long ago that humans are collectively creating the world that they are living in, if they want to live differently there is a need to change. Regenerative education means to develop practices that awaken potential in individuals, communities and places within the educational settings that have been dormant or depleted of life. Regenerative education has become of the utmost importance to reconnect with humanities evolutionary capabilities to create conditions conducive to life.

This testimonial is thought provoking as it shows a courageous attempt to create, design and implement new holistic social dynamics within formal and informal educational initiatives. It narrates the breakthroughs and breakdowns of everyday people interested in engaging in a healthier participation with learning and life to co-create an emergent praxis for the future of education.

2 The Birth of an Evolutionary Learning Ecosystem

Today's uncertain and complex time requires education to rethink its purpose, to reconsider the social forms and structures that were built to achieve it. What better way of learning than by observing how Nature behaves. Biomimicry (Vanaga and Blunberga 2015) is a nature inspired discipline that searches for solutions to human problems, the best examples of interbeingness and organized action have biological principles. The provincial school program was an unintentional sample of what could happen if these principles are used to design for psychological and social regeneration. If we define regeneration (Mang and Reed 2020) as the ignition phenomena of new life or energy to something that was depleted of its physical vitality or meaning, then exploring Nature's examples of living systems might bring new innovative and invigorating ideas to the educational field.

In 2020 a small team of professionals, influenced by the work done in such a program and led by Dr. Alexander Laszlo from the Vivir Agradecidos Foundation, started exploring practical ways to create a social container that allowed educational initiatives to mutually co-evolve its practice. The intention was to engage through nested systems by cultivating new ways of thinking, feeling and acting which were organic and connected to a larger whole. The principles of natural ecosystems, such as adaptability, connectivity, diversity and the ability to redistribute and recycle resources cater as fundations to create an Evolutionary Learning Ecosystem (ELE), an umbrella term to describe this larger whole a space to encounter one another (Pickett and Cadenasso 2002). The next step was Inviting the proliferation of initiatives that were already vibrant and present within the current system and many others outside who were working as informal options to grow together. To learn how to be, do and work as an educational ecosystem. This was a unique innovative idea at the time.

An Evolutionary Learning Ecosystem (Laszlo 2001, p. 319) was defined as a system that supported mutual learning, ignited collective intelligence, developed capabilities and evolved each educational initiative's practice. Similar to a biological ecosystem it was made up of a community of living organisms (the participants) and the physical environment (the virtual space) where they interacted with each other. The ELE embraced initiatives mainly in the Buenos Aires province, with a few spread out in other provinces. Its purpose was to be able to respond to the current challenges of the educational field through collective learning, to create a nutritious social field, an environment that allowed initiatives to connect, empower themselves and provide feedback, share experiences, reflect on their practices and their actions. To amplify resources towards the individual's self-development and to cultivate an interaction in resonance with the world we desired collectively to bring forth. The ELE lived for 2020 and 2021 in its entirety during COVID—lockdown restrictions. The ELE grew as a response to a complex and volatile time where separation and individuality became the norm, having an incalculable impact on the human psyche.

The selected initiatives were quite peculiar. There were different educational organizations representing public and private schools, educational support organizations, government, educational innovation initiatives, non-profit foundations, initiatives in the development of social-emotional skills, and health clinics for youth and adolescents. Each initiative was based and had action in their own territory. In order to understand the challenges that emerged in territory as learning opportunities there was a need to understand fully the potential that arose in the interconnection of the actors with each space:

- **The territory**: was the real field where the initiatives work for change, the context. This territory is unique in its composition and in its essence, it expresses itself in particular ways. It is a unique environment. It is an alive, always changing, a field of entanglements that continuously expresses itself. David Seamon writes that a sense of place (the essence of territory) can be "described as the specific character, atmosphere, and expressive energy of a particular environment or locale (2022, p. 1)." This sense of place shapes human relationships and interactions, shapes its communities and the larger society. It becomes a unique learning setting with its own demands and resources. It unfolds in nested networks that are in reciprocal engagement such as the neighborhood, the natural physical environment, the province, the region, and so on.

Fig. 1 This image creates a visual dynamic of the interplay between the territory, educational initiatives and the evolutionary learning ecosystem. The locality of the territory and the learning that emerges engages positive feedback loops in the other two dimensions, within the initiatives and its participants and collectively within the ecosystem

- **The initiatives**: were composed by the people that perceived, sensed and worked towards developing generative solutions for a particular educational challenge. Each initiative had a constellation of resources and potential that made it unique.
- **The Evolutionary Learning Ecosystem**: The space where the set of initiatives engaged in deep mutual learning and where collective intelligence emerged to fully understand the challenge that each one of them had to face (Fig. 1).

Learning took place on the periphery of the system, in the ultimate reality of the teacher, of the students, of the context in general. Due to the prevailing learning fragmented approach, it was decided to create a different pedagogical experience to read and make sense of each particular situation. For this reason, the territory was the one that informed about the changes, transformations, challenges and developments to the initiative and then to the ELE. The quality of each initiative participation in territory was crucial. Profound participation involves opening new organs of perception, to use all human capabilities such as sensing, feeling and intuition and not only thinking. The question that the project came to face was:

How can each initiative learn and participate fully in such a way that new meaning and possibilities of collective action open up?

As Bortoft (1996) writes, lived experience is intrinsically holistic. Developing consciousness about human experience gives greater chances to engage in relationships (being ourselves, others and the world) that allows the present situation to emerge more fully true to itself. In other words, when becoming conscious of experience there is a choice of true participation, with all human capabilities, and as a result new information comes up. There is a depth in the perception of life which is seen intuitively that is absent when students, teachers or principals only use their analytical mind. By making visible and deepening participation the initiatives were able to discover with curiosity, active listening and without judgment what was emerging in the territory, because it constituted the essential source of information from which the response sprouted. Participants and initiatives needed to engage with their analytical and holistic mind to profoundly comprehend the full meaning of each challenge. The ecosystem space became the place where crafting an adequate response for the initiative through collective wisdom was possible.

3 Organically Grown

Argentina had a 172 day lockdown, in 2020, and for almost the whole year students did not come back to in person classes. Most educational institutions were going through an unimaginable transformation, specially the public schools which did not have the technological resources to reach all their students. Inequality grew two completely different experiences of the lockdown for educational communities. One of the public schools where it was difficult to track most students, where sometimes whats app was the only technology teachers had to send homework and learning activities. Another aspect of it was the absolute reorganization of the in person classes to online classes sustaining long hours in front of a computer for higher social-economic status students. A lot of the space in the ELE gatherings was used to process emotionally what was happening within the educational settings due to PANDEMIC. It was about digesting the emotional turmoil in each individual, communities and the society at the time, so that the ecosystem too could collectively navigate wisely the waters of that present moment. In doing so the participants of the ELE brought their resources and practices into the gatherings and there were turns to lead and be led. The small leading team that started the ELE had a role to bring the first impulse to create movement, it was becoming a true self-generating ecosystem.

The ELE gathered weekly virtually for two hours. At the very beginning, the only technology used to create a safe space and tap into collective intelligence was through facilitated conversations. There was no clear vision on how to generate a powerful space, but with time, there were collectively established agreements to make the gathering space a trustworthy one, to set a holding container. The small leading group prepared the gathering agenda weaving in what emerged in the previous meeting, and adding new elements or new dynamics that felt would nurture the understanding of the experience in territory. The path of evolution of the ELE was organically brewed by the emergence themes within these gatherings, each encounter came out from a previous encounter. There was no intention of setting a prescribed direction or learning to the ELE, since it had a life of its own it could discover through its emergence its unique path. At the start, it was confusing, but it felt it was key not to control the outcome in these uncertain times, but to respect its particular essence. As Heidegger (Bortoft 1996, p. 74) wrote " to let that which shows itself be seen from itself in the very way in which it shows itself from itself", was exactly our approach (Fig. 2).

Diversity was encouraged to preserve the original expression of each individual, the immense variety of resources that were available within the ELE, which would

> **Agreements: How do we navegate within the Ecosystem?**
>
> In order to create the propitious space for substantial educational innovations to emerge through this approach, it is necessary for the participants to agree on the following points on how to behave within the ELE:
>
> - We bring ourselves into space with an open mind, an open heart, and an open will.
>   - Being cordial, taking care of the language and the ways of communicating.
>   - Respecting different points of view and personal decisions.
>   - Being receptive, deep listening and learning from mistakes.
>   - Accepting that perhaps the answers we seek are yet to be built.
>   - Seeking collaboration to co-construct new forms and models.
> - Being aware that we are part of an interconnected whole where each part is essential.
>   - Always speak in the first person.

Fig. 2 These are the agreements that were collectively laid to preserve the trust in the ELE. They become the principles of behavior

not exist if it was a homogeneous system. None of the participants alone possessed all the intelligence, inspiration, strength and perseverance required for the challenges that arose on a day-to-day basis in the ever-changing world. The key was the ability to organize the ecosystem, to understand that possibilities emerged as they were weaved by new combinations of relationships within it. From these combinations, endless opportunities were generated and sprouted new self-organizing gatherings. In the process of working collaboratively, there was a feeling of support and care for those who needed it. The benefits of collaboration were endless.

4 Financial Support

In the second trimester of 2020, the ELE faced a major challenge. To sustain itself and grow it further it had to find financial support. The team that launched the ELE worked mostly donating their time, most of their work within the Vivir Agradecidos Foundation was dedicated to other endeavors. The love and dedication to the ELE community was more time consuming than expected, so they started experiencing some struggle to sustain the dedication of promoting the ecosystem development and growth. There was no cost for the initiatives nor the individuals to participate, the ELE offered a free space for this new experiential way of working together. The foundation was exceeding their monetary capacity to sustain financially all their projects and the amount of money allocated towards the ELE needed to come from another source. Pressure started to emerge for the ELE to become financially independent.

At the time, the small group of professionals that were involved in launching the ELE took the task of looking for seed money or investments from contests or bigger organizations that would like to support such projects. None of the small group had knowledge of the funding or philanthropic world. As there were already stressful aggravating conditions due to the COVID situation there was no intention from the small leading team to generate a new concern within the initiatives so they kept the tension and challenge for themselves. Looking back this was their first mistake, not opening up and being honest about this main challenge the ELE faced at the time. If they would have acted truly collaboratively an emergent solution would have come up from the ecosystem collective intelligence. The fear of bringing another element to the strain of variables that they were already navigating felt too excessive.

The small leading team chose to find a strategy to solve this situation when they realized a lot of the work done by the ELE had qualitative results. It was noticed that in explaining and communicating the purpose of the ELE to other funding organizations there was a lack of understanding. There were some important conversations with people involved in different projects for educational innovation, to integrate a perspective on how to make the ELE financially independent while maintaining its essence. The realization after quite a fit of frustration was that what was successful of the ELE was its unique purpose. If there was an opportunity to teach others to have their own ELE, their own creative process to learn together and solve challenges, the ecosystem could continue offering a valuable space for initiatives in education. If the possibility of teaching others how to holistically and collectively face challenges new innovative responses would rise for them. The idea of achieving the ELE purpose and even making it scalable became very present. This is how the small leading team came up with the Circles of Community Response (CRC), a creative holistic process to collectively solve the challenges that were arising for each initiative that could be used in other Evolutionary Learning Ecosystems. If there was a probability to condense the process and offer it to other ecosystems as a new social container that could bring a desirable future, the original ELE could get funding.

5 The Circles of Community Response

Creating a systematical and practical holistic approach to problem solving was in itself a challenge. The design and condensation of the creative process that was happening within the Evolutionary Learning Ecosystem took a few months. The first movement was to become aware of how the ELE was held and facilitated, there was a reflection review on what was its essence and those ingredients that made it such a special space. The social dynamics and methodologies that were used were highly influenced by *The Presencing Institute* (presencinginstitute.org, 2023), *The Work that Reconnects (journal.workthatreconnects.org,* 2023*)* and *Delicate Activism (*www.proteusinitiative.org, 2023*).* All of these communities embrace collective creation of knowledge and have at their core an alive practice to approach social phenomena. The same intention that the ELE was incubating, to support a community that with trust and love could ignite collective intelligence to evolve learning practices and educational spaces. ELE was sustaining three premises at its core. These premises acted like the holistic frame by which the ecosystem engaged with the events and challenges that presented themselves to the initiatives. The premises were the following:

- **The response to a challenge lies within**: Only by diving deep into a situation and fully exploring it with fresh eyes will you be able to read and tap into the potential that lies within. In any situation, there are invisible forces that play a fundamental role in shaping it. Only by truly understanding and connecting with all that there is of the situation is that we can see what is asking of us.
- **The questions that the initiatives or ecosystem hold, focus the attention and guide their evolution**: This premise brings the power back to the soul's fertile soil. Within the individual, in his/her inner ecology lies a deep ancient wisdom that if they just sit with a question the answer will reveal itself for him/her or the community. This is true for the person but also for a group, asking the right questions allows life to reveal itself to us.
- **In order to gain a holistic understanding of a situation, there is a need to engage with all human qualities**: Humans ability to make meaning of life lies in their capacity to be open to all the information that comes up through their organs of perception. Thinking, feeling, sensing and intuition are some of the ways of knowing the world. Most of the time individuals engage in problem solving through their thinking capabilities and they miss relevant data about the quality of the situation. Only through profound participation in experience is that humans can make full meaning of a situation. Hence, the importance of developing the capabilities to bring all ways of knowing to the forefront of our understanding individually and collectively.

These premises lay the ground for all the work that collectively occurred in the ELE. At the beginning, groups of people got together around a certain challenge or topic they wanted to explore. When the small leading team condensed the creative process for problem solving called these groups the Circles of Community Response (CRC) which were formed based on the interest of the educational initiatives in relation to a particular challenge. This interest was a shared challenge, participants believed that they could contribute to the understanding of rooted causes, or because they saw that it was a challenge that affected many layers of the educational community, or because they chose the challenge they felt called to and so on. There was no specific reason as a rule to choose one challenge or the other but true motivation and commitment to contribute.

To create an easier path to engage with the learning process the design created four defined stages, which are the following:

- **Stop** and Explore, corresponds to bringing the participants to the present moment to explore with an open mind, heart and will to face a challenge. It is important to use all senses as a group to dive deep into the particular situation that has been presented. Through a series of questions, deep listening and generative conversations participants seek to discover more about the challenging situation, to be able to read it through with all participants' collective senses. In this phase, participants seek to learn from the situation, discover what makes it noisy, what brings new information, what is stagnant, what perspectives disconfirm pre-existing beliefs, and read the hidden patterns. It is important to enter this stage without pre-existing judgments and feeding our capacity for wonder.

- The second phase was called **Look**. In this phase participants deeply connected with what listening evoke for them related to the challenging situation, what this community needs and what are the greatest future possibilities that want to emerge from this situation. Through stillness, reflection and key questions, participants explore if something has shifted within the situation after the first phase. Participants marinate what has come up, what has moved, what tensions are still there and which ones have emerged. Participants hold questions close to their heart not to answer them right away, but to become aware that their inner wisdom can bring something to the circle that can ignite collective intelligence.
- The third stage indicates it is time to **Advance Prototyping**. In this phase, two fundamental aspects are sought: the first is to generate ideas, plans and action. The second has to do with putting them into practice to get immediate feedback on what works and what doesn't. Here participants organize themselves and enter into collaborative action. It is time for effervescence of ideas and possible implementation strategies. In addition, participants put their ears to the ground to hear what the terrain is communicating and what is coming up. Participants ask questions such as: Is what was offered what you needed?? Is there a need for adjustments? Which ones?
- The fourth stage is **Materialize**. This stage is the last of the stages. Here participants find out about the effects of the iteration of the prototype. The implementation of the plan that emerged in the third stage or a new flow of action created for this system. At this stage participants can trace the outcomes and effects that were initially aligned with what emerged. One more possible result is the formation of another team that carries out a project that emerges in response to the challenge presented (Fig. 3).

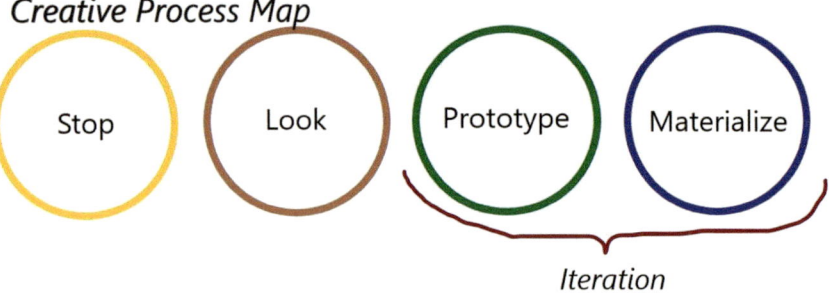

Fig. 3 One of the first drafts about the creative process map of the Circles of Community Response

From these four stages or instances participants could collaboratively create an innovative solution for the needs of each particular situation, listening deeply to the context from which it comes and providing an answer from the wisdom that is housed in the Ecosystem.

Each CRC had roles that were distributed between the participants to take care of the creative process as well as the knowledge that emerged due to collective intelligence. Each participating person would take a role. The roles were the Guardian of the Challenge, an individual who brought a challenge to the CRC space and represents an educational initiative, who is deeply interested in working, supporting and being proactive in relation to the challenge it brought. The Record Keeper, who kept notes, impressions, feelings, sensations, or whatever she/he thought was relevant to each encounter. This role may vary, but keeping records was very important for understanding the process as a whole and being able to connect together what was said and experienced from one meeting to another. The Guardian of Time helped find a healthy rhythm in shared spaces. Each process needed to have a beginning and an end. Her/His task was to guarantee that the steps are completed at the indicated times. The Facilitator of the process was the person who accompanied the group in each of the stages, guiding it to fulfill the objectives that were set and facilitating dynamics that were relevant to what was emerging within the circle. Circle Co-Creators were people who represented other educational initiatives interested in the challenge presented by the guardian of the challenge, and committed to moving the process forward.

One of the most valuable aspects when forming a CRC was its diversity. The more diverse its members were, the more systemic sourcing was available for understanding the challenge. Therefore, diversity and the inclusion of other perspectives were encouraged. A Circle was made up of a minimum of 5 people and a maximum of 12 people, with an ideal of 7 participants. It was discovered that this number allowed collaborative work in an agile and practical way. Each Circle had its own personality, its own characteristics. The members of the Circle determined the work dynamics, be it the tools to be used, the number and frequency of the meetings, their formats, the external resources that would be taken into account, as well as the requests with which the CRC would go to the Ecosystem, either to inform about the process or to ask for help. This structure was just an approximation of what the small leading group found was working the most within the ELE.

6 The Pioneer Team a Case Example

The pandemic affected the inner lives of most individuals, there were no more rituals or structure in the domestic flows. It was all mushed up together, the working environment for employees, the school space for children and youth, where everyone ate or exercised or rested. Work felt like a long sequence of exhausting virtual gatherings. Wellbeing had to become a priority to address the rapid decline of mental health in children, youth and adults.

A recurrent theme and pattern that came in many of the ELE gatherings was the concern for the mental health of teenagers. Within the initiatives quite a few noticed the rampant decline of it in schools. As this theme became one of the main drivers of the generative dialogues with the ELE, when the CRC methodology was created, participants decided to explore it more thoroughly. The ecosystem used the theme to explore and learn about the CRC, it was the first iteration to test what from the creative process worked and what needed to be improved.

The challenge that the CRC undertook was to explore the extent of the impact of youth mental health under quarantine. There was a sense of increased anxiety and depression in the adolescent population during the COVID context. Through careful research a crystallizing idea to measure what was the current state of the situation surface, in order to give visibility to what was happening. The CRC designed a survey with three already validated scales that measured anxiety, depression and resilience to test the hypotheses.

The survey was applied to 1216 teenagers between the ages of 14 and 18, from more than 40 educational institutions. These schools were scattered in five Argentinian provinces. The results of the survey showed an increase in discomfort during the quarantine for youth, and worked as an exploration of their mental health. In pre-pandemic times students of this same age were at a crucial stage of their development, needing to experience social relationships with peers and navigate their formative years for adulthood. The situation impacted the amount of exposure to real life moments that shape in positive ways their growth. All the experiences or the lack of them in quarantine forged their personality and contributed to their mental health. Due to the survey results most participants and educational institutions worked on developing emotional processing and wellbeing practices.

The results could not be measured as cause-effect, but it generated awareness within the educational institutions that participated in the research. Here is where Margaret Mead's phrase (goodreads.com, 2011) resounds deeply "Never doubt that a small group of thoughtful, committed citizens can change the world. Indeed, it is the only thing that ever has.". The pioneer team is just one of the examples of the rippling effects that the ELE and CRC's could have had within the psychological and social realm of the educational system. A group of individuals with the right support can engage and work together in collaboration to transform education. The challenges that surface in everyday life are opportunities of growth and transformation. Opportunities to regenerate the psychological and social fabric that the current fragmented system has damaged or depleted. The ELE and CRC's worked as incubators of community participation to create and innovate in ways that were beyond the fragmented current meaning-making capacities.

7 Conclusion

What emerged within the ELE and the CRC's was a new way of being and doing together, and it was on a continual process of creation. The financial support that the small leading team tried to find did not come in time, pressure was coming in from many fronts. Throughout this learning process there was a crack that grew slowly but steady and that was not addressed quickly enough, which was the power dynamics that emerged with the creation of the CRC. The CRC became an innovative solution for the initiative's challenges but fragmented the ecosystem, by not including and opening up the financial preoccupation to all the ELE's participants a sense of exclusion and mistrust germinated. Instead of taking the challenge as an opportunity to grow together, to be genuine and transparent, the small leading team made a wrong decision which cost the existence of the ecosystem. Some initiatives felt the idea of the CRC brought from outside as a crack in the horizontality dynamics of power, as an outside service to the organic emergence of the Evolutionary Learning Ecosystem.

This occurrence and the failed attempt to get funding from outside sources, sealed the ELE destiny. The ecosystem started to drain motivation and engagement from participants. The CRC's idea and creative process was offered to another project for schools at the Vivir Agradecidos Foundation. The participants of this other project created 290 CRC's in public schools spread out to the north and east of the country. All of those CRC's birth an innovative solution to their current challenges. The most noticeable aspect of this growth was the modeling and the space to engage holistically in collaboration with others. Learning how to live and engage together to create a more connected world. It is difficult to precisely calculate the extent of the effects of all those participants and schools that worked with the CRC's methodology in their contexts and lives.

Each individual is a world unto themselves, with their resources and their strengths, who learns in the encounter with another about their limitations and potentials. Martin Luther King Jr. said (mlkglobal.org/vision, 2021) "We are all caught in an inescapable web of mutuality, tied to a single garment of destiny. What affects one directly affects all indirectly." We are made to live together because of the interrelated structure of reality. It is only by looking deeply at our own past, present and future that we understand how we are linked to those of others. When we understand that we are part of an interrelated whole and that we influence its future, we understand that we are not only responsible for ourselves, but also for how we contribute to our context. Just as the individual is a living organism that has a narrative of itself (beliefs and values), a community is also a living organism, since it is made up of individuals.

As a new paradigm is emerging, in which the universe is experienced as a network of fundamentally interconnected relationships that make up life (Capra and Luisi 2014). Communities behave as a living organism embedded in a larger social system. Positioning ourselves and communities as part of a larger Evolutionary Learning Ecosystem requires some personal internal movements in relation to the perspective we had on what it means to work and learn with others. We need to cultivate other forms of leadership, generate more horizontal relationships and nurture trust in oneself, in the other and in the power of the ELE. As a living organism, each human being reveals themselves in interaction with their environment, with others and with nature. As initiatives within the ELE modified the environment and adapted to it in the process. We had an impact as an individual and a greater impact as an ecosystem. The CRC's seek to find answers by connecting with a higher level of consciousness in order to collectively discover sustainable solutions aligned with those results that we do want to have. With this case it was possible to have a glimpse of a way forward and practice to become the ecosystem we want to see in the world.

There are as many combinations of collective organization as people on the planet, if we want to collectively bring forth a world of beauty and health then we must be courageous enough to shift towards a new profoundly interrelated paradigm of being and doing together by developing new social containers. The ecosystem worked to develop an awareness of InterBeing (awareness of interdependence), making it easier for all participants to discover the impact they had as individuals, initiatives and as an ecosystem within the environment in which they were all immersed. Through exploration and iteration it was discovered that creating new regenerative social containers for powerful educational practices requires navigating personal and collective shadows. If evolution is how the world learns, and learning is how humanity evolves (Laszlo 2019), then what lies ahead is to become a transparent ecosystem with the intentional evolution of resources to learn and lead in order to co-create our healthy and desirable future.

References

Bohm D (1998) La totalidad y el orden implicado. Editorial Kairós
Bortoft H (1996) The wholeness of nature. SteinerBooks
Capra F, Luisi PL (2014) The systems view of life: a unifying vision. Cambridge University Press
Kaplan A, Davidoff S (2023) "The proteus initiative." Proteus Initiative – Home. www.proteusiniti ative.org/. Accessed 21 May 2023
Laszlo A (2001) The epistemological foundations of evolutionary systems design a. Syst. Res. Behav. Sci.: off. J. Int. Feder. Syst. Res. 18(4):307–321
Laszlo A (2019) Education for the future: the emerging paradigm of thrivable education. World Futures 75(3):174–183
Macy J (2023) "The work that reconnects." Journal.Workthatreconnects.Org, journal.workthatreconnects.org/. Accessed 21 May 2023
Mang P, Reed B (2020) Regenerative development and design. Sustain. Built Environ. 115–141
Maturana, H. R, Varela, F, Behncke, R (1984) El árbol del conocimiento: las bases biológicas del entendimiento humano (Vol. 1) Organización de Estados Americanos, OEA
Mead M (2011) Never doubt that a small group of thoughtful, committed citizens can change the world; indeed, it's the only thing that ever has. GoodReads, Quotes, 61107. www.goodreads. com/author/quotes. Accessed 12 December 2017
Pickett ST, Cadenasso ML (2002) The ecosystem as a multidimensional concept: meaning, model, and metaphor. Ecosystems 5:1–10
Scharmer O (2018) The essentials of theory U: core principles and applications. Berrett-Koehler Publishers

Scharmer O (2023) "Deep systems." Presencing Institute. presencinginstitute.org/. Accessed 21 May 2023

Seamon D (2022) Sense of place. International encyclopedia of geography: people, the earth, environment, and technology

"The Vision." MLK Global, 14 January 2021, mlkglobal.org/vision/

Vanaga R, Blumberga A (2015) First steps to develop biomimicry ideas. Energy Procedia 72:307–309. https://studycrumb.com/alphabetizer

A Critical View on the Strengths and Challenges of Outdoor Preschools and Nature-Based Education in Spain

Katia Hueso-Kortekaas

Abstract Nature-based education is growing throughout the world. Although it already existed at the beginning of the 20th century, WWII in Europe and the Civil War in Spain put an end to it. The first outdoor preschool in Spain started in 2011, now ca 60 exist. Nature-based education includes family-led playgroups, outdoor preschools and forest school sessions provided by conventional schools. From 2022 this approach is taught at the university and academic literature on the topic has grown. However, it has not been officially recognized by the authorities, leading to various challenges for families and practitioners. One priority is obtaining legal recognition by, among others, establishing quality criteria, covering pedagogical, logistical, and environmental aspects, to reassure decisionmakers. Another priority is the expansion of training for practitioners. These key aspects stimulate an environmentally sound, sustainable, and fair access to nature-based education. In summary, this contribution offers a critical view of the strengths and challenges of nature-based education in Spain. This work adopts a qualitative, (auto-)ethnographic perspective, drawing on the experience of the author, co-founder of the first outdoor nursery in Spain, as well as on semi-structured interviews with key stakeholders in the field.

Keywords Nature-based education · Outdoor education · Spain · Weaknesses · Lessons learned

1 What is Nature-Based Education? a Conceptual Framework

Outdoor and nature-based education is a broad concept that has been attempted to define by many scholars and practitioners over the past decades, usually in a descriptive manner. Versions abound in different contexts across Europe and America, e.g., Del Rosso (2010, pp. 40–42), Donaldson and Donaldson (1958, p. 17), Forest School

K. Hueso-Kortekaas (✉)
Department of Mechanical Engineering, ICAI, Comillas Pontifical University, Madrid, Spain
e-mail: khueso@icai.comillas.edu

Association (2022), Larimore (2016, pp. 33–34), MacEachren (2013, p. 90), Michek et al. (2015, p. 739), Waite et al. (2016, pp. 5–6), Wilson (1984, pp. 5–6). Significant conceptual and practical overlaps exist with environmental education, while they do not always address the same public, methods, or targets. Environmental education is goal-oriented; experiential, process-oriented, and nature-based initiatives are typically place-based (Adkins and Simmons 2002). The three of them, however, benefit from the direct experience of nature, which is the core of nature-based education.

It is worthwhile to take a closer look at the definition of "nature education" proposed by the now-inactive Spanish National Federation of Nature Education (EDNA). This organization aimed at supporting nature-based preschools in Spain and training professionals in this field. It defines nature education as "pedagogical approaches aimed at anyone interested that offer regular, direct and permanent contact with wild nature, in which nature is the source of materials and sites and is the driving force of a respectful lifestyle with it" (Hueso Kortekaas 2022, p. 97). Despite being far from flawless, it will serve as the conceptual framework in the context of this contribution.

This definition makes it clear that nature is not only "used" as an educational resource, but that the essence of this approach is to (re-)establish a profound connection with it, that will ensure learning and development outcomes beyond specific curricular aims. Significant emotional and personal hands-on experiences in nature will be firmly and lastingly engraved in the participants' psyche, reinforcing the acquisition of skills and competencies associated with them. There is a strong association between learning and place. Since these place-based experiences will become part of the personal history of the participants, the desire will arise to protect the places where these have occurred. Ultimately this will result in pro-environmental attitudes and behaviors that will contribute to a better planet (Chawla 2007, p. 145; Hueso Kortekaas 2022, p. 97).

Aside from this long-term effect of nature-based education, spending time outdoors has multiple added benefits that are perceived almost straight away. Direct learning, acquired through immediate contact, through the senses, with the natural environment, is much more effective than when obtained from vicarious experiences, as usually happens in a classroom. Thus, we speak of a "cognitive benefit" that consists not only in acquiring skills and competencies but also in their foundation. Likewise, spending time outdoors in a context of respectful accompaniment and active listening, another of the main features of nature education, allows the development of social and emotional skills that are very useful for life in general (Hueso Kortekaas 2022, p. 97; Hueso Kortekaas and Mayorga Martín 2020, pp. 37–40]. In addition, there is a broad scientific consensus on the benefits of staying in the natural environment for physical and mental health, also in the educational context, e.g., Cudworth and Lumber (2021, p. 75), Lovell and Roe (2009, pp. 51–52), Miller et al. (2021, p. 1135), Mygind et al. (2019). Free movement, the irregularity of the terrain, the contact with natural elements, and the changing environment to which it is necessary to adapt, all constitute a source of strength and resilience that are precursors of a healthy lifestyle (Brussoni et al. 2015, p. 6447; Janssen and LeBlanc 2010, p. 13; McCurdy et al. 2010, pp. 109–111).

Thanks to the positive and significant experiences they generate in natural settings at an early age, nature-based education and mainly forest/outdoor nurseries, preschools, and kindergartens are also seen as precursors of a pro-environmental lifestyle in later life. This is not only because of the positive affect to place but also thanks to the social and emotional skills that lead to proactive leadership in adult life (Chawla 2007, pp. 152–153). They can thus be considered a more sustainable way of educating and regenerative towards the environment, as they actively contribute to nature preservation nature and include indigenous worldviews, in the long run (Haraway 2016, pp. 99–110; Johnston 2019; Mycock 2020, p. 3).

2 The Historical Context of Nature-Based Education

Despite the apparent importance of contact with nature from an early age, it is not something to which formal education has paid much attention. Education in Western cultures has long ago distanced itself from the vernacular knowledge of nature and children grow away from the innate relationship of biophilia (Louv 2008, p. 43–44; Waite et al. 2016; Wilson 1984). Despite this, many scholars emphasized nature as a pedagogical environment, even in Antiquity, with Aristotle defending empirical learning in nature. Perhaps best known in this context is Jean Jacques Rousseau, who laid the foundations for considering nature as a pedagogical resource, for scientific study, physical education, and health promotion. In this last aspect, he was a precursor of the hygienist movement proposing staying outdoors as a tool for health improvement in the mid-19th century. At that time, in 1839, Friedrich Fröbel opened his first *Kindergarten* with the idea of connecting children with nature, albeit from a more religious perspective, so that they would experience nature as the work of God. Shortly afterward, in 1910, Margaret and Rachel McMillan founded the first open-air nursery in London for disadvantaged children. Maria Montessori did the same in Rome in 1907. With a similar objective, although more akin to Catholic doctrine, Father Andrés Manjón created the *Ave María* Schools in Spain, reaching 300 centers throughout the country. Inspired by Rousseaunian thought, the New Schools arose in Switzerland, whose ideologist, Adolphe Ferrière, considered it essential to be in the countryside and one of whose principles is the protection of and respect for living beings (Bernal Martínez 2000, p. 176–179; Hueso Kortekaas 2017, pp. 158–160).

In interwar Europe, open-air schools emerged with certain notoriety, following the example of the *Waldschule* of Charlottenburg (Germany), founded in 1903 and the pioneer of this approach. The Catalan pedagogue Rosa Sensat promoted the creation of the *Escola del Bosc* and the *Escola del Mar* in Barcelona, and similar initiatives arose in Madrid and Andalusia. The main objective of these schools was to improve children's health by spending time outdoors. This idea of nature as a setting for learning and a precursor of healthy development was also promoted by Francisco Giner de los Ríos, founder of the *Institución Libre de Enseñanza* (ILE), who also applied it in higher education with adult students. His work was continued by his disciple and successor, Manuel Bartolomé Cossío, founder of the Pedagogical Missions, an initiative that brought cultural resources to isolated, poor rural areas in Spain (Canes Garrido 1993, pp. 150–153; Esteban Frades 2016, pp. 262–264; Hueso Kortekaas 2017, pp. 158–160).

After a period of latency, strongly marked by World War II in Europe, the Civil War in Spain, and the costly post-war periods, the interest in nature as a pedagogical setting arose again at the end of the 20th century. While the primary motivation for outdoor education in its origins was respiratory health, at the turn of the millennium it is seen as a response to overindustrialisation and a need to regenerate the connection with nature. It is considered that the first modern nature-based school was founded in 1952 in Denmark when Ella Flatau started taking her own children and others to the forest to play. In Germany, Ursula Sube created the first private *Waldkindergarten* in Wiesbaden in 1968. However, it was not until 1993 that this model was standardized, with the center started by Petra Jäger in Flensburg. This led to a wave of others opening nationwide, with a number that today could exceed 2,000.[1] In the rest of Europe, the model truly expanded at the beginning of the 21st century and can be found in most countries, albeit with different degrees of implementation. Similar schools can be found outside Europe in North America, the Far East, Australia—the so-called *bush kinder*—and New Zealand. Canada and New Zealand have shown a keen interest in recovering and teaching indigenous knowledge in their settings (Hueso Kortekaas 2017, pp. 160–162; Hueso Kortekaas 2021, pp. 137–139). Fuelled by the pandemic, the interest in the outdoors as a safe and healthy setting turned the attention of the public at large towards this thus far marginal approach. In 2022, almost three years into the pandemic, no figures have been found to measure the increase in nature-based schools, although, in the author's experience as a consultant in the field, it seems to have grown.

3 Nature-Based Preschools and Similar Initiatives in Spain

Even though Spain was well aligned with the rest of Europe at the start of the 20th century with respect to outdoor schools, modern ones took much longer to start. The post-war period was long and cumbersome, and education policies were steered by the right-wing, Catholic thinking of the Franco dictatorship, which left little room for innovation. In addition, rural areas were seen as backward and significantly underdeveloped with respect to urban areas, and nature was not considered a valuable educational resource. In the first decades of Spain's democracy, at the end of the 20th century, a solid socioeconomic and industrial development pushed society further away from its natural roots. At the turn of the millennium, with an ever-growing interest in the environment and sustainability, an interest in the outdoors started to grow, first as a setting for leisure and adventure sports, later more holistically as a

---

[1] https://www.bvnw.de/ [Retrieved in November 2022].

lifestyle option. Only in 2011, the first outdoor preschool project,[2] the Nature Play Group *Saltamontes*, was created, inspired by its counterparts in northern Europe. In 2015, when about a dozen initiatives existed in the country, the National Association of Education in Nature, the forerunner of the abovementioned Federation, was founded (Hueso Kortekaas 2022, p. 99). A decade after *Saltamontes* opened, circa 60 outdoor preschools exist,[3] most running on a full-week, full-day basis for the children enrolled. Aside from the main nature-based (pre-)school activity, many initiatives offer additional services, such as early years playgroups, extracurricular activities in nature, holiday camps, family gatherings, thematic workshops, etc. (pers. obs.). Less common are forest school-like sessions, i.e., groups of students from traditional schools who regularly visit nature-based schools. More often these sessions are organized by the schools, with the help of specialized staff (pers. obs.).

Aside from outdoor nurseries/preschools, traditional schools gradually incorporate outdoor teaching, using schoolyards and nearby parks, gardens, and beaches. Thanks to nationwide initiatives such as *Patios x Clima*[4] and *El Nou Safareig*,[5] many of them are undergoing a process of greening schoolyards, and authorities are providing funds to transform these areas from concrete-filled, straight-lined sports grounds into more organic designs with sand, vegetation, and shade, e.g., Baró et al. (2021). The benefits of these schoolyards are not only those related to being outdoors but also are known to reduce inequalities and mitigate abusive behavior (Lanza et al. 2021, p. 9; Luís et al. 2020, p. 3; Puhakka et al. 2019, pp. 12–13). In addition, these benefits reach a much larger population of children, compared to those attending outdoor nurseries and preschools, who are indeed a minority. On the other hand, many schools have undergone superficial changes and do not fully embrace outdoor teaching beyond a cosmetic level. During the pandemic, photos of traditional classrooms, furniture, and all, moved outdoors were seen on social media. Others even consider themselves "nature schools" while offering limited sessions in nature, not more than a few hours per week (pers. obs.). This can perhaps be interpreted as a sign that this approach is gaining momentum in society, at the risk of commodifying it, as has already been critiqued in the case of forest schools in the UK (Leather 2018, pp. 11–12).

---

[2] In Spain, nurseries are aimed at ages 0–3 and preschools to ages 3–6. From 6 years of age, children attend primary school. Schooling is compulsory from the age of 6; below that, schooling is optional.

[3] An unofficial listing can be found here: https://escuelainnatura.com/blog/2021/05/16/directorio-escuelas-en-la-naturaleza-espana/ Accessed 25 November 2022.

[4] https://patisxclima.elglobusvermell.org/ Accessed 25 November 2022.

[5] https://elnousafareig.org/ Accessed 25 November 2022.

4 Challenges to Nature-Based Education and Lessons Learned

The expansion of nature-based education in Spain is relatively recent and limited in geographical and pedagogical scope. This option still faces numerous challenges that must be addressed before it is consolidated. It can provide a natural alternative for those interested in offering an education to their children or pursuing a professional career in the field. The main challenges, albeit related, are individually outlined below. These observations are based on the personal experience of the author as the cofounder of the first nature-based preschool in Spain, trainer, and consultant, as well as interviews and informal conversations conducted with four specialists and pioneers in nature-based education in Spain. The interviewees were Emma Camina (herein forth E.C.), co-founder of the first nature preschool in Spain; Noelia García (herein forth N.G.), founder of *Bosqueko* in the Canary Islands, one of the officially recognized nature preschools and academic coordinator of the expert course on nature education at the University of Las Palmas de Gran Canaria; Paz Gonçalves (herein forth P.G.), founder of the nature-based center *Amadahi* in Galicia and adjunct professor at the University of A Coruña and Jara Loira (herein forth J.L.), practitioner and trainer at *Bosquescuela* in Madrid.

4.1 Equal Access to Education

Nature-based education in Spain is still a marginal option open only to a minority of children, for various reasons. The most obvious is the small number of initiatives, concentrated mainly in areas with higher population density within Madrid, Galicia, Catalonia, Basque Country, and Andalusia. Geographical accessibility is therefore limited. Also, these projects are privately run and need to cover all the costs by themselves, hence the families need to cover them with their monthly fees. Subsidies and grants are difficult to obtain because this activity does not match most of the options available in funding calls (which typically cover leisure activities, environmental education, or formal education activities). This is a significant issue for all parties involved because projects need to sustain the activity, pay the staff, and keep the fees as low as possible, a delicate balance act. In addition, most of these initiatives have shorter opening hours than regular schools. Therefore, they are unattractive for full-time working parents, especially when commuting (E.C. pers comm.). The lack of official recognition prevents this approach from reaching other families, who may be willing to join but do not have the economic resources to do so.

The lack of apparent formality of most projects seems to be a deterrent for reluctant families, who—despite their interest in the benefits of nature—worry about the curricular contents (N.G., pers. comm.) or the transition to the following stages of education, such as primary school (P.G., pers. comm.). Access to desired schools may need early registration to secure a place and families may prefer to enroll their children in traditional schools from the start.

4.2 Official Recognition

The main obstacle to the consolidation of nature education in Spain is administrative authorization as formal schools, which must be complied with at national, regional, and local scales, depending on the aspects involved. The first step is authorization by local authorities to perform an economic activity, compulsory for all, regardless of age or activity. The second is to comply with regional and/or national regulations, depending on the activity the initiative formally offers and to whom. This is why most of the projects are focused on the early years (0–6 y.o.) stage, which does not require formal schooling and is more flexible in choosing activities and aims. To obtain recognition as a school, they must qualify in the pedagogical project and the infrastructure. In the first case, this means that the programming of the activities in schools that are applying for recognition needs to match the official curriculum of the corresponding region (autonomous community, in official terminology). Often play-based projects cannot fit in this scheme, because curricula are often content-based and not competence-based, leaving little room for adaptation and creativity. In the second case, outdoor schools do not need a part or most of the infrastructure required, as they would hardly use it (such as paved schoolyards or multipurpose rooms, for instance). The law is very precise with the technical requirements for school buildings and nature-based initiatives would need to make an essential and possibly useless investment. Given that authorities have the capacity to interpret the regulations to a certain extent, this adds uncertainty to project proposals and risks unfair recognition processes (Hueso Kortekaas 2022). Only two nature-based preschools have been officially acknowledged, *Bosquescuela* in Madrid and *Bosqueko* in the Canary Islands, both as regular schools. In the first case, it has been considered an "innovative" project and has been freed of specific formal requisites (pers. obs.), while in the other case, all formal requirements have been met (N.G., pers. comm.). These two cases show how serendipitous the process can be. Aside from these difficulties, most initiatives have found alternative ways to acknowledge their activity, be it at a local or regional scale, by being registered, for example, as a nature club or a leisure center. Others are still in the early stage of operations and are yet to comply with even the most basic authorization, namely as economic activity, granted by each municipality (pers. obs.). A few consciously choose not to engage with "the system" and prefer to go underground, arguing civil disobedience as their primary motivation (pers. obs.).

4.3 Quality Assurance

Despite the small number of projects engaged in nature-based education, there is an excellent variety of pedagogical approaches, management styles, and formal structures among them. Most projects are run by associations formed by school community members (teachers and parents, sometimes also external participants). The management of the projects varies from horizontal decision-making to a more top-down, hierarchical style. Pedagogical decisions are sometimes part of this participatory process, but most often are left in the hands of the practitioners. Given the economic fragility of the project, salaries can be precarious, with unpaid holidays or sick leaves. Although the law requires staff in regulated schools to have formal teacher training, nature-based projects often prefer other professional profiles, related to social work, environmental education, or creative professions such as arts or yoga instructors. This may cause issues with the pedagogical quality of the work. Related to this is the need to document and assess the performance of these projects and provide evidence of the benefits of nature-based education at all levels, which is not often done (E.C. pers. comm.). There is a need to research and generate tools for teachers to provide this evidence to families and authorities (N.G. and P.G. pers. comm.). This has also been identified as one of the challenges to outdoor teaching in a study performed in Wales (Davies and Hamilton 2016, pp. 6–8).

Other problems related to the lack of funding or irregular infrastructure are the logistics of daily life at school. Examples are providing food, transportation of children, safety, or hygiene. The education laws are also very clear on these aspects. Therefore, one of the challenges to be solved before requesting official acknowledgment is to comply with the law and provide a minimum of quality at all levels (pedagogical, logistical, and managerial). The Federation EDNA prepared a set of quality criteria (Hueso Kortekaas and Mayorga Martín 2020, unpublished work) that should have served as a reference framework for families, nature-based schools, and authorities. In the first case, to reassure families that could be hesitant about the well-being of their children in projects considered as "alternative". The second is to help benchmark projects and serve as a guide for their daily functioning. In the latter case, to inform the authorities to be more flexible with interpreting the law (e.g., understanding the outdoors as the "schoolyard", instead of requiring a paved one, N.G. pers. comm.). However, due to the unforeseen closure of the Federation EDNA in 2022, these quality criteria were never applied.

There is a fine line between establishing quality criteria while leaving room for pedagogical creativity and style among nature-based preschools and creating a one-size-fits-all model. One of the established initiatives tried to offer a franchise model for nature preschools, but the costs have been too high for entrepreneurs to engage in it, and it has been unsuccessful. Leather (Leather 2018, pp. 11–12) warned of the risk of the *McDonaldization* of forest schools in the UK. Commodifying nature-based education is already underway in the greening of schoolyards (see above) and is likely to happen with full nature-based preschools at some point.

4.4 Training and Professional Opportunities

Such a young and innovative pedagogical approach has an obvious flaw: the lack of adequate staff training. Given the specificities of the setting (outdoors instead of a classroom) and the pedagogical style (e.g., free play as a learning tool, active listening), specific training for nature-based education professionals is considered essential by different authors (Knight 2009, pp. 49–54; Martin 2008, pp. 33–34; Michek et al. 2015, p. 742; Miklitz 2001, pp. 78–81). The Association EDNA offered short introductory trainings on nature-based education between 2017 and 2021, run by instructors who were Spain's first nature-based preschool founders and practitioners. Other entities, also nature preschools themselves, earn a significant income providing introductory or thematic training, typically over a weekend, although some are extended over several weekends (aka modules). Many of these trainings were inspired by foreign nature-education training models, such as Forest Schools in the UK or *Waldkindergärten* in Germany. In 2015, the first more extended training (120 contact hours) was created thanks to an agreement between *La Violeta*, a private entity with a consolidated training scheme in active education, and the University Rey Juan Carlos in Madrid, discontinued due to lack of profitability. In 2022, the first proper university-led expert-level training (over 500 h) was created at the University of Las Palmas de Gran Canaria and is still running. In the last few years, there is a constellation of trainings of different lengths, approaches, and complexity levels that caters to different needs and profiles. However, none are endorsed by any external entity and an objective comparison between them is hard to find. The depth and pedagogical quality of these trainings are very diverse and not always apparent to the public, with some seeming to follow a commercial motivation (Leather 2018, p. 13). Participants are left to decide by reading and interpreting the publicity of each training, which is not always clear. This is also a problem for the projects themselves, as they do not always know what trainings are valid for their needs or not when selecting staff (E.C., pers. comm.). In addition, only official degrees, such as those offered by the university, give credits in accreditation processes, especially relevant for civil servants (basically, all staff working in public schools). At the time of writing, there seems to be a certain market saturation, and some of the trainings are likely to disappear (pers. obs.). On the other hand, lagging somewhat behind, are trainings specifically aimed at families, to allow going deeper into their reasons to choose a nature-based project and better understand nature's benefits for their children (N.G. and P.G., pers. comm.).

A related issue to the professionalization of nature-based educators is the need to join forces. It is essential to have an organization that represents nature-based education, provides training and advice to projects and practitioners, disseminates, generates knowledge, and creates links with society. Especially relevant is to provide science-based evidence to policymakers and thereby contribute to a more straightforward, more equitable recognition process. The former Association/Federation EDNA had these goals in mind but did not manage to solve internal issues and is no longer operating. There seemed to be diverging priorities, caused by the different degrees of experience and the management styles and ethos of each project (E.C., pers. comm.). In the meantime, half a dozen regional and national organizations and informal networks are interested in nature-based education. However, none of them share the focus and scope that EDNA planned to have.

4.5 Lessons Learned

In the past few years and especially after the start of the pandemic, nature—and outdoor-based education has been perceived as safe, healthy, and effective. There is an increasing demand for this model by both families and practitioners. In the first years, families were not well aware of the pillars of this pedagogical approach and simply chose it as a means to create distance from mainstream education. While yet far from being a real option in equal terms to other pedagogical alternatives, nature-based education has reached the media regularly. Today, most families and practitioners have basic knowledge and reasonable expectations (E.C., pers. comm.).

Some of the most veteran projects are now consolidated, with a clear identity beyond the obvious—spending most of the time outdoors—and offering diverse approaches beyond this aspect. Examples are organic food growing, (land) art sessions, connection to local traditions and folklore, etc. These projects are genuinely regenerative in their nature, as they promote a healthy, resilient, and sustainable lifestyle for both the children and the school community as a whole and positively impact the local community. In general, the public reacts positively when told what nature-based education is about, even when they are unfamiliar with its benefits (J.L., pers. comm.).

Training of professionals is also improving, with a broad offer that caters to those interested in practical aspects and those who aim at management positions, entrepreneurs, and researchers in the field. Teachers in all settings gradually encounter nature-based teaching and move their classes to nearby nature (J.L., pers. comm.). There seems to be less fear of being outdoors. Both popular science books and academic works on the topic are being published (see the reference section for some of them), although some gaps still need to be filled. A deeper discussion on this goes alas beyond the scope of this contribution.

5 The Future of Nature-Based Education in Spain

All respondents coincide in highlighting the spatiotemporal window of opportunity that the pandemic has given Spanish nature-based education initiatives to demonstrate their validity and strength. The diversity of Iberian cultures, climates, and landscapes is another crucial advantage that favors the visibility, the perception of feasibility, and subsequent consolidation of outdoor education projects.

Soon, nature-based education will likely experience significant growth—both in geographical terms and in-depth—in traditional schools. Families and professionals are exerting important pressure on education authorities to transform schoolyards into more healthy, equitable settings. This approach is also likely to percolate to upper educational levels, such as primary and secondary stages, or even in higher education. Society is gradually accepting, celebrating, and demanding nature-based education. The last step is to convince policymakers, hopefully with the assistance of scholars, and disseminate sound evidence of its benefits. In the end, nature-based education is a regenerative approach to improve the quality of society and nature that will contribute to better people for a better planet.

Acknowledgements I am greatly indebted to my colleagues Emma Camina, Noelia García, Paz Gonçalves, and Jara Loira for their time and insight during the interviews and informal conversations. Any inaccuracies in the transcription and interpretation of their words are solely attributable to me.

References

Adkins C, Simmons B (2002) Outdoor, experiential, and environmental education: converging or diverging approaches? ERIC Clearinghouse on Rural Education and Small Schools, Charleston, WV

Baró F, Camacho DA, Del Pulgar CP, Triguero-Mas M, Anguelovski I (2021) School greening: right or privilege? Examining urban nature within and around primary schools through an equity lens. Landsc Urban Plan 208:104019

Bernal Martínez JM (2000) De las escuelas al aire libre a las aulas de la naturaleza. Áreas. Rev Internacional De Ciencias Sociales 20:171–182

Brussoni M, Gibbons R, Gray C, Ishikawa T, Sandseter EBH, Bienenstock A, Chabot G, Fuselli P, Herrington S, Janssen I, Pickett W, Power M, Stanger N, Sampson M, Tremblay MS (2015) What is the relationship between risky outdoor play and health in children? A systematic review. Int J Environ Res Public Health 12(6):6423–6454

Canes Garrido F (1993) Las Misiones Pedagógicas: educación y tiempo libre en la Segunda República. Rev Complutense De Educación 4(1):147

Chawla L (2007) Childhood experiences associated with care for the natural world: a theoretical framework for empirical results. Child Youth Environ 17(4):144–170

Cudworth D, Lumber R (2021) The importance of Forest School and the pathways to nature connection. J Outdoor Environ Educ 24(1):71–85

Davies R, Hamilton P (2016) Assessing learning in the early years' outdoor classroom: examining challenges in practice. Education 3–13

Del Rosso S (2010) Waldkindergarten: ein pädagogisches Konzept mit Zukunft? Diplomica Verlag, Hamburg

Donaldson GW, Donaldson LE (1958) Outdoor education a definition. J Health Phys Educ Recreat 29(5):17–63

Esteban Frades S (2016) La renovación pedagógica en España: un movimiento social más allá del didactismo. Tendencias pedagógicas 27

Forest School Association (2022) History of forest school. https://www.forestschoolassociation.org/history-of-forest-school/. Accessed 10 Nov 2022

Haraway DJ (2016) Staying with the trouble: making kin in the Chthulucene. Duke University Press, Durham, NC

Hueso Kortekaas K (2017) Somos naturaleza. Plataforma Editorial, Barcelona

Hueso Kortekaas K (2021) Educar en la naturaleza. Plataforma Editorial, Barcelona

Hueso Kortekaas K (2022) El futuro de la educación en la naturaleza. In: Artero M, Calzada P (Coords) Dosier Graó Renaturalizar los espacios y los tiempos educativos, Barcelona, pp 97–100.

Hueso Kortekaas K, Mayorga Martín M (2020) Guía de escuelas en la naturaleza. La Traviesa Ediciones, Málaga

Janssen A, LeBlanc AG (2010) Systematic review of the health benefits of physical activity and fitness in school-aged children and youth. Int J Behav Nutr Phys Act 7:40

Johnston L (2019) Forest school pedagogy and indigenous educational perspectives: where they meet, where they are far apart, and where they may come together. In: Trifonas P (ed) Handbook of Theory and Research in Cultural Studies and Education. Springer International Handbooks of Education, Springer, Berlin, pp 1–14

Knight S (2009) Forest schools and outdoor learning in the early years. SAGE, London

Lanza K, Alcazar M, Hoelscher DM, Kohl HW (2021) Effects of trees, gardens, and nature trails on heat index and child health: design and methods of the Green Schoolyards Project. BMC Public Health 21(1):1–12

Larimore R (2016) Defining nature-based preschools. Int J Early Child Environ Educ 4(1):32–36

Leather M (2018) A critique of "forest school" or something lost in translation. J Outdoor Environ Educ 21(1):5–18

Lovell R, Roe J (2009) Physical and mental health benefits of participation in forest school. Countrys Recreat 17(1):20–23

Louv R (2008) Last child in the woods: saving our children from nature-deficit disorder. Algonquin Books, Chapel Hill, NC

Luís S, Dias R, Lima ML (2020) Greener schoolyards, greener futures? Greener schoolyards buffer decreased contact with nature and are linked to connectedness to nature. Front Psychol 11:567882

MacEachren Z (2013) The Canadian forest school movement. Learn Landsc 17(1):219–233

Martin P (2008) Teacher qualification guidelines, ecological literacy and outdoor education. J Outdoor Environ Educ 12(2):32–38

McCurdy LE, Winterbottom KE, Mehta SS, Roberts JR (2010) Using nature and outdoor activity to improve children's health. Curr Probl Pediatr Adolesc Health Care 40(5):102–117

Michek S, Nováková Z, Menclová L (2015) Advantages and disadvantages of forest kindergarten in Czech Republic. Procedia Soc Behav Sci 171:738–744

Miklitz I (2001) Der Waldkindergarten. Dimensionen eines pädagogischen Ansatzes. Neuwied Kriftel, Berlin

Miller NC, Kumar S, Pearce KL, Baldock KL (2021) The outcomes of nature-based learning for primary school aged children: a systematic review of quantitative research. Environ Educ Res 27(8):1115–1140

Mycock K (2020) Forest schools: moving towards an alternative pedagogical response to the Anthropocene? Discourse Stud Cult Polit Educ 41(3):427–440

Mygind L, Kjeldsted E, Hartmeyer R, Mygind E, Bølling M, Bentsen P (2019) Mental, physical and social health benefits of immersive nature-experience for children and adolescents: a systematic review and quality assessment of the evidence. Health Place 58:102136

Puhakka R, Rantala O, Roslund MI, Rajaniemi J, Laitinen OH, Sinkkonen A, ADELE Research Group (2019) Greening of daycare yards with biodiverse materials affords well-being, play and environmental relationships. Int J Environ Res Public Health 16(16):2948

Waite S, Bølling M, Bentsen P (2016) Comparing apples and pears? a conceptual framework for understanding forms of outdoor learning through comparison of English Forest Schools and Danish *udeskole*. Environ Educ Res 22(6):868–892

Wilson EO (1984) Biophilia. Harvard University Press, Cambridge, MA

The Art of the Possible. Imagining and Modelling Regenerative Cultures Through Participatory Experience

Joanna Crowson, Jorge Gallardo, and Andrew Zionts

Abstract Against the backdrop of the global environmental emergency and growing levels of climate and eco-anxiety, this chapter explores narratives and methodologies that, without denying the extent and complexity of the problems and solutions, motivate citizens to act. The three authors present the results of their qualitative research undertaken in community projects, including a training course for youth workers, regenerative agriculture initiatives and activities blending art and ecology, showing how a focus on shared processing of the emotional impact of the huge biodiversity loss we are experiencing can empower participants to become change agents. They describe their use of eco-literacy approaches and permaculture principles to develop deeper conversations around the issues with people of all ages and share some results from their on-going action inquiry into radical tenderness, decolonial futures and taking collective thinking from sustainability to regeneration, from gratitude to reciprocity, and from denial to acceptance of our radical interdependence with nature.

Keywords Eco-literacy · Permaculture · Decolonial futures · Deep ecology · Art and community · Action research

---

J. Crowson (✉) · A. Zionts
Almenara Culturas Regenerativas, Vejer, Cádiz, Spain
e-mail: joannacrowson@gmail.com

A. Zionts
e-mail: info@almenara.online

J. Gallardo
Párpado - soñando en el umbral, Vejer, Cádiz, Spain
e-mail: parpado.net@gmail.com

J. Crowson · J. Gallardo · A. Zionts
Be.Time SCA, Vejer, Cádiz, Spain

© The Author(s), under exclusive license to Springer Nature Switzerland AG 2025
C. Cobreros et al. (eds.), *Regenerative Design*, Cities and Nature,
https://doi.org/10.1007/978-3-031-76890-3_19

1 Introduction

La Janda is one of the six regions of the province of Cádiz (Andalusia), in the south of Spain. A large part of the region is included in the Parque Natural de los Alcornocales, which covers 173,619 ha of protected land, home to the largest cork oak forest on the Iberian Peninsula and a variety of landscapes, fauna, flora and Mediterranean and riparian woodland. The other natural park in the region is La Breña y Marismas del Barbate, located in the municipalities of Barbate and Vejer de la Frontera. This maritime-terrestrial park covers 5077 ha, 1152 marine and 3925 terrestrial. It protects the pine forest of La Breña, between Los Caños de Meca and Barbate, together with the cliff and the maritime strip that borders it, as well as the marshes of the Barbate River, which extend from La Barca de Vejer to the mouth of the river, bordering the hills of the Sierra del Retín. The area that corresponds to the former Laguna de La Janda is still an important natural area. Once one of Europe's most important wetland sites due to its position on the main bird migratory route between Europe and Africa, it was drained under an illegal concession in the 1960s during Franco's dictatorship, and since then has been used for intensive agriculture, including growing rice. After a lengthy campaign, a recent court decision has led to a project to recover part of the wetland. La Janda is a region with a markedly agricultural and livestock farming economy, which in recent years has come to depend increasingly on intensive coastal tourism, with the construction boom, inflated rental costs and water problems that come with that model. The other characteristic economic activity is traditional fishing, with the historic *almadrabas* of Conil, Barbate and Zahara de los Atunes being of particular importance.

We (the three authors of this article) either have strong family connections to the area or have been living and working here for over 30 years. We each have our own permaculture project, and collaborate in different ways in cultural, activist and permaculture activities. However, in 2019 our shared work began to feel more urgent. The publication of the IPCC's *Special Report: Global Warming of 1.5 °C* (2022) and Jem Bendell's *Deep Adaptation* paper in (2018) led us to initiate a series of conversations with our neighbours and other collaborators that somehow felt unsatisfying, quickly polarising into stories of collapse, feelings of hopelessness or what seemed like overconfidence in technological solutions giving rise to business-as-usual responses. Our research had shown that during the last glaciation (in the early Neolithic period) the southern part of the Iberian Peninsula was an oasis of biodiversity from which life could re-emerge to repopulate Europe as it was freed from the ice. We started to wonder how we might preserve the natural resources found here as a tool against mass extinction while also tackling the realities of the need for mitigation and adaptation in the face of the climate crisis. This was the context of the initial research question that we focused on as we began this new phase of collaboration: **How can we have deeper conversations with our neighbours about the gravity of the climate and ecological emergency that motivate us all to act?**

2 Setting the Scene

As change agents, we have an idea of what we are trying to achieve, but often the change we are pursuing is systemic and long term, and this makes it difficult to assess the impact of specific actions. This is clearly the case in the context of the climate and ecological emergencies. If you devise a business marketing plan, you set your aims (so many clicks, so many enquiries, so many conversions, etc.), and depending on whether you meet your target—your KPIs—you evaluate your campaign and either tweak, change or repeat it. Of course, we too set some objectively measurable targets—so many people attending an event, so many talks given to local groups and so on. However, we wanted to achieve better **quality** conversations, which is essentially a subjective aim, in other words, qualitative not quantitative. We also wanted these conversations to motivate action in the context of the climate and ecological emergencies, but action can mean different things to different people. Add to this the fact that a society is a complex system, something "organic, adapting, becoming, and emerging", not something mechanical, "predictable, where change goes to plan and in which the cogs in the machine, including people, are unchanging and controllable" (Boulton et al. 2015). In the reality of the complex world, it is difficult to research social interactions and processes objectively, but if we are to identify what works and what doesn't, then we need to pay particular attention to this complexity. Boulton et al. (2015: 112) ask:

> What would a research process able to deal with the complexities of the world have to do? It would have to try to give room for, and not constrain, the emergence of something new during the investigation, and it would need to make more allowance for the unexpected. It would need to follow the process of the emergence of new factors and outcomes. It would need to pay attention to the systemic nature of the situation and not require too narrow a path of inquiry, but allow the taking of detours and the consequent widening of the factors of interest. It would facilitate the noticing of emergent contextual features or of new entrants into the situation. It would recognize the importance of the detail—the particularity, the granularity of the conditions and events, and the particularity of the order in which things occurred, the sequence of events.

In the world of social science much work has been done on qualitative research techniques and we have mainly drawn from the fields of action research and narrative research. For us an important aspect of action research is that it makes no pretence to being objective, in fact it aims to work in the messiness of real life, connecting:

> …practice and ideas in the service of human flourishing … engaging those who might otherwise be subjects of research or recipients of interventions to a greater or less extent as inquiring co-researchers. Action research does not start from a desire of changing others 'out there', although it may eventually have that result, rather it starts from an orientation of change *with* others… Action research primarily arises, however, as people try to work together to address key problems in their communities or organisations—some of which involve creating positive change on a small scale and others of which affect the lives of literally millions of people. (Reason and Bradbury 2008: 1)

Narrative research and case studies are methods that are particularly relevant to researching complex situations (Boulton et al. 2015) since they help to collect multiple causes and multiple outcomes. They follow the time sequence of events, often from multiple perspectives bringing in diverse voices, highlighting what has happened and what is beginning to emerge around a particular initiative. They also allow for the recognition of an extended epistemology (Heron and Reason 2008), valuing the different ways of knowing of participants in the inquiry, and place value on the full range of human experience, including some of the difficult emotions that come up in the uncertainty that arises in the context of the climate and ecological emergency.

An important area of work for us has been how to deal with the emotional response that naturally appears when we stop denying the scale and depth of the eco-social crisis we are facing. This can feel like a big responsibility because we can't make everything OK. Our approach has been to make space for feelings of fear, grief, anger and so on, to recognize them in ourselves and others and to accept them and work with them, seeing them as important information and the other face of the love we feel for the Earth. Joanna Macy's work has been fundamental to this approach, and we have drawn on tools from deep ecology and The Work that Reconnects methodology (Macy and Brown 2014), including the Council of All Beings[1] (Seed 2007). This has been a rich area of exploration and growth, and exercises drawn from our personal mindfulness practices teach us how to be aware and pay attention in difficult moments, staying present and allowing time and space for feelings to be processed.

Our permaculture practice has had an important influence too, in that it has led us to aspire to apply the ethics, principles, tools and practices afforded by permaculture design to every area of life. Although the three of us each initially started with our personal domestic or land design projects, we noticed a shared growing resentment of the diluted idea of sustainability as sustaining a given situation over time (including what may be an impoverished and degraded landscape and/or community). Collectively, we aspire to create our own local version of a regenerative culture, as defined by Wahl (2016):

> A regenerative human culture is healthy, resilient and adaptable; it cares for the planet and it cares for life in the awareness that this is the most effective way to create a thriving future for all of humanity.

In this chapter, we follow three different case studies: Catalina regenerative agriculture project; BarbaT arts, ecology and participation project; and our local activism through collaborations with the Mesa Redonda de Permacultura (Permaculture Round Table) and Extinction Rebellion (XR) La Janda. We will use a chronological approach that will braid together these three threads which demonstrate three different aspects of a regenerative culture in combination with permaculture's triple ethic (Holmgren 2002): regenerative agriculture (earth care), regenerative community culture (people care), and regenerative activism (fair shares or future care).

---

[1] The Council of All Beings is a communal ritual in which participants step aside from their human identity and speak on behalf of another life-form.

Known as 'Catalina' in homage to a local 15th century woman landowner, the first case study is a land-based regeneration project located near the town of Vejer de la Frontera (see map in Fig. 1) that is exploring different forms of community life and collaboration with nature, based on the principles of permaculture and developing specific practices in relation to the location and condition of the land. A number of groups, volunteers and specialists have been working on the project since 2014, led by Andrew (one of the authors of this chapter). The work has involved implementing water retention techniques, reforestation, regenerative agriculture and composting, as well as deep ecology and group management through sociocracy. In 2019 the original 4-hectare project expanded to include a further 9 ha of land. Although this included some native forest on the western slope of Vejer, much of the land had been deforested, was overgrazed, and was at risk of severe erosion. The project is based on restoring the functioning of the ecosystem, albeit on a small scale.

BARBA-T, the second case, is a multidimensional programme of cultural actions organised by Be.Time, the cooperative founded by Jorge Gallardo (again, one of the authors of this chapter), Karmit Evenzur and Polina Stoynova, in which the other two authors participate, among other agents and citizens. The programme connects scientists with artists and inhabitants in the basin of the Barbate River. Its aim is to generate love for the territory and make its wildest and most tender side visible. Its first two iterations took place in Vejer over 3 months in 2020 followed by a further 3 months in the town of Barbate at the beginning of 2021. The programme continues to breathe, almost in secret, longing for an encounter with the next town in the shared basin and its coasts and hills.

The third case study is related to the local activism in which the three of us are involved. This goes back to 2011 when Joanna (the third author) co-created Encuentro Gaia Spirit, a three-day festival focused on three pillars: care for the environment, care for emotional and physical well-being, and arts and culture to cover the vital role of shared fun and celebration. This event gave rise to the first Permaculture Design Certificate course in the area, which ran one weekend a month for 9 months on Joanna's land. When the course came to an end, Joanna, Andrew and other graduates founded the Mesa Redonda de Permacultura de Cádiz, a local group with a small core of around 10 permaculture practitioners in the area. Our initial work was with permaculture, mostly through outreach activities designed to share knowledge and ideas, carrying out volunteering days, permablitz[2] events, talks and workshops. It was during this time that our thinking moved from a focus on sustainability to regeneration—to the aspiration to leave the world better than we found it. At the same time, an increased sense of urgency led us to scale up our activities with a focus on the climate and ecological emergency. However, as stated at the start, we were very eager to create deeper conversations around the issues: conversations that could invite curiosity about the situation and a desire to explore and develop potential solutions, rather than polarising and ending in denial,

---

[2] An informal gathering in which a group of volunteers come together to learn and share skills related to permaculture and sustainable living.

Fig. 1 Map showing the area, with the arrow indicating the location of Catalina, one of the case studies discussed in the chapter (© OpenStreetMap)

feelings of impotence or a false sense of security. As Peter Hawkins shows in his Eco-Phase model (Whybrow et al. 2022), curiosity is a great place to start a conversation or process, opening the way to exploration and the seeking of information, before developing awareness in everyday life, building into engagement and finally action. Since our aim was to generate local action to mitigate and adapt to the climate and ecological emergency, this was the approach we decided to experiment with.

This chapter is an invitation to accompany us through a series of action learning experiments that we carried out in the three areas mentioned above between 2019 and 2022. We will then share a series of conclusions and lessons learned that we

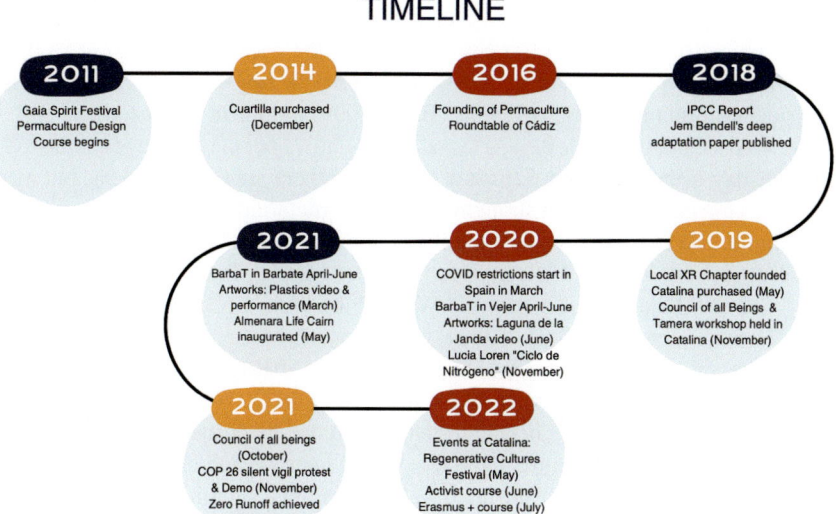

Fig. 2 Chronology of events (Andrew Zionts, 2022)

will be feeding back into our on-going work. For an overview of the chronology of events, please see Fig. 2 Timeline.

3 Reset: From Sustainability to Regeneration (2019)

As stated earlier, our starting question for this initial phase is: How can we have deeper conversations with our neighbours about the gravity of the climate and ecological emergency that motivate us all to act? Our permaculture group—Mesa Redonda de Permacultura—is an experiment in horizontal organisation that in 2019 has over 200 members. However, it has experienced a loss of purpose and a narrowing of vision: our meetings tend to be social gatherings, and there is often no focus on permaculture content. Many new members have no knowledge of permaculture but are attracted by its association with organic gardening and the idea of living in the countryside. At the same time, Andrew's acquisition of neighbouring land leads to the expansion of the Catalina project, so we take this as an opportunity to refocus our outreach on permaculture design tools. Our own inner work at this time has led us to a shared sense that we need to connect to the wider community of life to nurture our action. In addition to our shared desire for deeper conversations, we also have the following specific question:

- How can we broaden the understanding of permaculture as a design framework to help respond to today's crises and bring permaculture tools into use in designing a resilient culture locally?

After the purchase of the land is finalised, we decide to hold two events with a double objective: one is to offer our neighbours the opportunity to engage in the on-going design process, and the other is to bring the voices of the land itself and the wider community of living beings (Berry 2009) to inform the design.

We hold a Council of All Beings (Seed 2007) and 41 people of all ages come to the free half-day event, held on a Saturday morning—there are familiar faces, but many are new, and the playful but thought-provoking methodology provides a fun and meaningful encounter that widens the focus of the project. The following voices are represented: Soil, Thyme, Flower, Tree (3), Life, Mother, Pain, Salamander, Eagle, Ether, Subsoil, Bird (3), Underground Water, Pine Tree, Nocturnal Animal, Mastic Tree (2), Sun, Tree Bark, Bird of Prey, Owl, Organic Matter, Water, Moss, Factory Pig, Snake, Waterfall, Fig Tree, Life Cycle, Spirit of Raptors, Hen, Star, Canary, Hollow Tree Trunk, Earth, Snail and Gorse.

Andrew has been training in water retention landscape design with Tamera Peace Research and Education Centre and a few weeks later we hold a workshop during a visit from Tamera's experts featuring a walk that connects intuitively to the land and its needs. Some 20 people participate in this. There is a small amount of overlap, but again we see new faces.

We're pleased with the level of engagement for both these events, and we notice that two elements come up repeatedly: the issue of the role of water in the much-needed regeneration of the soil and the importance of local participation in the Catalina project. This leads us (the authors) to frame two new questions:

- Can we restore the functionality of our ecosystem and see springs come back in our lifetime?
- How can we continue to include more voices in the design of Catalina, including those from the non-human community?

We set our first objectives: a zero-runoff design and the creation of Almenara, which is both a place, located on the hill on the edge of Catalina, where the land meets public road, and an outreach project. *Almenara* is the Arabic-Spanish word for the beacon fire that can both raise the alarm in warning and be a beacon of hope calling us home, and for a traditional irrigation system—in time it will be the name of the legal body we plan to set up as an umbrella organisation for our activities.

In mid 2019 we create a local Extinction Rebellion group, XR La Janda, under the slogan "*Somos naturaleza defendiéndose*" (We are nature defending itself) and focus on the first demand "Tell the Truth". We set the objective of holding conversations with 100 people around the climate and ecological emergency before Christmas 2019. We hold large meetings in our local towns of Vejer, Barbate and Conil and achieve reasonable attendance (approximately 90, 20 and 30 people respectively). As a result of these, we are also invited to give other talks, including in our local high school (for about 100 12-year-olds), to a gathering of rural women's associations (some 200 women), a 3rd-age conference (about 50) and in a regional aviation company (another 50 or so).

People need more information about the issues, but most of the people who attend are already familiar with much of the science, and we observe that what they really want is to discuss what's happening. We change the format of our talk over time because we see that making space for grief and connection takes the conversation deeper and makes the problem feel more personal. We still provide data but leave increasing space for feeling and responding. We identify the need to create more informal spaces for conversation and so we design and get partial funding for a series of community events to be held in 2020 around regenerative cultures. At this time, Be.Time gets funding for a its hyper-local project—BARBA-T · Art, Science and Participation in the River Barbate Basin—a Citizen Expo-Laboratory.

Summarising our achievements in this first phase, we identify some clear quantitative outcomes—we reach many more people through our talks than expected, exceeding our target, and new people are attracted to our permaculture activities, some of whom join the group. In terms of content and processes, the three of us develop our skills around creating space for dealing with emotional responses. In qualitative terms, we are satisfied to observe that conversations slow and deepen as we are better able to hold the space.

4 The Pandemic Slows Us Down (2020)

We are in full flight at the start of 2020 with a calendar packed with activities and actions scheduled until the end of May, but the lockdown brings uncertainty, cancellations and changes to plans. Flexibility is highlighted as an important aspect of resilience. We want to continue with our work to deepen conversations through acknowledgment of the emotional impact of the ecological crisis, and art is the vehicle we are best able to work with in this tricky time.

The first stage of BarbaT in Vejer has to go mostly online. Even so, some collaborations can happen:

- A collaborative ephemeral artwork with a focus on eco-literacy and permaculture design is created outdoors involving 18 volunteers over one week—El Ciclo del Suelo Vivo by the artist Lucia Loren. It forms the centre of Catalina's mandala garden (see Figs. 3 and 4).
- Be.Time and XR La Janda collaborate with 18 local artists and activists to make the video: I Dream of Being What I Was, in defence of the Laguna de La Janda.[3] This is produced in English and Spanish, has a place in the Be.Time exhibition and is uploaded to Be.Time's YouTube channel, as well as the channels of XR UK, XR Spain and XR Cádiz. It has had approximately 4000 views online.

---

[3] BEE TIME · PROYECTOS DE INVESTIGACIÓN EN ARTE Y ECOLOGÍA. (2021, August 17). *Sueño con ser lo que era—BARBA-T · Arte, ciencia y participación en la cuenca del río Barbate*. BARBA-T · Arte, Ciencia Y Participación En La Cuenca Del Río Barbate. https://barba-t-expo.beetime.net/proyecto/sueno-con-ser-lo-que-era/

Fig. 3 Lucía Loren's ephemeral artwork, part of BARBA-T Art, Science and Participation in the River Barbate Basin—a Citizen Expo-Laboratory (Fig. 3 Alberto Omiste, Be.Time SCA, 2020)

Fig. 4 Andrew Zionts, 2020

Reforestation work and water retention design and implementation continue in Catalina. Although we continue our work, and we reach people through participation in the creation of the artworks, volunteering days for tree planting, school workshops and through views of the video, this also feels like a period of gestation. On-line networks are very important in this period and surprise us by their power to bring diverse voices into our work as well as many new ideas. Joanna particularly highlights the importance of her participation in CliFiWri (a climate fiction writing group with eight members that meets online on the second Tuesday of every month) and in A School Called HOME, "a gathering place and learning community for those who are drawn to the work of regrowing a living culture" and which offers a Mighty social network and monthly conversations on Zoom led by Anna Bjørkman and Dougald Hine. These two virtual communities play a vital role as sources of inspiration and learning in the context of sitting with and responding to the eco-social crisis. Jorge's participation in Cubo Verde, a national network, involves research dealing with the relationships between institutions and art projects in the field, led by long-term cultural mediators and brings an important regional context to our work. The support Andrew receives from the Tamera team helps keep the Catalina project on track and consolidates the learning experience at a distance.

Summarising our achievements in this difficult second year, we create and disseminate two works of art that deal with the eco-social crisis and eco-literacy, as well as holding small scale actions and activities that reach small groups. This feels like a big achievement, given the pandemic. In qualitative terms, we continue to develop our own exploration of the dimensions of the eco-social crisis through our networks, using the spaciousness suddenly created by the cancellation of most group activity to seek learning opportunities beyond our hyper-local situation.

5 Gathering Pace: From Gratitude to Reciprocity (2021)

In the early part of the year restrictions come and go, but we manage to increase our level of organised activity as the year goes on. The three of us add new questions to our shared exploration:

- How can we extend the work to feel grief in connection to the climate and ecological crisis to the younger generation?
- How can we handle this information with others without generating paralysis?

The second stage of Be.Time's BarbaT project takes place over 3 months in Barbate during spring, this time with an exhibition that can be visited throughout and a series of workshops that involve working with information about the eco-social crisis with school children from both localities (Barbate and Vejer). Focusing on the water management of the River Barbate, students explore how climate change is affecting the population. In order to generate bonds of union and affection with the territory, one of the workshops is dedicated to the collective creation of an emotional map. These workshops, each with three sessions, are attended by 769 students.

Fig. 5 Almenara Life Cairn installation and inauguration ritual (Fig. 5 Polina Stoynova, Be.Time SCA, 2021)

The second phase of BarbaT comes to a close with a collaboration between Be.Time and XR La Janda through the creation of Almenara—Monumento a la Vida, a life cairn that is both an art installation and focus for ritual (see Figs. 5 and 6). A text by a local poet is used on the plaque, and this attracts the interest of many townspeople, with 60+ attending the inaugural outdoor event. Placed on liminal land on the unmarked border between public space and private land (Catalina), it thrives on the uncertainty of ownership and receives many visits, including from mysterious carers who bring new stones, tidying up the space and leaving stones in beautiful patterns. During this time, Be.Time collaborates with XR La Janda, XR Bielefeld and the Barbate group of Nasty de Plastic in an activist performance around plastic pollution, a beach clean-up and an information stall in Barbate's fish market (see Figs. 7 and 8). Due to the restrictions that come and go, the first live performance is seen by very few people, but the video "We're Drowning in Plastic[4]" has achieved around 2000 views. A handful of people participate in the beach clean-up and the information stall is visited by a smaller number of shoppers than we'd hoped, probably due to social distancing regulations in force at the time.

The work in Almenara with XR, and in the school workshops offer us spaces in which to integrate imagination and mourning in processes that help manage disturbing information derived from scientific research on climate change and the predictions that it produces for the territory. With COP26 on the horizon, we start

---

[4] Extinction Rebellion UK. (2022, February 4). *PROTECT OUR OCEAN!|Extinction Rebellion UK* [Video]. YouTube. https://www.youtube.com/watch?v=QZ-v72t0uQE

The Art of the Possible. Imagining and Modelling Regenerative Cultures … 455

Fig. 6 Alberto Omiste, Be.Time SCA, 2021

Fig. 7 "We're Drowning in Plastic" performance (NUDO, Be.Time SCA, 2021)

to wonder what we can do to encourage our neighbours to engage further. We hold another Council of All Beings activity at Catalina in October. This time some 27 people of all ages participate, representing the following voices: Stick Insect, Root, Bald Ibis, Ant, Wild Olive, River, Sap, Wolf, Owl, Cliff Rock, Mastic Tree, Bee, Wind, Oil, Holm Oak, Caterpillar, Sun, Water, Fungus, Waterfowl, Stone Pine, Decomposition, Bald Mountain, Thistle, Unity in Diversity, Tree and Earth's Crust. We use the recording to produce a communique from the community of living beings

Fig. 8 "We're Drowning in Plastic" info stall in Barbate's fish market (Polina Stoynova, Be.Time SCA, 2021)

of La Janda with a message for the humans attending the summit. The document goes out to our local schools, as well as being read in Glasgow at several fringe events—it even travels to Cornwall (UK), and is read out in an arts and ecology event in a town square. We're especially delighted when several teachers contact us to say they plan to work with it in their classrooms.

With our local XR group we decide to hold a Silent Vigil in the square outside Vejer's Town Hall throughout the entire COP26, from November 1st–12th. The idea is to have a single person dressed in Red Rebel costume sitting/standing in silence from dawn till dusk, looking up at the Town Hall. Our central demand is for our Town Council to take action by drafting the local climate adaptation plan they are legally required to produce, and our campaign slogan is COP26: *Inacción = Muerte* (Inaction = Death). The Red Rebel costume symbolically represents the community of living beings of La Janda, expanding the notion of agency and affected parties, and also means we can take it in turns to play the role, so we can sustain the action for the full 12 days. We convene a demonstration for the final day.

This poetic, symbolic action inspires the participation of 19 people who take it in turns to stand in the square. The silent figure has power to inspire curiosity and being solitary, is not too intimidating, so every day many different people approach to ask why we're there. Silence in the face of a direct question feels almost violent, so we give a simple answer based on our feelings and hand out a leaflet with more information. The constancy of the action over time builds momentum, and we achieve some presence in the press (a first). All four classes of the town's 11/12-year-olds make animal masks at school and come to the demo to participate in the name of species at risk from climate change and habitat loss, taking it in turns to read out sections of the communique (see Figs. 9 and 10). We are delighted by their engagement (it's the best activity of the year, we're told) but disappointed by the lack of adult presence.

The year closes in December with torrential rain that causes damage and flooding in the area, however observation at Catalina shows zero runoff has been achieved.

Summarising our achievements in this third phase, we can again identify some quantitative outcomes. Zero run-off is achieved at Catalina, which translates to the capture and infiltration of 104 million litres of water in a year of average rainfall. We also create and disseminate further works of art and performance that deal with the eco-social crisis. We reach many new people in ways that allow for deeper dialogue around the ecological emergency, including 769 schoolchildren through workshops, and at least another 100 12/13-year-olds through their response to the Communique from the Beings of La Janda. Last but not least, we finally get some press coverage for our activism, with articles appearing in at least four local and regional newspapers. In terms of content and processes and qualitative outcomes, we explore and confirm the impact of ritual and poetic symbolism on our communication with our neighbours, discovering that it inspires others to participate in our actions, and excites curiosity so people approach us to find out more. Both adults and children seem moved to (symbolic) action when the starting point is gratitude and connection to the wider community of living beings.

Fig. 9 Masked children during COP26 silent vigil protest (NUDO, Be.Time SCA, 2021)

Fig. 10 Red Rebels during COP26 silent vigil protest (NUDO, Be.Time SCA, 2021)

6 New Organisational Capacity: Radical Interdependence (2022)

After two years exploring small-scale actions, we can now begin to weave our learning into more complex projects.

Andrew and his team celebrate the achieving of zero runoff at Catalina by formulating a new question:

- How can the succession to a closed canopy forest be accelerated to change our microclimate in our lifetime?

Since Catalina can now open to visitors and host events, we start planning. The first thing we design is a series of small permaculture events with the Mesa Redonda de Permacultura, culminating in a Regenerative Cultures festival—around 125 local people attend this free, one-day event in May during which 25 people from local environmental projects present their response to the ecological emergency through workshops and a human library. We gather feedback from participants and the overwhelming feeling is positive (see a collection of the feedback in Table 1). We plan to repeat the experience annually.

We follow this a month later by running a three-day course for activists called "Creating Regenerative Cultures". Over the long weekend, we facilitate a Work That Reconnects (WTR) spiral (Macy and Brown 2014) for 17 activists from all over Spain. We use an online survey to gather feedback on infrastructure (it's the first time we've carried out a residential event at Catalina) and responses in general are positive with some specific suggestions that are useful for us, since we are programming another

Table 1 Feedback from participants in Regenerative Cultures festival

Examples of feedback received from participants in the Regenerative Cultures Festival on 7th May 2022:
"It feels very comforting to know that so many people are doing their little bit to make this planet a better place."
"Prior to attending, I knew nothing about permaculture or sustainability. It was incredible to see how many people came together to learn about this mission."
"These kinds of encounters give meaning to our business and life project (which go hand in hand), as an encouragement and support from the community to continue with a vision of sustainability and resilience in its broad sense, through design techniques, group organisation mechanisms, and ways of communicating and giving each other warmth. In our case all these inputs are of particular importance, as we are developing an eco-tourism camp with a nature classroom, in which we want to prioritise raising awareness of the state of the environment we live in at different scales, and what we can do to seek healthier alternatives."
"I distinctly remember saying … "This has been the best day of my life!" My favourite part of the whole day was dancing with everyone as the sun went down over the mountain. It's one of my happiest memories."
"Joy, to be with people who have many things in common and care for the environment. Curiosity, to know the concrete actions that are being carried out in the area."
"… a very beautiful day, of meetings and reunions with dear, appreciated and unknown people, but all united by our purpose of improving ourselves and our planet."

residential activity within a month. The feedback on content is helpful too, with participants asking for a reduction in pace:

> a little more relaxed schedule to be able to integrate the contents and meet the people

and more bodywork:

> I think that some exercises focused on the body and movement interspersed with those that were being done could help to integrate the mental and conceptual aspects of the work…I missed at some point leaving the head and being more in the body…to express myself through the body instead of words…

Participants also expressed their appreciation of the process with comments like this:

> "When I see the generosity in front of me it feels so good in my body that I want, need and cry out to life to more consciously give more back as well, thank you I am grateful with an open heart."

However, our later attempts to get wider feedback on impacts are met with no response, and we identify the need to improve our practices for gathering narrative feedback and impressions from participants.

In July we launch New Eco-Narratives for European Youth[5]—an intensive one-week course funded by Erasmus+ during which four facilitators and 16 participants from eight European countries working with young people and other groups around the climate and ecological emergency meet to explore the huge ecological and social predicament we are facing using varied lenses that might help us deeply understand

---

[5] https://www.parpado.net/eco-narratives/

where we are and where we are headed as well as dealing with the feelings and emotions this brings up. The facilitation draws from permaculture, deep ecology and WTR, as well as environmental constellations, mindfulness and bodywork practices, art and ecology and work on decolonial futures. Through fostering a deep sense of radical interdependence with nature and belonging to the world, we hope to mitigate against denial of the scale of the problem and/or burnout. We gather initial feedback through an improved online survey and three months later, narrative feedback on more lasting impacts.

The feedback immediately after the course is generous and very helpful. It's clear that the process has been useful and nurturing for participants, but also challenging, and for future iterations our programme design will require the inclusion of even more bodywork and spacious presence for some people. Interestingly, as was the case in the previous weekend course, the deep ecology activities receive a slightly lower valuation overall (see Fig. 11), and we wonder whether for certain profiles they may prove more challenging in terms of the current paradigm, or whether there is an issue with framing here, since WTR continues to be highly valued in spite of large overlaps.

The feedback received shows that participants especially valued the chance to get to know each other through the presentation of their personal narratives. This participant's extensive feedback is fairly typical in tone:

> Without wishing to sound dramatic, I feel that sharing my Personal Narrative has, in a really profound way, changed my life. Never in my life have I been given such an open floor to speak about my experiences … I think that the environment I was in, and some of the WTR activities we'd already done earlier in the week helped lay the groundwork for this. The other participants on the programme also deserve so much credit in the process of the Personal Narratives experience—I knew before I started speaking that I was in such a safe space, and that nobody in the room was going to pass any negative judgement on anything I might say…The love we all received and were able to share in the moments and days after sharing our narratives was also really beautiful, and I feel that the opportunity to listen to each other helped us truly understand and connect with one another in a way that we might not have fully been able to otherwise. Certain things that other people said in their narratives really shaped me too…

As we collaborate, we have been aware that our ultra-local focus has advantages for building community, and one of our on-going aims has been to visibilise our work so that our neighbours can see what we are doing, in an effort to create inspiring replicable interventions that have been tested and shown to work. However, one of the profound realisations that came from this international activity was that by building our skills as we work together, we can create situations that can create a ripple effect and have impact far beyond our immediate reach. This is important to us, and in fact is one of the reasons for writing this chapter to summarise our research. To test our perceptions about the impact of the course on praxis, we asked participants for feedback on the lasting impact three months later and we were delighted with both the number of replies and what participants had to say.

In answer to the question about whether participants had applied what we learned together in their work with young people or in their work in general, one participant described facilitating connecting to nature in a children's summer camp, while

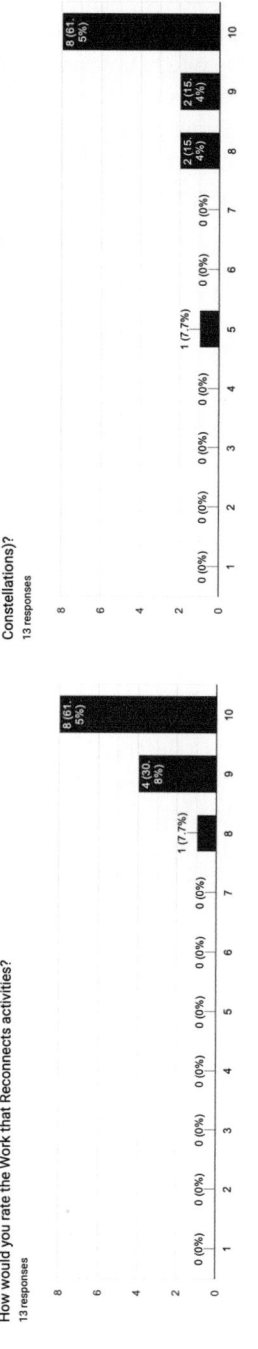

Fig. 11 Comparison of feedback on WTR and deep ecology activities

another was using ideas around listening to ancestors in their work with businesses, graduate recruits and schools. One described using some of the ideas presented in her reading group and in a series of talks she had felt inspired to deliver, while another described creating a video tutorial called Drawing Inspiration from the Wisdom of Woodlice, for the Irish Schools Sustainability Network's Climate and Nature Summit which includes elements of the training course.

When asked about whether the learning had ongoing value and what they had taken back to their personal life, one participant had discovered:

> ... the pain that I feel may not just be my individual pain, due to my personal history and emotions. It might very well be a pain that is mine, because I am an aching limb of a wider body which is deeply in distress. **I am not on the Earth, I am not of the Earth, I am the Earth.** The Earth suffers, I suffer. As simple as that. We are nature defending herself. ... Pain can very well be a healthy response to a distressed situation. Realising this has been extremely liberating for me.

Others described creating new mind maps, experiencing paradigm shifts and learning to develop the capacity to sense deeply through the body and thus listen to the wider system. Here's one participant on her experience:

> [It was] deeply empowering. What I took home with me (and still prevails) is a sense of 'emerging transformative communities', but also the urgency to involve the majority of society in transformations that will reach us whether we like them or not. The question is no longer whether these transformations will come but how they will find us and how 'well-equipped' we will be, practically and emotionally. The 'eureka moment' that every single person will undergo as a member of our societies with regard to the eco-social crises may entail pain and solitude, and thus destruction, or gratitude and community, and thus life.

One thing that seems significant is a shift participants experienced in their attitudes to themselves. It's not uncommon for activists to suffer from burn-out, often because they feel guilty about stopping and taking a break. In the face of so much destruction, it's easy to feel that time out is wasted time. One participant said the week had helped her to reframe the concepts of illness and wellness, and to understand better what she needed for her own healing. Another said:

> ...in one of the closing activities we were asked to formulate some kind of commitment to ourselves and to the world ... I had a hard time narrowing it down to a single sentence, but in the end, I came to the word: 'courage'. And yes. Within my limitations I think I am trying. What's more, sometimes I think that to continue living and militating in the face of so much migrant loneliness and in the context of so many interconnected crises is already courage.

One of the most satisfying outcomes for us has been the answer to our early question around broadening the understanding of permaculture as a design tool with an important role to play in creating a regenerative culture. We were pleased to hear how the work with Holmgren's 12 permaculture principles and the permaculture flower (Holmgren 2002) led several participants to adopt these tools in their professional work. In her paper summarising her "R-Rural" action-research project funded by two public universities in Andalusia which included Be.Time's work as a case study, Rocío Nogales highlighted that "the 12 permaculture principles were an interesting compass that allowed me to explore their possible application in other areas, such as art and culture" (Nogales Muriel 2022), drafting a table that offered a possible adaptation of these to artistic and cultural praxis. Another told us:

> I have introduced the seven domains of permaculture[6] in the way I re-asses and re-design the work I do. Taking into account what we learned about the domains during the course has helped me to have a more holistic view on the work I do with designing spaces for systems change and I now offer this view for people to see whether their projects have blind spots and which domains need more attention.

Summarising our achievements in this last phase of our research project, we again have some clear quantitative outcomes. Our one-day Regenerative Cultures Festival saw the participation of 125 people plus activists; 16 activists and 3 facilitators participated in our three-day Creating Regenerative Cultures residential workshop; and 16 youth workers and 4 facilitators participated in our week-long New Eco-Narratives for Youth international residential course. In qualitative terms, we identified the need to improve how we collect feedback—professionalising here will enable us to increase our impact, access to funding and the replicability of our actions. We have identified the power of certain activities to increase our impact: the Regenerative Cultures festival provided support for other groups to expand and consolidate their activities (even leading to a similar event in the northern part of the province following our model), and the international course showed clear evidence for its ongoing ripple effects.

7 Conclusions

This chapter is an exploration of a three-year window on our ongoing work. We have paused to summarise our progress several times already, and this seems an appropriate place to close and to share some tentative conclusions that will be informing our next steps. Although our research has largely had a local focus, and is therefore very context specific, we believe that it provides insight into the results of working with a variety of tools that can easily be adapted to other contexts, feeding into a variety of approaches towards building diverse regenerative cultures everywhere.

We started this chapter by describing the move from an aim of achieving sustainability to one seeking regeneration, and as our research continued, we also identified other attitudinal shifts needed in our current paradigm, in particular the need to move from gratitude to reciprocity, fully exploring the ramifications of interdependence and interbeing in our work with others. It has become clear to us that when people feel connected and that they belong, they see the value in their actions as a part of the system. In the face of such huge, system-wide problems, feeling a part of something bigger takes on a vital role.

In terms of the first of permaculture's three ethics—'earth care'—we see that communicating the urgency and importance of water retention landscapes and

---

[6] The participant is referring to the Permaculture Flower that features in David Holmgren's book (2002).

working to restore the broken water cycle is an important first step in restoring the functionality of ecosystems in our area. Implementing this work and creating a zero-runoff landscape has been a fundamental base for our work, however it is incomplete without our next stage of accelerating the succession to closed canopy vegetation systems. In our efforts to inspire people by sharing a scalable and adaptable/replicable model, we have also found that creating a landscape where the improved health of the landscape and ecosystem is clearly visible to others is the most effective way to create impact.

There is an increasing desire and demand for people moving to rural spaces to find resources to support this transition; at the same time there is a lack of an understanding of permaculture as a holistic framework which addresses this demand. Our local group needs more training to understand and apply the design methodology so that we can have a much greater impact beyond being a local group providing mutual support. It's also valuable to share the tools of permaculture with people who are not necessarily interested in a back-to-the-land solution, since as a 'pair of glasses' the permaculture principles have applications in every sphere of community design.

In this area, our reliance thus far on narrative and visual feedback is a limitation to our work; we also need to work with scientific indicators to have a more universally recognizable way of evaluating our impact. We need to do more work on eco-literacy and citizen science, to improve our own skills as well as those of others so they can assess their own impact. Furthermore, in a world dictated by economic markets, our work needs to be economically viable to be valued and implemented by early adopters. Though we understand the 'invisible values' such as ecosystem services, increased biodiversity, etc., these don't readily translate into economic benefits for agriculture projects. Creating economically productive models of regenerative agriculture will be an important step for growing regenerative cultures, and we will be focusing more on this in future.

In the areas linked to the other two permaculture ethics—'people care' and 'fair shares' or 'future care'—we have found it difficult to separate our conclusions because they overlap so much. One tentative reason to explain this is the fact that so much of the work we have done through our arts and ecology work in the second case study has involved deepening our connection to nature and developing radical interdependence—in short it aligns very much with the activist work we carried out in the third case. For this reason, we present our remaining conclusions together.

We have observed that a focus on shared processing of the emotional impact of the huge biodiversity loss we are experiencing can empower participants to become change agents. Connecting to love is an important key to this. Our work with deep ecology practices connects the various threads of our regenerative work and helps us as individuals bridge inner landscapes to the outer landscapes and find purpose and meaning which informs both. Working with our neighbours to become citizens who belong and take responsibility rather than simple consumers is a powerful way to create change (Alexander et al. 2022).

In the course of our research, we repeatedly identified our own need to commit to our inner work to be able to facilitate and hold the space for others exploring a shift to a regenerative culture. One particular issue for us has been around the decolonial aspects of this work, and we identified the need to support each other in our exploration of radical tenderness. This led us to set up a small action inquiry group based on the work of the Gesturing Towards Decolonial Futures Collective using the Global Education Otherwise materials (2019) to better ground our future work.

We also recognise that we must take care of ourselves first. If we can't sustain our own activism, then having any positive impact on anyone else is unlikely. To quote Bayo Akomolafe (2022): "What if the way we respond to the crisis, is part of the crisis?" We see the need to take a completely new approach to the issues, in particular applying a different time scale to our expectations of results—thinking in terms of 'within our lifetime' and beyond. Collectively it will be helpful to expand our notion of agency and work with nature, placing trust in the Earth's capacity to heal over time. As Akomolafe says "the times are urgent; let us slow down" (2022), and for us this is reflected in our self-care. This means that we choose to produce events which are manageable and meaningful for us first and foremost, assuming that we are fractals of society, and that our work will resonate with others. We recently observed in our second edition of the Regenerative Cultures festival that we can organise these events and enjoy them with little or no stress and pressure if the design is right.

Lastly, finding creative ways to engage in conversations with people who otherwise would be unreachable has been a key to our growing impact. And while our hyper-local focus is vital for paradigm shift by focusing on moving from 'mile wide inch deep' movements to 'inch wide mile deep' movements (Brown 2017), taking on larger scale projects makes our impact more far-reaching. We see that our local impact can be amplified if we share our practices through funded, target, more complex activities which spread beyond our local networks, and we find this wider impact encouraging.

8 What's Next for Us?

Catalina's closed canopy question is evolving into a permaculture and citizen science project to test two approaches to reforestation adapted for our climate: syntropic agriculture and Miyawaki forests. We are working on the design for a local arts and ecology project around eco-literacy and the restoration of the water cycle under the umbrella of Almenara. We have obtained Erasmus + funding for another international one-week course in 2023 entitled "Living on the frontline of climate change in Europe: Tools and practices for working with European youth." And we are planning or participating in further courses, performances, arts interventions and actions with our neighbours and the wider community of living beings in our territory.

Acknowledgements Our research on regenerative cultures has been developed in several blocks of actions and programmes. One of them was Barba-T, funded by the Daniel and Nina Carasso Foundation (Fondation France) and with the collaboration of Santander Foundation, Diputación de Cádiz, Universidad de Cádiz and Junta de Andalucía. https://barba-t-expo.beetime.net/ (2020-2021).

Extinction Rebellion International provided funding from their 2019 Regenerative Cultures funding cycle for the video "I Dream of Being What I Was" and the weekend course for activists "Creating Regenerative Cultures". The EU provided funding for ongoing research and knowledge sharing with other European organisations (June 2022, April 2023). https://www.parpado.net/econarratives/.

References

Akomolafe B (2022) A slower urgency • writings – Bayo Akomolafe. www.bayoakomolafe.net, https://www.bayoakomolafe.net/post/a-slower-urgency. Accessed 18 May 2023

Alexander J, Conrad A, Eno B (2022) Citizens: why the key to fixing everything is all of us. Canbury Press, London

Bendell J (2018) Deep adaptation: a map for navigating climate tragedy. http://www.lifeworth.com/deepadaptation.pdf

Berry T (2009) The sacred universe: earth, spirituality, and religion in the twenty-first century. Columbia University Press, New York

Boulton JG, Allen PM, Bowman C (2015) Embracing complexity: strategic perspectives for an age of turbulence. Oxford University Press, Oxford

Brown AM (2017) Emergent strategy: shaping change, changing worlds. Ca Ak Press, Chico

Gesturing Towards Decolonial Futures Collective (2019) Global citizenship education otherwise. Gesturing towards decolonial futures, 18 May 2019. https://decolonialfutures.net/portfolio/global-citizenship-education-otherwise/. Accessed 2 Nov 2022

Heron J, Reason P (2008) Extending epistemology within a co-operative inquiry. In: Reason P, Bradbury H (eds) The Sage handbook of action research. participative inquiry and practice, 2nd edn. Sage Publications Ltd.

Holmgren D (2002) Permaculture: principles & pathways beyond sustainability. Holmgren Design Services, Hepburn, Vic.

Intergovernmental Panel on Climate Change (IPCC) (2022) Global warming of 1.5°C: IPCC special report on impacts of global warming of 1.5°C above pre-industrial levels in context of strengthening response to climate change, sustainable development, and efforts to eradicate poverty. Cambridge University Press, Cambridge. https://doi.org/10.1017/9781009157940

Macy J, Brown MY (2014) Coming back to life: the updated guide to the work that reconnects. Gabriola Island, Bc: New Society Publishers

Nogales Muriel R (2022) Sembrando Para Quienes Nos Sucederán: Resignificación de 'Lo Rural' a Través Del Arte Y La Cultura En Un Contexto de Crisis Ecosocial (R-Rural). Revista Para El Análisis de La Cultura Y El Territorio 23. https://doi.org/10.25267/periferica.2022.i23.18

Reason P, Bradbury H (2008) The SAGE handbook of action research. SAGE Publications Ltd, UK. https://doi.org/10.4135/9781848607934

Seed J (2007) Thinking like a mountain: towards a council of all beings. New Catalyst Books, Gabriola Island

Wahl DC (2016) Designing regenerative cultures, Triarchy Press, Axminster, England

Whybrow A, Turner E, McLean J, Hawkins P (2022) Ecological and climate-conscious coaching. Taylor & Francis

Conclusions

There is Hope. Better Possible Futures Through Regenerative Design

Carlos Cobreros, Emanuele Giorgi, and Tiziano Cattaneo

Abstract Life on the planet has continuously regenerated and sustained itself for billions of years. The functioning of ecosystem services has been affected, risking their resilience. In this situation of converging multiple crises, an urgent and radical rethinking of our degenerative and destructive model of development is needed. Sustainability is no longer enough, we need to move towards regenerative sustainability and regenerative development, which aims to heal our damaged world, enabling it to evolve into thriving living systems in which the health and well-being of the entire system are continuously increasing. To move to a regenerative paradigm, new goals, purposes, narratives, practices, and standards need to be recognized. This closing chapter of the book "Regenerative Design. New Contexts, new visions, Emerging Practices and Perspectives" aims to review and reflect on the main conclusions of new approaches, experiences and narratives, discussions around our necessary affiliation with nature, around Regenerative Thinking, Regenerative Development, and Regenerative Design, from the urban scale to product design scale, from rural to urban contexts, from the application in communities to the importance of generating a suitable framework for learning, development and resonance, to extract guidelines for future regenerative design approaches and for a regenerative, better and possible living future.

Keywords Sustainability · Regenerative thinking · Regenerative development · Regenerative design · Hope

C. Cobreros (✉)
School of Architecture, Art and Design, Tecnologico de Monterrey, Querétaro, Mexico
e-mail: ccobreros@tec.mx

E. Giorgi
School of Architecture, Art and Design, Tecnologico de Monterrey, Chihuahua, Mexico
e-mail: egiorgi@tec.mx

T. Cattaneo
Department of Civil Engineering and Architecture, University of Pavia, Pavia, Italy
e-mail: tiziano.cattaneo@unipv.it

1 Introduction

Life on the planet has continuously regenerated and sustained itself for billions of years (Capra and Luisi 2014). We are currently experiencing the transition from the Holocene epoch, the only state of the planetary system that we know with certainty can sustain human societies (Steffen et al. 2015), to the Anthropocene, defined by overwhelming human activity on the natural world (Zingoni de Baro 2022). The Anthropocene is committed to constant growth under a model of industrial development, which sees the planet more as a factory from which to extract resources, rather than a living system with which to integrate. This leads to an unprecedented planetary-scale socio-ecological and climate crisis (Salk 2019; Masson-Delmotte et al. 2021).

The functioning of ecosystem services has been affected (Zingoni de Baro 2022), risking their resilience (Steffen et al. 2015). Natural processes have interfered (Cardinale et al. 2012), climate stability (Zari 2018; de Rooij and van Hattum 2022), the integrity of the biosphere, biochemical fluxes and the Earth system and its biodiversity has been affected (Salk 2019; Servant-Miklos 2020; de Rooij and van Hattum 2022), to the extent of generating irreversible changes (Rands et al. 2010). Future environmental conditions, and threats to the biosphere, and all its life forms, are far more dangerous than previously thought (Bradshaw et al. 2021). The prevailing economic system does not respect planetary biophysical limits (Wahl 2016), causing ever-increasing degradation leading to a state of planetary emergency (WWF 2016; Lenton et al. 2019), accompanied by armed conflict, growing economic inequality (Goodall and Abrams 2021), disease, and pandemics (Salk 2019).

In this situation of converging multiple crises (East 2020), we attack symptoms rather than causes, which prevents us from achieving states of neutrality or thriving and flourishing living systems (Du Plessis 2012; Russell 2013; Kopnina 2015). An urgent and radical rethinking of our degenerative and destructive development model is needed, transforming destructive patterns of thinking and being into patterns that nurture all life holistically (Du Plessis 2012; González-Márquez and Toledo 2020). Will, social learning and action are urgently needed to make living systems thrive (Mang and Reed 2012a; Reed 2007). It is needed for the health and well-being of all species (Salk 2019), to weave the natural basis of our actions towards a resilient and thriving future (de Rooij and van Hattum 2022). It is necessary a paradigm shift, mindset and framework (East 2020), breaking with mechanistic worldview and reductionist thinking (Abson et al. 2017; du Plessis 2012), operating within the planet's biophysical limits and under the need for economic slowdown, incorporating social justice, respect for nature's regenerative capacity, integrity, and resilience of earth systems (Zingoni de Baro 2022).

2 Sustainability is not Enough

Sustainability is based on a mechanistic, reductionist, deterministic, dualistic, anthropocentric worldview, where nature must be controlled and managed (Benne and Mang 2015), which considers humans and the rest of life to be separate and environmental resources to be at the service of human consumption (Gibbons 2020a). The term is unspecific in its definition, unambitious in its aims (Du Plessis 2012; Russell 2013; Mang and Haggard 2016; Bastianoni et al. 2019), focuses on symptoms rather than causes, superficial leverage points, and supports unsustainable patterns of action and thinking (Du Plessis 2012; Caradonna 2022; Gibbons et al. 2018; González-Márquez and Toledo 2020; Clark and Harley 2020).

Sustainability, as an anthropocentric paradigm in crisis, offers no guidance on how to arbitrate between the competing drivers of economic growth, planetary boundaries, and social justice (East 2020) and fails to achieve its own goals (Stafford-Smith et al. 2017; Kroll et al. 2019; González-Márquez and Toledo 2020). The Sustainable Development Goals offer goals, but do not support thriving living systems in an integrated and holistic manner (Gibbons 2020a, b), and are difficult to integrate synergistically (Stafford-Smith et al. 2017; Kroll et al. 2019). More holistic, inspiring, ambitious, and motivating goals should be proposed (Jørgensen et al. 2015; Mang and Haggard 2016; Ives et al. 2020; González-Márquez and Toledo 2020; Gibbons 2020a). Sustainability promises not to affect future humanity but does not envisage the possibility of co-evolving with nature. Sustainability is no longer enough (Wahl 2016).

We need to move towards regenerative sustainability (Lyle 1996; Reed 2007), which aims to heal our damaged world, enabling it to evolve into thriving living systems in which the health and well-being of the entire system are continuously increasing (Gibbons 2020a; Iyer et al. 2021; Mang and Haggard 2016). To move from a sustainability-based paradigm to a regenerative paradigm, different goals, practices, and standards need to be recognized (du Plessis 2022). The need for resilience provides a new definition of sustainability, which is based on a stable state that can regroup and move forward when equilibrium has been disrupted (Mang and Haggard 2016), but maintains certain inconsistencies to meet immediate local human needs, may sacrifice the integrity of larger living systems (Mang and Reed 2012b).

It is increasingly recognized that human endeavours must be aligned with life principles that centre on sustaining life. A transition is needed towards futures where socio-ecological systems are designed to balance anthropogenic activity and the integrity of natural ecosystems (Wals 2019; Tafuni and Heß 2019), where humanity sees itself as part of nature and living systems and is (re)integrated into the biosphere processes that maintain health and sustain life (Wahl 2006).

Advances in biology, ecology, neuroscience, systems thinking, and complexity theory, now offer us a systemic view of life (Capra and Luisi 2014; Benne and Mang 2015), enabling us to engage with it, to use the health of ecological systems as a basis, to shift our fragmented worldview to a whole systems model and thus understand the interrelationships of living systems (Reed 2007). Humanity must assume complexity, becoming familiar with uncertainty and ambiguity, as fundamental characteristics of our lives and the natural world (Wahl 2016). Since living systems are constantly changing in ways that are never completely predictable or certain, it must be learned to accept uncertainty, allow time for regenerative processes to develop, and focus more on developing regenerative capacity than on achieving fixed goals (Meadows 1999; Wahl 2016; Gibbons et al. 2020; Gibbons 2020b).

3 Regenerative Thinking, Development and Design Framework.

We need to create a positive future through an ecological worldview based on living systems (Du Plessis and Brandon 2014). We need transformative, fundamental, radical changes in our understanding of our relationship with each other and with nature (Hawken 2007), from the local to the global, in education, in the economy and financial system, governance, industry, transport, infrastructure, energy systems, water control and provision, agricultural and food systems, health systems and social systems (Wahl 2016).

Today we can speak of a Regenerative Development (RDe) Thinking Framework and approach based on living systems theory, based on an understanding of place as something unique that enhances the capacities necessary for living systems to increase in complexity, diversity, capacity to sustain all life and potential for change to provide options for future health and well-being (Gibbons et al. 2020). Although Regenerative thinking has been developed and used in practice for over 20 years, only recently started gaining the attention of theorists, academics, and practitioners (Du Plessis 2012; Cole 2012b; Du Plessis and Brandon 2014; Robison and Cole 2015; Gibbons et al. 2018).

Regenesis Group explores the process of place-based RDe in depth, through underlying premises: a new role for humans; working within a development process; regeneration and exploration of development alongside design (Mang and Reed 2012b). It is vitally important to understand place as a unique multi-layered network of living systems within a geographical region resulting from complex interactions, over time (Mang and Reed 2012a).

Regenerative Design (RDi) helps us translate RDe understanding into applications (Zhang et al. 2015). Du Plessis (2012) identifies and develops the theoretical foundations of regenerative design: Human systems are integral parts of ecosystems; Human activities contribute positively to the function and evolution of ecosystems; Human efforts must be a function of the specific context of the place; Participatory and reflexive processes are necessary for the design and development of regenerative places.

RDi builds on recognized core concepts of regenerative approaches: Systems thinking, place-based and oriented towards contributing to positive outcomes (Svec et al. 2012); It allows for increased social and natural capital (Clegg 2012); It looks for the co-evolutionary relationship between socio-cultural and ecological systems (Cole et al. 2013). The RDi framework, rather than presenting sets of specific design strategies and guidelines that might inhibit creative solutions, is populated with provocative questions that change and can be adapted to different situations, from community engagement, to detailed design explorations (Cole 2012a).

RDi is beginning to demonstrate that it is possible to revitalize existing degraded precincts and design new ones as quality living environments capable of regenerating natural and urban ecosystems (Zingoni de Baro and Macedo 2020). Regenerative design includes ecological techniques and design technologies (e.g. living buildings, permaculture, biophilic design) that can be used in service of a larger RDe trajectory (Du Plessis and Brandon 2015; Gibbons et al. 2018).

While RDe provides the framework for identifying patterns within a living system, RDi is integral to the process of shaping patterns and actions (Mang and Reed 2012a) and the practice of engaging the natural world as a medium and generator, drawing on the living and natural systems that exist in a place (Littman 2009).

RDe builds on RDi (Lyle 1996), ecological design and ecological planning (Van der Ryn and Cowan 2007), biophilic design (Kellert et al. 2008); in the challenges of buildings and living communities (Yeeles et al. 2023), in biomimicry (Benyus 1997), permaculture (Holgrem 2002) and ecovillage design (Christian 2003; Litfin 2014). RDi transcends at different scales, from the organism, the product, the building, the community, the city, the region and the planet. RDi can open pathways towards guiding principles of positive thinking, positive humans, positive living buildings, positive living communities, positive cities, positive climates, and positive influence on nature, overarching and/or cross-cutting themes related to identity, form and function of a regenerative environment that integrates them (Thomson et al. 2022). The next phase of regenerative impact will be powerfully led by cities, as they have the resources, growth and innovative capacity to make such changes (Thomson and Newman 2018; Thomson et al. 2018) and from there their influence can spread to the rest of the territory and scales.

4 There is Hope

Civilisation is and always will be under threat, but we must opt for a hopeful future (Diamond 2011). There is hope because change can happen (Newman 2022). Hope as active hope, as Goodall defines it, is necessary to find enabling pathways for humanity and the rest of the planet's inhabitants, based on our intellectual capacity, the resilience of the natural world, the power of youth and the indomitable human spirit, which needs to focus on how we can live in greater harmony with nature (Goodall and Abrams 2021). The idea of planetary health provides a unifying concept on which efforts towards remediation and survival can focus, with evidence derived from the history of the natural world suggesting that there is reason for optimism for a transition to a new era of symbiotic flourishing (Salk 2019). Hope is reinforced by success in achieving beneficial changes, as well as beginning to see a different vision of the future emerging, bit by bit (Meyer and Newman 2020; Newman 2018).

Humanity is not doomed to a future dystopia. Regenerative Development and the design of Regenerative Cultures make us visualize a better and possible future. We can design and organize our culture, our Economy, our societies to protect, improve and celebrate life, from questioning current paradigms, living the questions in greater depth, towards equality, sustainability and justice, where delving into the "why" will guide us to the "what" and the "how" (Wahl 2016).

Before a regenerative approach is widely accepted, several challenges need to be overcome, such as the need to abandon the lure of "expertism" and the belief that technological solutions, alone, have the power to solve all our complexes. Challenges as humanity (Benne and Mang 2015) and rather cultivate creativity that, from systemic thinking, from the understanding of the complexity and changing nature of the world, from the potential rather than the problem, guides us towards that better future. It can occur from the development of the capabilities of living systems to evolve in a co-evolutionary mutualism process, a process of supporting the inherent capabilities of self-organization and improvement of life, of living systems (Gibbons 2020b), with humanity always understood as part of that living system where it is inserted.

There is hope, already many things have changed and we have new socio-technical systems that are much more sustainable and even regenerative, so we can now move towards their generalization, but we have to choose this future and not return to the exploitative approach to development so deeply rooted in modernity (Newman 2022).

Regenerative thinking and development posits humanity as a powerful agency of change that can recognize the interrelationship with all life and align our worldview, thinking, and actions with the principles of living systems to catalyse prosperity, with the principles of nature at all scales (Gibbons 2020a).

To make this qualitative leap as humanity, new narratives are needed in the way we think and develop our cities, from the application of biophilic design, as a first step of affiliation with nature, from a multi-species mind, rethinking and reinventing the way we produce our products, from an idea of Advanced Design, Responsible Innovation and Circularity that can be a course of action to address the needs of the complex systems around us. We need to change and evolve consciousness from presence, from our inner ecology to care for and transform our outer ecology.

The regenerative paradigm can and should contribute from the capacity of all natural, cultural and economic systems of a place. The way we bring nature into our cities needs to be rethought and strengthened, as a first step towards an affiliation with nature, the development of an ecological awareness for further regenerative development by being nature. The different aspects that enable and make regenerative development possible must be strengthened, from the natural territory, from the rural territory to the urban territory, where more attention needs to be paid in terms of its positive impact.

Work for educational and training programs for professionals, scientists, students and community members, as further theoretical development is needed, along with case studies, design experiments, and adaptive design (Felson et al. 2013). Although in this book we can already see interesting experiences, at different scales and levels, of transformation towards more sensitive learning systems that enable change towards the regenerative.

To make this qualitative leap as humanity, we need hope and courage to commit to change, courage to work for greater awareness and sensitization about regenerative paradigms, principles and practices. Above all, it is necessary to cultivate greater sensitivity and ecological awareness, greater spirituality, understood as that consistent being and that being present, present for change, for evolution as a professional, as a person, as a society, as humanity, a change that allows for that future. better and possible, that regenerative future, that living and prosperous future where we all fit, humans and the rest of the species that cohabit in this treasure of a world that has been inherited to us after billions of years of evolution.

References

Abson DJ, Fischer J, Leventon J, Newig J, Schomerus T, Vilsmaier U, von Wehrden H, Abernethy P, Ives CD, Jager NW, Lang DL (2017) Leverage points for sustainability transformation. Ambio 46:30–39

Bastianoni S, Coscieme L, Caro D, Marchettini N, Pulselli FM (2019) The needs of sustainability: the overarching contribution of systems approach. Ecol Ind 100:69–73 [CR]

Benne B, Mang P (2015) Working regeneratively across scales—insights from nature applied to the built environment. J Clean Prod 109:42–52

Bradshaw CJA, Ehrlich PR, Beattie A, Ceballos G, Crist E, Diamond J, Dirzo R, Ehrlich AH, Harte J, Harte ME, Pyke G, Raven PH, Ripple WJ, Saltré F, Turnbull C, Wackernagel M, Blumstein DT (2021) Underestimating the challenges of avoiding a ghastly future. Front. Conserv. Sci. 1:615419. https://doi.org/10.3389/fcosc.2020.615419

Capra F, Luisi PL (2014) The systems view of life: a unifying vision. Cambridge University Press

Caradonna JL (2022) Sustainability: a history. Oxford University Press [CR]

Cardinale BJ, Duffy JE, Gonzalez A, Hooper DU, Perrings C, Venail P, Naeem S et al (2012) Biodiversity loss and its impact on humanity. Nature 486(7401):59–67

Christian DL (2003) Creating a life together: practical tools to grow ecovillages and intentional communities. New Society Publishers, Gabriola Island, BC, Canada [CR]

Clark WC, Harley AG (2020) Sustainability science: towards a synthesis. Annu Rev Environ Resour 45. http://nrs.harvard.edu/urn-3:HUL.InstRepos:42660129 [CR]

Clegg P (2012) A practitioner's view of the 'regenerative paradigm.' Build Res & Inf 40(3):365–368

Cole R (2012a) Regenerative design and development: current theory and practice. Build Res Inform 40(1):1–6

Cole R (2012b) Transitioning from green to regenerative design. Build Res Inform 40(1):39–53

Cole RJ, Oliver A, Robinson J (2013) Regenerative design, socio-ecological systems and co-evolution. Build Res & Inf 41(2):237–247

de Rooij B, van Hattum T (2022) Design for regeneration – a nature-based future perspective on cities. In: Roggema R (eds) Design for regenerative cities and landscapes. Contemp Urban Des Think. Springer, Cham. https://doi.org/10.1007/978-3-030-97023-9_6

Diamond J (2011) Collapse: how societies choose to fail or succeed: revised edition. Penguin [CR]

Du Plessis C (2012) Towards a regenerative paradigm for the built environment. Build Res & Inf 40(1):7–22

Du Plessis C (2022) The city sustainable, resilient, regenerative – a Rose by any other name? In: Roggema R (eds) Design for regenerative cities and landscapes. Contemporary urban design thinking. Springer, Cham. https://doi.org/10.1007/978-3-030-97023-9_2

Du Plessis C, Brandon P (2014) An ecological worldview as basis for a regenerative sustainability paradigm for the built environment. J Clean Prod (2014). https://doi.org/10.1016/j.jclepro.2014.09.098

East M (2020) The transition from sustainable to regenerative development. Ecocycles 6(1):106–109

Felson AJ, Bradford MA, Terway TM (2013) Promoting Earth Stewardship through urban design experiments. Front Ecol Environ 11:362–367 [CR]

Gibbons LV (2020a) Regenerative—the new sustainable? Sustainability 12(13):5483

Goodall J, Abrams D (2021) The book of hope: a survival guide for trying times. Celadon Books

Gibbons LV (2020b) Moving beyond sustainability: a regenerative community development framework for co-creating thriving living systems and its application. J Sustain Dev 13:20–52

Gibbons LV, Cloutier SA, Coseo PJ, Barakat A (2018) Regenerative development as an integrative paradigm and methodology for landscape sustainability. Sustainability 10:1910. https://doi.org/10.3390/su10061910

Gibbons LV, Pearthree G, Cloutier SA, Ehlenz MM (2020) The development, application, and refinement of a regenerative development evaluation tool and indicators. Ecol Indic 108:105698.[CR]

González-Márquez I, Toledo VM (2020) Sustainability science: a paradigm in crisis? Sustainability 12:2802 [CR]

Hawken P (2007) Blessed unrest: How the largest social movement in history is restoring grace, justice, and beauty to the world. Penguin

Holgrem D (2002) Permaculture: principles and pathways beyond sustainability. Holmgren Design Services, Hepburn, Australia

Institute, Hartland, VT

Ives CD, Freeth R, Joern F (2020) Inside-out sustainability: the neglect of inner worlds. Ambio 49:208–217 [CR]

Iyer HS, DeVille NV, Stoddard O, Cole J, Myers SS, Li H, Elliott EG, Jimenez MP, James P, Golden CD (2021) Sustaining planetary health through systems thinking: public health's critical role. SSM-Popul Health 15:100844

Jørgensen SE, Fath BD, Nielsen SN, Pulselli FM, Fiscus DA, Bastianoni S (2015) Flourishing within limits to growth: following nature's way. Routledge [CR]

Kellert SR, Heerwagen J, Mador M (2011) Biophilic design: the theory, science and practice of bringing buildings to life. John Wiley & Sons

Kopnina H (2015) Neoliberalism, pluralism and environmental education: the call for radical reorientation. Environ Dev 15:120–130

Kroll C, Warchold A, Pradhan P (2019) Sustainable Development Goals (SDGs): are we successful in turning trade-offs into synergies? Palgrave Commun **5**:140 [CR]

Lenton TM, Rockström J, Gaffney O, Rahmstorf S, Richardson K, Steffen W, Schellnhuber HJ (2019) Climate tipping points—too risky to bet against. Nature 575(7784):592–595

Litfin KT (2014) Ecovillages: lessons for sustainable community. Polity Press, Cambridge, UK [CR]

Littman JA (2009) Regenerative architecture: a pathway beyond sustainability. Masters theses, 303

Lyle JT (1996) Regenerative design for sustainable development. Wiley international Press, Hoboken, NJ, USA. ISBN 978-0-471-17843-9

Mang P, Haggard B (2016) Regenerative development and design—a framework for evolving sustainability. Regenesis Institute, Santa Fe, NM, USA. ISBN 978-1-118-97286-1

Mang P, Reed B (2012a) Designing from place: a regenerative framework and methodology

Mang P, Reed B (2012b) Regenerative development and design (Chap. 303). In: Encyclopedia sustainability science & technology

Masson-Delmotte VP, Zhai P, Pirani SL, Connors C, Péan S, Berger N, Scheel Monteiro PM et al (2021) IPCC, 2021: summary for policymakers. In: Climate change 2021: the physical science basis. Contribution of working group i to the sixth assessment report of the intergovernmental panel on climate change

Meadows D (1999) Leverage points: places to intervene in a system. The Sustainability

Meyer K, Newman P (2020) Planetary accounting. Springer [CR]

Newman P (2018) Sustainable earth begins its journey. Sustain Earth 1:2. https://doi.org/10.1186/s42055-018-0005-2[CR]

Newman P (2022) Regenerative development: the hope-based future. Curr Adv Geogr Environ Earth Sci 8:109–130

Rands MR, Adams WM, Bennun L, Butchart SH, Clements A, Coomes D, Vira B et al (2010) Biodiversity conservation: challenges beyond 2010. Science 329(5997):1298–1303

Reed B (2007) Shifting from 'sustainability' to regeneration. Build Res Inf 35:674–680

Robinson J, Cole RJ (2015) Theoretical underpinnings of regenerative sustainability. Build Res & Inf 43(2):133–143

Russell JM (2013) Thrivability: breaking through to a world that works. Triarchy Press, Devon, UK [CR]

Salk JD (2019) Planetary health: a new reality. Challenges 10(1):7

Servant-Miklos V (2020) The love at the end of the world. Towards an existential ecological ethic. https://www.degruyter.com/document/doi/10.1515/jbpa-2020-0009/HTML?Lang=en

Steffen W, Richardson K, Rockström J, Cornell SE, Fetzer I, Bennett EM, Biggs R, Carpenter SR, de Vries W, de Wit CA, Folke C, Gerten D, Heinke J, Mace GM, Persson LM, Ramanathan V, Reyers B, Sörlin S (2015) Planetary boundaries: guiding human development on a changing planet. Science 347:1259855. https://doi.org/10.1126/science.1259855

Stafford-Smith M, Griggs D, Gaffney O, Ullah F, Reyers B, Kanie N, Stigson B, Shrivastava P, Leach M, O'Connell D (2017) Integration: the key to implementing the sustainable development goals. Science 12:911–919 [CR]

Svec P, Berkebile R, Todd JA (2012) REGEN: toward a tool for regenerative thinking. Build Res Inf 40(1):81–94

Tafuni V, Heß J (2019) Education for sustainable development as a catalyst and the role of students in the future management of HEIs. In: Proceedings of the keynote at the Bologna process conference, Bologna, Italy, 24 June 2019

Thomson G, Newman P (2018) Urban fabrics and urban metabolism: from sustainable to regenerative cities. Resour Conserv Recycl 132:218–229

Thomson G, Newman P, Hes D, Bennett J, Taylor M, Johnstone R (2022) Nature-positive design and development: a case study on regenerating black cockatoo habitat in urban developments in Perth, Australia. Urban Sci 6:47. https://doi.org/10.3390/urbansci6030047

Van der Ryn S, Cowan S (1996) Ecological design, 10th anniversary, 2nd edn. Island Press, Washington, DC

Van der Ryn S, Cowan S (2007) Ecological design, 10th anniversary. Island Press, Washington, DC

Wals AEJ (2019) Transgressing the hidden curriculum of unsustainability: towards a relational pedagogy of hope. Educ Philos Theory 52:825–826

Wahl DC (2006) Bionics vs. biomimicry: from control of nature to sustainable participation in nature. In: Design and nature III: comparing design in nature with science and engineering, vol 87, pp 289–298

Wahl DC (2016) Designing regenerative cultures. Triarchy Press, Bridport, UK

WWF (2016) Living planet report 2016: risk and resilience in a new era. WWF: Gland, Switzerland [CR]

Yeeles A, Sosalla-Bahr K, Ninete J, Wittmann M, Jimenez FE, Brittin J (2023) Social equity in sustainability certification systems for the built environment: understanding concepts, value, and practice implications. Environ Res: Infrastruct Sustain 3(1):015001

Zari MP (2018) Regenerative urban design and ecosystem biomimicry. Routledge

Zhang X, Skitmore M, De Jong M, Huisingh D, Gray M (2015) Regenerative sustainability for the built environment–from vision to reality: an introductory chapter. J Clean Prod 109:1–10

Zingoni de Baro ME (2022) The anthropocene and the urbanising world. In: Regenerating cities. Cities and nature. Springer, Cham. https://doi.org/10.1007/978-3-030-90559-0_2

Zingoni de Baro ME, Macedo J (2020) The role of regenerative design and biophilic urbanism in regional sustainability. The case of Curitiba. In: Fanfani D, Matarán Ruiz A (eds) Bioregional planning and design, vol II. Springer, Cham. https://doi.org/10.1007/978-3-030-46083-9_13

GPSR Compliance

The European Union's (EU) General Product Safety Regulation (GPSR) is a set of rules that requires consumer products to be safe and our obligations to ensure this.

If you have any concerns about our products, you can contact us on ProductSafety@springernature.com

In case Publisher is established outside the EU, the EU authorized representative is:

Springer Nature Customer Service Center GmbH
Europaplatz 3
69115 Heidelberg, Germany

Batch number: 09530254

Printed by Printforce, the Netherlands